APPLIED OPERATIONS
RESEARCH IN FISHING

NATO CONFERENCE SERIES

I   Ecology
II   Systems Science
III Human Factors
IV Marine Sciences
V   Air—Sea Interactions
VI Materials Science

II SYSTEMS SCIENCE

APPLIED OPERATIONS RESEARCH IN FISHING

Edited by
K. Brian Haley

University of Birmingham
Birmingham, England

Published in cooperation with NATO Scientific Affairs Division by

PLENUM PRESS · NEW YORK AND LONDON

Library of Congress Cataloging in Publication Data

Nato Symposium on Applied Operations Research in Fishing, Trondheim, 1979.
  Applied operations research in fishing.

  (NATO conference series: II, Systems science; v. 10)
  Sponsored by NATO and the Norwegian Fisheries Research Council.
  Includes index.
  1. Fisheries—Mathematical models—Congresses. 2. Fish trade—Mathematical models
—Congresses. 3. Fishery management—Congresses. 4. Operations—research—Congress-
es. I. Haley, Keith Brian. II. North Atlantic Treaty Organization. III. Norges fis-
keriforskningsråd. IV. Title. V. Series.
SH331.5.M48N37 1979   333.95'6'.072   80-27780
ISBN 0-306-40634-9

Proceedings of a NATO Symposium on Applied Operations
Research in Fishing, held August 14—17, 1979, at the
Marine Technology Centre, Trondheim, Norway

©1981 Plenum Press, New York
A Division of Plenum Publishing Corporation
227 West 17th Street, New York, N.Y. 10011

ORGANISING COMMITTEE

- Dr. Emil Aall Dahle, Norwegian University
  of Fisheries (chairman)

- Mr Torbjorn Digernes, Institute of Fishery
  Technology Research

- Mr Nelvin Farstad, Norwegian Fisheries
  Research Council

- Mr Arnold Hansen, The Ship Research Institute
  of Norway

EDITORIAL AND PROGRAMME COMMITTEE

- Professor K.B. Haley, University of Birmingham,
  United Kingdom (chairman)

- Mr. T. Digernes, Institute of Fishery
  Technology Research, Norway

- Mr. N. Farstad, Norwegian Fisheries
  Research Council

- Mr. D.J. Garrod, Fisheries Laboratory Lowestoft,
  United Kingdom

- Mr. K. Haywood, White Fish Authority,
  United Kingdom

- Dr. L. Mathiesen, Norwegian School of Economics
  and Business Administration

FOREWORD

Arnold Hansen

Director
Marine Technology Centre
Trondheim
Norway

Norwegian fisheries are presently facing serious problems, but also some promising challenges.

Most important is the fact that nearly all the major fish-stocks have been over-exploited, either by an overall too large fishing effort or a too large effort on wrong year-classes, resulting in stock-sizes reduced well below an economically optimum level or even nearly depleted. The atlanto-scandic herrings, for instance, has been below an exploitable level for several years. The recommended total allowable catch of Norwegian-Arctic cod for 1980 is 390,000 tons compared to more than 800,000 tons a few years ago. The Norwegian industrial fisheries are today mainly based on capelin. The Soviet Union has successfully claimed an increased share of this resource, resulting in an accordingly reduced catch quota for the Norwegian purse seining fleet. As a result of this resource situation the excess catching and processing capacity is great. Maintaining this excess capacity means high production costs. Both short term and long term planning for a better capacity adoption to the resources are necessary, as are means of policy to obtain this goal. (In other words fishery management is a necessity.)

Generally speaking the fishing industry is energy intensive. Fuel prices have increased dramatically the last year. Further increases may be expected as we gradually change from the present politically based situation of a non-realized excess crude oil production capacity into a real shortage of oil fuels.

The fuel oil consumption in fishing can be considerably
reduced by technical improvements like optimum design of vessel and
machinery including implementation of waste heat recovery systems,
reduced towing resistance by new net designs etc. There are also
ample room for technical improvements in the processing industry.
The greatest potential, however, for reducing the fuel consumption
lies in optimum choice of gear and processing methods and in new
fishery systems solutions. Conceptual studies have to be performed
and operations research will be one important tool in this work.

One promising challenge for the Norwegian Fisheries is the use
of both more of the total catch for production of human foodstuffs
and improved yield in the processing. Traditionally 70 - 75% of
the total Norwegian catch has been reduced to oil and meal for
animal feed. A higher degree of utilization corresponds well with
the generally accepted nutritional goal of lowering human consumption
of animal fat as fish products have a fat content both lower and
of a more acceptable quality than animal fat. A great effort on
market research, and in product and processing development is
necessary for reaching these goals. Again operations research will
be important both in feasibility studies and at the implementing
stage.

In concluding this part of my speech I would like to alter
somewhat and make some of Cyrus Hamlin's words mine: "The
Norwegian Fisheries needs Operation Research. The complexity,
the increasing political and nutritional pressures on our fisheries,
will cause the rational and orderly techniques of operations
research to become of crucial importance in solving the difficult
problems lying ahead in the best possible way."

Operations research is a stepwise and time consuming process
including:
- problem definition and simplification
- model construction and validation
- data collection and analysis
- model quantification/solution and/or experimentation
- analysis and interpretation of results
- implementation of results
and finally
- continuous validation and updating of models.

It is of vital importance that the development does not stop
after model construction which often seems to be the most interesting
work, but that the steps are carried all through.   The developer
himself or the development team must get system models implemented
as tools of analysis for decision makers in the industry, or get
the knowledge acquired through the development work made available
to them.   Only then will the fishing industry benefit from the
OR-efforts which in turn is the only measure of OR-project success.

It is therefore promising to see the symposium papers stressing the application aspect.

Fishery Management requires integrated models or linked models of different sub-systems: biological models, bio-economic models and industrial models. Both approaches mean an interdisciplinary development activity requiring people with different educational background ranging from marine biology to marketing. The present and future OR-problems of fisheries cannot be solved by one scientist or discipline alone. Inputs from several scientific areas are required to analyse and therby acquire necessary knowledge of the complex relationships that exist. A forum for exchange of knowledge and views is of vital importance for the future develop-ment of operations research to a powerful and useful tool for fishery administrators, fishermen and industrial managers.

PREFACE

These proceedings represent the papers given at a NATO
Symposium on Applied Operations Research in Fishing jointly
sponsored by NATO and the Norwegian Fisheries Research Council
(NFFR) and organised by the Norwegian University of Fisheries,
Trondheim Division and the Institute of Fishery Technology
Research.   It was held at the Marine Technology Centre in
Trondheim, Norway, August 14-17, 1979.   There were over 100
participants and the organising committee is to be congratulated
on the highly successful venture.   Although it is perhaps
invidious to select particular individuals I claim the right
to personally thank Dr. Emil Aall Dahle and Mr. Torbjorn Digernes
for their untiring efforts.   The conference was the best organised
that I have attended and the hospitality was unforgettable.

K.B.H.

CONTENTS

FISHING OPERATIONS

FISH INDUSTRY -

SCOPE, PROBLEMS AND APPROACHES

COMPREHENSIVE MODELLING OF FISHERIES:

COMMENTS AND A CASE STUDY

Leif K. Ervik, Sjur D. Flam and Trond E. Olsen

Department of Science and Technology
The Chr. Michelsen Institute
Fantoftvegen 38, N-5036 Fantoft, Norway

INTRODUCTION

In addition to supplying all our food, biological systems provide virtually all the raw materials for industry with the exception of minerals. Biological production is therefore a main determinant of economic wealth.

The increasing pressure on these systems due to human population growth and technological change has in many areas reached an unsustainable level; a point where the productivity of biological systems is being impaired.

Oceanic fisheries, a principal source of high quality protein is no exception to this pattern as can be seen in figure 1.

The problem is international and will remain to be so. Typically several nations harvest the same stocks of fish which migrate across national economic boundaries to spawn or to search for food. All species considered in our project have this attribute.

The division of total catch between nations has most often been guided by historical data. Because of this a single country might purposely overfish to rise its future quota.

Thus, the common property problem will be with us even after the law of the Sea Conference has reached its conclusion. In order to alter this situation one has to think about entirely new concepts. Concepts like:

3

- general experiments that produces fish that stay within a
  boundary,
- fish boundaries created by ultra sonics, electrical current,
  smell, etc.,
- international corporations owned by several states that have
  all fishing rights for one or more species.

Such technical and organizational new developments might change
the nature of the fish industry. However, the organizational
solutions might well prove impossible. The fishing industry is also
a part of society at large.   Efficient strategies for catching
fish with least effort are often deemed sub-optimal. Typically,
modern regulations have a tendency to reduce the effectiveness of
the catching system.   The typical situation in the "old" fishing
nations is thus one of overcapacity and an effort to reduce this
overcapacity and distribute the limited catch equitably.

The situation in the pelagic fisheries in Norway fits this
general description. The reduction in total catch has sharpened
the underlying conflicts of interest in distributing a limited
resource, internationally as well as nationally.

These developments have created a need for a holistic view of
the fishing system. This is not to say that partial analysis is of
no interest. Within each part of the industry there are numerous
problems that need to be solved.

We may mention the problems concerning the biological production
capacity of fish populations, efficient management of fishing fleet
and processing plants, and powerful marketing operations. Analysis
of problems of this kind is probably best conducted through a
partial approach, concentrating attention on the relevant part of
the system.

In a partial approach, the perspective is narrowed down, and
one addresses those aspects of a problem that are relevant to a
specific sector. Typically then conflicts due to interactions with
other sectors of the system may be neglected.

The solution envisaged by a partial model will most often
consist of a modification of one or several of its exogenous
variables. The problem is thereby "passed on" to another part
of the system.   This may be of little help to a decision-maker
who also has responsibility for this other part. Because of
conflicting objectives he may not be free to modify those variables
in the prescribed way.

For example, from a biological perspective it may be concluded
that a certain species is heavily overexploited, and that fishing
effort must be reduced.   The decision-maker with wide responsib-

ilities is then faced with a number of new questions: What actions should be taken to reduce effort?   What will be the direct consequences on employment and the industry's economic situation? What will the secondary effects be?   Are there compensating mechanisms in the industry influencing the resource base?

From our point of view the question of secondary effects is important. These effects indicate the strength of interaction between the system parts. If these effects in reality are not negligible, then a partial approach in addressing the present problem is too narrow.

In keeping a broad perspective on the system, interactions between the parts becomes essential. We are here faced with problems more related to how the overall system operates, that is, how the parts of the system interact. In a national planning perspective, these questions become important. How does the system as such react to perturbations in one part?   What will be the consequences on employment of a certain reduction in a national quota?   What are the likely consequences on fishing effort and thereby on fish stocks from a subsidy on fish products?   In national planning there will also be normative aspects: How should the industry as such be organized and managed to satisfy the needs of society as fully as possible?

Modelling the whole system sounds ideal, but it has its own set of problems and trade offs. One important purpose of modelling is to gain understanding of how the system works.

"In the extreme case, a simulation leads to "Bonini's paradox". A model is built in order to achieve understanding of an observed causal process, and the model is stated as a simulation program in order that the assumptions and functional relations may be as complex and realistic as possible. The resulting program produces outputs resembling those observed in a real world and inspired confidence that the real causal process has been accurately represented. However, because the assumptions incorporated in the model are complex and their mutual interdependencies are obscure, the simulation program is no easier to understand than the real process was." (Dutton and Starbuck[1]).

To avoid this trap, the model has to be kept relatively simple and transparent. If we at the same time want a broad outlook, we are forced to operate on a highly aggregated level, leaving out details that could have been included in a partial approach.

In a broad outlook, we maintain a systems perspective on the fish industry. However, to model a system as such is a totally unstructured task, not likely to be of much value. Since everything "in reality" is interrelated and since one can never capture all

aspects of a real world system in one model, one faces the problem
of limitation everywhere.

In order to be able to select which mechanisms and variables
to include in a model, a focus for the modelling effort must be
identified. Without this, modelling degenerates into an attempt
to give "a true description of the system".   Since no single
description of "true reality" is valid in all contexts, the attempt
to "map reality" is doomed to fall.

As an example of what could possibly be included, consider
the problem of quotas.   Reduced quotas may show the fleet to be
inefficient, but if a large partially employed fleet is a powerful
argument in negotiations, it may, because of distributional effects
pay to keep a number of more or less idle vessels. If overcapacity
at the same time tends to rise total quotas, this may threaten the
resource base.

Choosing one aspect is part of problem definition.   The
problem definition phase is very important in all modelling
efforts, but even more so in models designed to cover a broad basis.

In an effort to learn from our and other experience concerning
the need for a clear and crisp problem definition, we devoted quite
a long time to this phase of the project we are going to talk about.

But before we go into details on the problem focus, we will
first outline the basic trends in Norwegian Industrial Fisheries.

NORWEGIAN INDUSTRIAL FISHERIES, A CASE STUDY

The Norwegian industrial fisheries exploit pelagic fish
populations in the North Sea and the Barents Sea. Some 450 vessels,
270 purse seiners and 180 trawlers presently take part in these
fisheries. The catch is processed to fishmeal and oil in some 40
factories spread out along the Norwegian coast from Vadsø in the
north to Egersund in the south.

The recent development in the purse seiner fleet is pictured
in figure 2.

Total catch delivered from meal and oil processing has varied
considerably during the last 20 years as shown in figure 3.

The increase in catch reflects the growth in productivity.
This development is not, however, without a discomforting aspect.
Up to now, it has been possible to increase total catch by
exploiting stocks previously unfished.   Possibilities of continuing
this trend appear limited. The old stocks have been heavily

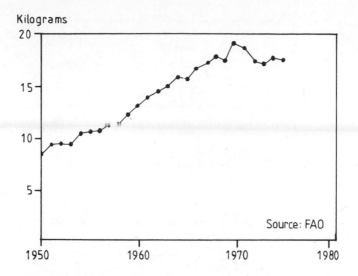

Figure 1. World Fish Catch Per Capita, 1950-75

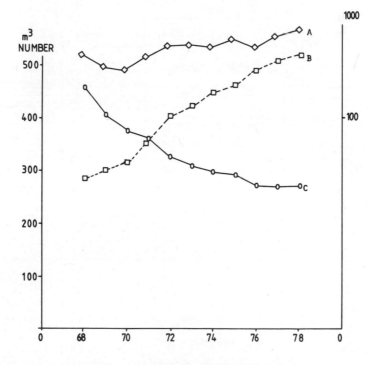

Figure 2. A = Total Cargo Capacity
          B = Average Cargo Capacity
          C = Number of Vessels

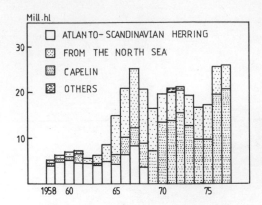

Figure 3. Quantity Delivered to Meal and Oil Processing

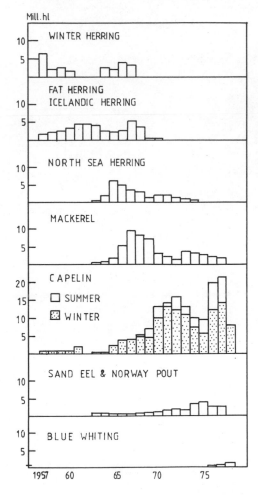

Figure 4. Yearly Catches. Composition of Quantity Delivered to Meal
           and Oil Processing

decimated and sometimes threatened by extinction (see figure 4).
It seems likely that the industry in the future will have to adjust
to yearly catches of approximately 15 million hl, a significant
reduction from the 20 million hl level of the early seventies.

Problem areas

This short historical sketch points to some important problems
in this industry.   The main problem is that of capacity adaption.
This problem is enlarged by the oscillations in various fish
population levels, and by the spatio temporal variability of the
catches (see figure 5).

The major issues are:

1. How should the fleet be composed and scaled over time?
2. What should be the pattern of development in the location,
    scale and number of factories?
3. How should excess capacity be evaluated?
4. How should conflicting goals regarding revenue, employ-
    mcnt and rural development be united?

To analyse these problems a bio-economic model of the Norwegian
pelagic fisheries is being developed at The Chr. Michelsen Institute
with financial support from the Norwegian Fisheries Research Council.
The overall intention is to explore how the industry should be
managed in order to have a fleet and factory structure which is
consistent with expectations on the development in markets and resources, and at the same time avoids awkward consequences in rural
areas.   The final model should therefore describe the likely
developments in markets and resources, how industry reacts to those
developments, and how these reactions reflect on resources and
markets.

Model boundaries

The relevant aspects of reality to include in the model have
to be chosen in the light of the problem definition.   However, for
a generally wide range of problem definitions figure 6 might give
a representation of the potential mechanisms that are at work.   At
any particular application, for a given problem definition many of
the interactions can safely be neglected.   Which are important and
which are not depend on both the real world and the problem
definition.   Using figure 6 to structure the discussion of the
included and excluded mechanisms, we will start with markets.

1. The products from the Industrial Fisheries are basically oil
    and meal.   The oil is used in the margarine production while

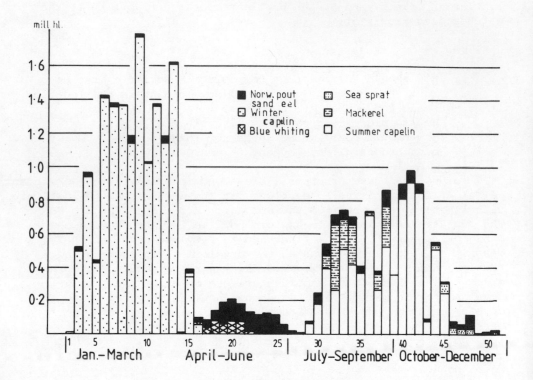

Figure 5. Weekly Deliveries in 1977

meal goes to feed for poultry and livestock.

The products from the rest of the industry go to human consumption.   These two groups are so completely different that there
is no possibility of substitution and we therefore assumed no
interaction between the two markets.

2.   USSR, Iceland and the EEC-countries do to some extent harvest
     the same pelagic atocks as we do.   Their catch is not considered
     large enough to influence the price of meal and oil significantly.
     The main exporter of fishmeal is Peru.   In the period 1970 -
     1973 Peru's fraction of total export market was around 50%.   In
     the same period Norway's export fraction varied between 6 and
     11%.

3.   Investment and competion for credit grant is not included in the
     short-run model.   Vessels participating in industrial fisheries
     may also in some cases engage in catching for direct consumption.
     The extent to which this happens is to a large degree controlled
     by the concession system.

     The content of other species in trawl catches has presently not
     been modelled.

4.   Biology.   The ecological dependencies between the species
     harvested may be important.   A substantial build-up of the
     Atlanto-Scandian herring might increase competion with the
     capelin on common grounds (and also cause a revived pelagic
     fishery to be included in the model).   There may be an import-
     ant predator-prey relation between the cod and capelin population.
     However, because of limited existing knowledge of these and other
     relations, and the need to have a manageable model, these have
     been neglected.

System boundary for the pelagic project

     Figure 7 illustrates the system boundary of the project:
"Economic Model of the Pelagic Fisheries".

     The feedback from catch to markets is for most scenarios very
weak.   Thus we see a picture of an industry squeezed between a
market and a resource base.   In such a situation there may be very
little relationship between a physical capital adjustment (one
that is physically able to catch say maximum quotas) and an economic
one.   Furthermore, the capacity resulting from private economic
adjustment under free access conditions will differ from the capacity
resulting from a strategy of maximum net income for the whole
industry.   Given high prices we might easily have a catch capacity
several times what is physically needed, and there would still be
pressures to expand.

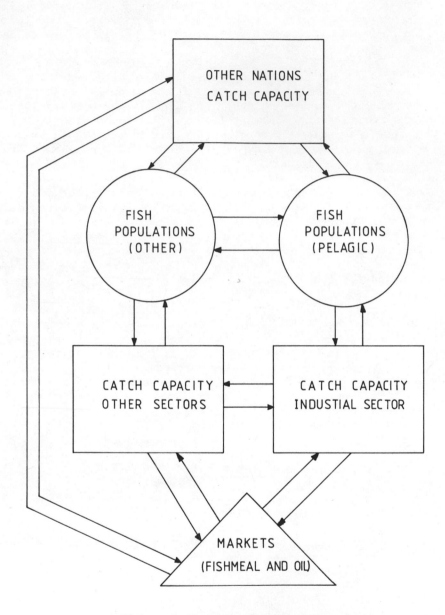

Figure 6. A General System Boundary

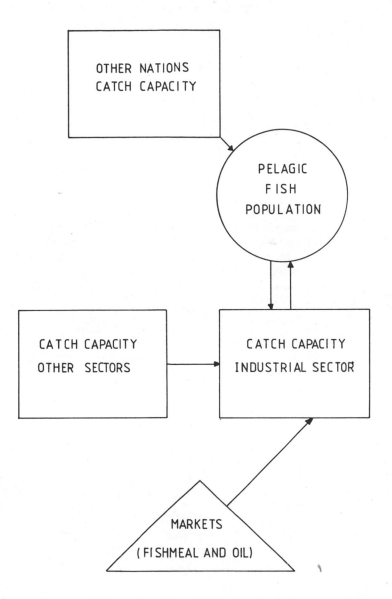

Figure 7. Final System Boundary

The interaction between capacity and fish populations is a
complex one. In the first year of the project we devoted quite a
lot of time on this subject. We developed a bio-economic model of
the capelin fishery, containing a detailed population model. This
work is still ongoing, but at an early stage we were able to conclude:

1. The trade-off between exploiting summer or winter capelin is
    a joint biological and economical question. With the present
    political situation in the Barents Sea it is also a question
    for political science to explore.

2. The uncertainties in the population model were enormous. We
    developed one model that behaved reasonably, given the available
    data. Several other models would have been equally good.

3. Present catch capacity far exceeds the quotas for the species
    considered here. At the same time there are social and
    political obstacles to a rapid reduction in capacity. If
    quotas are set autonomously without a view towards present
    economic and capacity conditions prevailing in the industry,
    then this is a strong case for decoupling the biological
    model from the economic one, making each exogenous to the other.

Early on we, for the above reasons, concluded that it would
be worthwhile to make a fairly detailed analysis of the industry,
taking biological variables as exogenous inputs. At the same
time we would be working independently with the question of
population/catch interactions.

A short-run capacity model for interdependent fisheries

Following the approach outlined above, effort has so far been
concentrated on the catch sector, with biological variables
represented at a rather aggregated level. The short-run dynamics
of this industry is illustrated in figure 5, which shows weekly
total landings to plants over a year (1977). The main fisheries
are intensive and concentrated over rather short periods of the
year. Processing capacity of plants near to fishing grounds then
become a limiting factor, and the catches have to be transported
to plants further away.

The total capacity of the system depends mainly on:

    - number and size of vessels and plants,
    - distances from plants to fishing grounds,
    - vessel's speed and the chosen transport strategy,
    - length of fishing seasons and availability of fish.

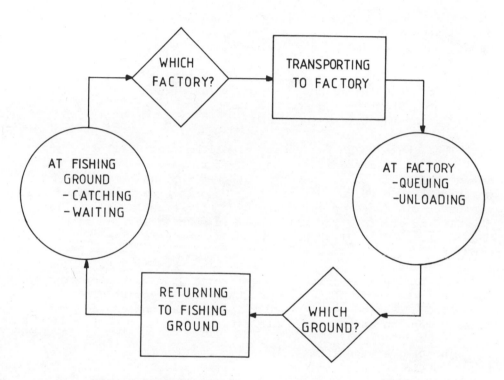

Figure 8. The States of a Vessel's Round-Trip

The interplay between all these factors presents a rather complex picture.   To get a thorough understanding of the capacity concept, a detailed simulation model has been developed.   The model gives a detailed description of the performance of up to 500 vessels taking part in six fisheries with up to 18 fishing grounds and distributing catch to some 40 plants along the coast. In a typical run it handles some 6000 vessel trips.   The model operates on a yearly basis.   It is a convenient tool for studying the performance of the system under various assumptions on the above mentioned factors.

The model is an event oriented simulation model, each event occurs when a vessel leaves a fishing ground.   Generally, a vessel may be in any of the following states (see figure 8):

    - at a fishing ground fishing or waiting,
    - transporting catch to a processing plant,
    - at a plant queing or unloading,
    - returning to fishing ground.

There have to be two choice algorithms included, one for choice of plant for delivery and one for choice of fishing ground. Given those choices, we need not at every moment keep track of which state the vessel belongs to.   We only have to register when it leaves one state to enter the next, for example when it leaves the fishing ground.   The time spent in each following state is then determined by distances, vessel's speed and plant's unloading capacity.

Any time a vessel leaves the fishing ground, the program up-dates all the relevant variables, the vessel's total catch, travelled distance, queuing and waiting time; the chosen plant's total reception and reception this week.

The choice algorithms simulating the way these choices are taken may be modelled in two ways; either as an imitation of the way they are taken in reality, or as exogenous controls.   In this model we have both methods represented.   The vessel's choice of plant for delivery is modelled close to reality, while choice of fishery is exogenously given by a preference table.   (This will be replaced by endogenous choice algorithms in a further development of the model.)

In reality, vessels are only partially free to choose a plant for delivery.   During the most important fisheries, vessels are directed from a central office.   The direction strategy followed by this office has significant impact on the system's capacity. The strategy followed at present has an equalizing effect with regard to the burden of transport requirements, placing the heavier part on the larger vessels.   This strategy is implemented in the

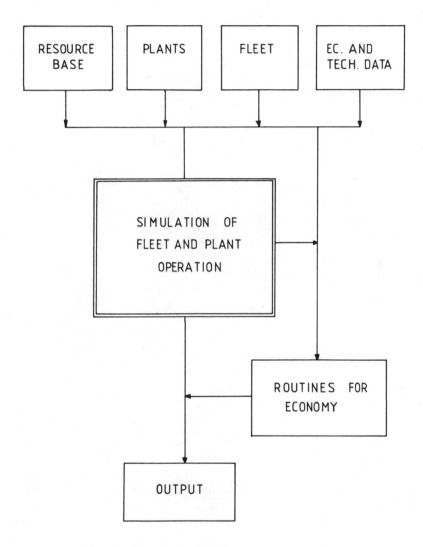

Figure 9. The Pelag 6 Program System

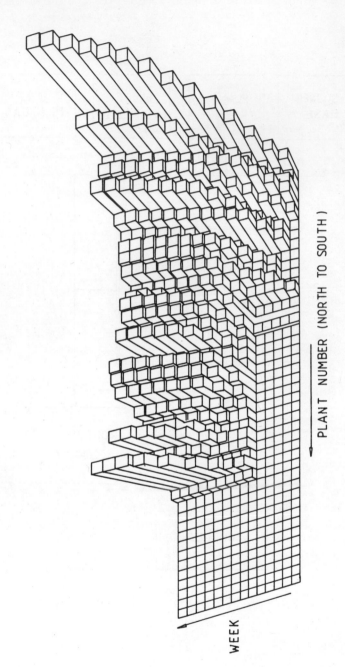

PLANT NUMBER (NORTH TO SOUTH)

WEEK

Figure 10. Accumulated Deliveries to Plants Over Time

model.   The model also has other strategies, giving the user
opportunity to experiment with various policies. The simplest
of these is to give the vessels freedom to choose the nearest open
plant for delivery.

A strategy using price discrimination as suggested in a study
of the Icelandic industry (Jensson[2]) has not yet been considered.
Such a strategy could, however, easily be incorporated. In fact,
there is an element of price discrimination in the present official
strategy in as much as transport costs are equalized through
subsidies.

So far, we have only considered capacity in physical terms.
Efficient capacity use is determined by costs and product prices.
A sub-model for economic evaluation of vessels and plants is linked
to the simulation model. This sub-model has prices and cost data
as exogenous inputs. Cost data for vessels are based on surveys,
and specify costs for the typical vessel in five cargo groups.

Individual cost data have been gathered for factories.

The organization of the program system is shown in figure 9.
Technological and economic data are stored on files which are
referred to several times during the simulation. The results from a
run, in the form of more or less detailed tables, may be inspected
on an output file.   Some graphical display of results is also
available. Figure 10 shows the generated distribution over time
(weeks) and space (plants from north to south) of delivered capelin
catches.

Result

As mentioned the study is centered around capacity adjustments.
The emphasis has been on studying the consequences of:

- establishing new factories or closing old ones.
- reducing the fleet and altering its size distribution,
- using alternative allocation rules for landing the catch,
- using alternative "sharing rules" between employed
  fishermen and owner and between vessels and factories,
- maintaining a given capacity by substituting between
  factories and fishing vessels.

Emphasis has been placed on working closely with the
representatives of the various parts of the industry.   In our
twice-yearly-seminars with the industrial representatives, the
results have been put forward and new problem areas discussed.
We would be glad to report these results to you, but we consider
it outside the scope of this paper.

Conclusion

The model constructed so far simulates the economic activities within the sector in a myopic way since the time horizon is one year ahead. This will be extended while suppressing some details, to a model simulating changes in industrial structure and populations. The results will of course heavily depend on the priority put on economic efficiency.

In simulation emphasis is put on displaying the consequences of various strategies. This approach is attractive when there are conflicting objectives involved, and no reasonable objective function can be agreed upon.

The question of optimality should, however, not be totally avoided. In particular, knowledge of the strategy giving maximum discounted net flow of income would be valuable. This strategy would specify the time evolution of capital and effort.

With respect to capital it is important to note that in the fisheries sale of capital goods often cannot be accomplished at the price of purchase. There are installation or dismantling cost added to the purchase price that cannot be recovered on scale. Sufficiently specialized machinery may have little value to others. It has recently been shown (Clark et al.[3])that "irreversible" investment significantly influences the optimal pattern of adaption in capacity and effort over time. In this context it is well known that studies based on control theory more often than not end up with a recommendation of rather bang-bang policies.

Such a policy has to be modified in the real world because of inertia and the costs of rapid transitions. It would be of particular interest to trace the effect of different policies with respect to subsidized scrapping, since such a program has lately been started.

REFERENCES

1. I.M. Dutton and W.H. Starbuck, Computer Simulation of Human
       Behaviour. New York, John Wiley (1971).
2. P. Jensson, A Simulation Model of the Capelin Fishing in
       Iceland. (This volume)
3. C.W. Clark et al. "The Optimal Exploitation of Renewable
       Resource Stocks: Problems of Irreversible Investment."
       Econometrica, Vol. 47, No.I (January 1979).
4. L.K. Ervik and R.C. Roman, "The Capelin Population and
       Exploitation - A Dynamic Analysis". CMI-ref. No. 75056-1 and
       2 (1975 and 1977).

5. L.K. Ervik and S. Tjelmeland, Capacity Adjustment in the
     Capelin Fisheries. CMI-ref. No. 75056-5 (1976) (in
     Norwegian).
6. L.K. Ervik and S. Tjelmeland, SIMPEL - A Simulation Model for
     Pelagic Fisheries. CMI-ref. No. 75056-9 (1977) (in
     Norwegian).

TOWARDS A UNIFIED STRUCTURE OF THE FRENCH FISHERY SYSTEM

Alain Fonteneau and Denez L'Hostis

VARECH - Institut Oceanographique

195, rue St. Jacques - 75005 Paris, France

"Fishing, from resource to market"

"In a fishing system, the dynamics of suicide is not fatidical": that could have been the motto of the French Working Group called VARECH.

Initiated by young specialists from different origins, as state-sponsored research institutions (CNEXO-COB, ISTPM, ORSTOM, INRA, Universities), administrations or private companies, based on voluntary adhesion through cooperation, the group was created at the end of 1977.

Its purpose was to propose a new multidimensional approach to the issue, a better self-regulation and monitoring of the system.

VARECH tried to elaborate the basis for new decision-making processes, through multicriteria and pluridisciplinary methodologies, independently from existing institutions. In a first step, six working committees were created, in a second step a general conclusion integrating the whole fishing system was attempted. This conclusion corresponds to the present paper.

In their work, the committees created by the VARECH Group have tried to outline the means for a harmonious functioning of the different parts of the French fishery system. This analytical approach was necessary at first, but could not deal with the two basic aspects of the problem which was at the origin of the VARECH Group: the relationships between the different parts of the fishery system, on the one hand, and its dynamics on the other.

The synthesis will be based upon "fishery simulations" which will be stressed according to our prospective goals.

THE RELATIONSHIPS IN THE "FISHERY SYSTEM"

Three types of relationships can be distinguished in the framework of a monitoring of the fishery system:

-       the commodities - type structure
-       the information - type structure
-       the decision   - type structure

A simplified organisational chart (figure 1) shows the various parts of the fishery system that need relating. Figure 2 indicates the relationships which seem compulsory to secure a good functioning of the system and particularly to achieve its rapid adaptation to constraints which are and will be modifying the French fishery system.   Finally, figure 3 indicates the present state of the structures within the French fishery system.

Some of the relationships have a restrictive character; without them the fishery system or some of its parts will very often reach a crisis or some kind of paralysis. The most critical relationships have retained the group's attention (Figures 1, 2, 3)

Relationships between producers and research

These relationships are of paramount importance. An exchange of information is essential for research workers in order to guide research programs according to the evolution of problems in the fishing sector and the state of stocks.

In the same way, the agents of production need to get information and forecasts about stocks, technology and markets if they want to make pertinent decisions.

There is no doubt that halieutic research in France has worked for a long time relatively isolated from fishing realities (producing agents) which has been prejudiced to its efficiency and has curbed its development.

The relations between research and producers could be developed along the following axes:

-       permanent supply by the production system of a flow of crude statistical information (biological and economic) which will give the basis for population dynamics and economic studies. In return, it is expected that synthetic and complete statistical information

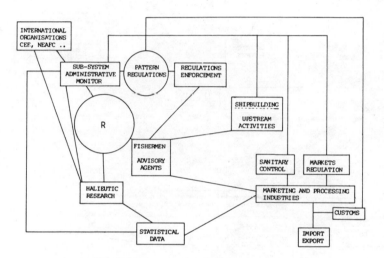

Figure 1. Organical Factors of the Fishing System

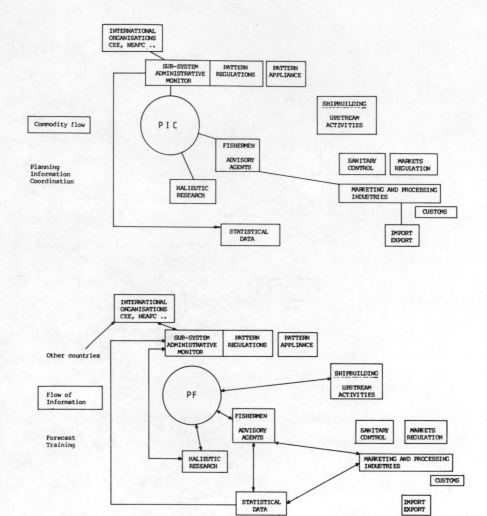

Figure 2. Graph 2: Optimum Situation in the Fishing System

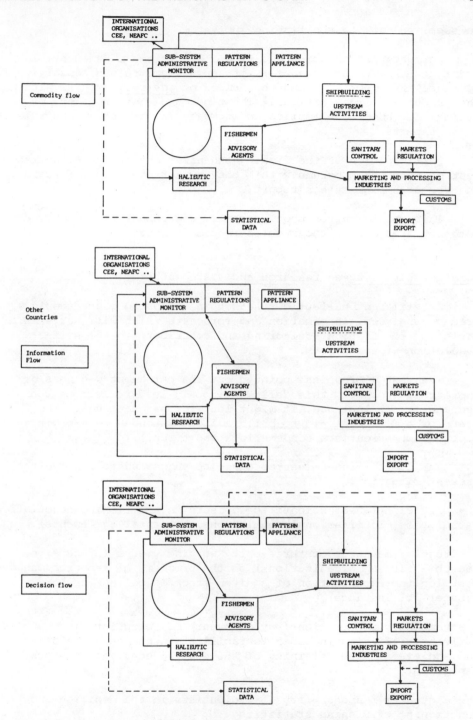

Figure 3. Graph 3: Present Situation in the Fishing System

be supplied back to the fishing subsystem,

-       permanent information and training in stock management. In
order to achieve this transfer, it seems absolutely necessary to
increase the importance and the number of <u>advisory experts</u> within
the producers' organizations. The high number of producers (a
maximum few hundreds) implies an increase in the number and role of
these experts,

-       participation of the fishing organizations in the establishment of research programs within the new structures put forward in
the research committee's report.

This participation should greatly contribute to the choice of
research axes and of the species to come first.

Relationships between research and state authorities

An efficient halieutic research in the broad sense should
produce analyses, information and forecasts which will serve as a
basis for the fishing monitoring and contribute to the self
regulation of the system.

The VARECH Group has pointed out the permanent weakness of
those relationships; this deficiency could lead to serious consequences: for instance, it might lead to the granting by the
state of a financial aid to build a fleet intended to exploit a
stock which scientists condemn in a short run.

Various concrete measures could be suggested to improve the
present situation:

1 - participation of scientific advisors in all fishing negotiations
(statistics, international commissions, bilateral agreements),

2 - regular meetings of three-party working groups (state - research - fishing organizations) on the problems of resource management (on the exploitation of a given stock of pollack or Norway
lobster for instance),

3 - temporary integration (say one year) of scientists into public
offices for fishing in order to facilitate exchange of information
and achieve a better training of the state executives in stock
management,

4 - a greater responsibity for scientists in the decision and
functioning of fishing statistics,

5 - a greater participation of public authorities in research by

formulating research targets, supplying sufficient means, and controling the achievement. If the halieutic research targets are those examined by VARECH, an important restructuring is necessary.

More financial means will sometimes be necessary in particular for research workers' salaries and research equipment (computers). A well-adapted halieutic research can be an economic investment for the community provided it is given appropriate means.

All these structures involve an increase in the number of staff. This apparent "surplus" of research workers will in fact prevent the research from becoming isolated.

Relations between fishing organizations and public authorities

Such relations already exist. They are however too often conflicting and exist mainly when a crisis occurs. A constant relation through the "forecasting, planning, coordination level" is necessary. This would avoid the creation of perfect fishery development schemes which in fact do not take into account the maximum tolerable rate of exploitation of the stock under consideration.

"Monitoring" relations of the system

The notion of "monitoring" of the system deserves a certain attention. The present crisis in French fishing really seems to stem more from various conjunctural phenomena such as resource shortage (for biological or political reasons), the skyrocketing energy costs, etc.

However, the deficiency of the information flow between the different agents in the fishing sector and the lack of forecasts appear to be other structural causes which have contributed to the present crisis.

In order to eradicate this deficiency, the fishery system will have to be thoroughly reorganized and the public fishing authorities will have to take some authoritarian measures.

Sooner or later, the system will probably regulate itself through the existence of information and forecasting flows between the agents of the system. During this period, the monitoring function of the public authorities should be progressively reduced to certain fundamental responsibilities only, such as arbitration between conflicting private interests or between private and public interests.

The works of the different committees have amply shown that the present fishing system has no satisfactory readjustment capacity. It appears particularly incapable of adapting itself to the new multiple heavy constraints that fishing faces today:

1 - constraints related to preservation of resource (quotas, licenses, meshing ...),

2 - economic constraints for the producer due to the rise of operating costs (oil, yield decrease),

3 - marketing constraints due to the internationalization of fish markets and the diversification of marketing methods,

4 - legal constraints due to the new International Sea Law and the creation of E.E.C. regulations.

But an adequate information flow between the agents should help to achieve self regulation of the fishing system at least partially. However, it would not be sufficient to modify the system in a satisfying way for the whole of the community.

A monitoring of the system is therefore essential and can be achieved at two levels:

Level A - At the level of each part of the fishing "subsystem". For instance, a research program can (and must) be impulsed by the research workers themselves; fishing agreement can be negotiated directly between a producer and a foreign country. The advantage of such a decentralized monitoring is obvious but it will have two major drawbacks:

1 - the possibility of any part of the system to act freely will be often limited by legal, administrative or financial constraints. Each time a monitoring "at the bottom" will be possible, it would be a good idea to support it by adequate measures for decentralized responsibility.

2 - Separate monitoring of the parts of the fishing system may prove awkward since conflicts and antagonisms are likely to appear soon between the policies of the different subsystems. A complementary monitoring at the top is therefore essential. This will be the level of public fishing authorities.

Level B - Public authorities have potentially at least the _information_ and the _monitoring commands_ at their disposal. These actions can be classified in the following manner:

- financing actions,
- actions of incentive, coordination and arbitration,

- actions as regards fixing and enforcing regulations.

These three types of action which constitute the control levels of the monitoring of the fishing system deserve a closer examination.

1 - The financing actions are by essence limited by the existing budgets that are hardly expandable.

The authorities in charge of the fishing policy should harmonize and coordinate the investment between the different parts of the system. It would be absurd for instance to invest hundreds of millions of francs in order to enforce a regulation on the conservation of the resource (control fleet, planes) without investing the five millions necessary to the creation of an efficient statistical system which is the "sine qua non condition" for the management of halieutic resource.

A potential activity in the field of financing deserves a particular attention: the management of the resources which are characterized by rapid collapse followed by rapid restocking. In this type of fishery, it would be possible to create a compensation fund in order to capitalize during the period of high exploitation by levying taxes (which could be proportional to fishing profits thus yielding important amounts in the periods of abundance). These sums would be allocated during the fishing stoppage decided before complete extinction of the resource and until exploitation can be resumed.

2 - The actions of incitement, coordination and arbitration. These actions do not necessarily imply important financing means or normative actions. Among them, a more adequate research planning can be an example for the French halieutic research organizations (I.S.T.P.M., C.N.E.X.O., O.R.S.T.O.M., I.N.R.A., Universities, fishing organizations).

One of the most important functions will be that of arbitration. It seems preferable to separate this action from the regulations it generates.

A rational management of the resource will lead to a considerable increase of the economic efficiency for the individual fisherman by limiting the number of persons having access to the resource. This surplus value (which may be important) granted to the fishermen comes directly from public funds (invested in the research and control of sea resource for example).

The share of this surplus value stemming from the utilisation of a public property will necessitate some sort of arbitration in order to select those "happy fews" which will have access to resource.

The profits of these optimized fisheries could be considerable in certain cases: this privileged situation could be counterbalanced by a system of taxation in order to amortize the public funds invested in the management of resource. There again an arbitration by the public authorities will be necessary.

3 - The actions of fixing and enforcing regulations can vary from a simple administrative memorandum to laws at an European level. The scope of action is vast and complex: security at sea, harbours, fish auctions, sanitary controls, foreign trade, preservation of resource, vocational training, statistical research, shipbuilding, social security schemes for fishermen, etc.

It will often be a good thing for the public authorities to decentralize and to share the regulation activity with the parties concerned every time it is possible.

For example, as far as resource management is concerned, regulations can only be decided and enforced efficiently through a greater responsibility of the producers. There again, a sufficient number of advisory agents and a better training of the fishing organizations seem essential.

THE DYNAMICS OF THE SYSTEM

French fisheries are rather unique in world fishing, but heterogeneous.

On a biological level, the heterogeneity will largely depend on how the stock will respond to steady pressure; we will thus distinguish between very fragile stock (of the herring or sardine type) often quite productive and the stocks which resist a constant fishing pressure (of cod or tuna type). A related criterion is the possibility for the scientist to foresee (in the short or intermediate run) the evolution of the resource which will influence the fishing and marketing strategies.

As far as production structures are concerned, a heterogeneity immediately appears between "industrial" and small-scale fishing. Even fishing industries are very heterogeneous depending on fishing systems and regions. In each case, the functions of these fisheries on a national level can be quite different and need to be determined.

From a legal point of view, the possibilities of rational exploitation and management of the resource are different according to the location of the fishery, whether it is a coastal stock under national jurisdiction, an E.E.C. resource, a resource situated in the economic zone of a third state or a highly migratory resource (tuna fish for instance) which is often captured in international waters.

Thirdly, marketing will mean taking into account the more or less international character of the market, present and potential preservation methods, the elasticity of demand, ...

These characteristics determine the real - time dynamics of resource, fisheries and markets. They constitute the essential information for the monitor in order to define a fishing policy.

For this purpose, our group has elaborated a basic analytical framework which sums up the main variables for the monitoring of a fishery, constraints, limiting factors and potentialities of the fishery (table 1).

As an example, the group has then tried to outline the characteristics of four stocks (scallops, North Sea herring, Bay of Biscay hake, tropical tuna) by using this framework.

A few indications on the most important relationship for the monitoring of each stock are then given in figure 4.

Finally, for each of the first three stocks, the figures of monitoring means require a few remarks.

The scallop system

The main characteristic of the scallop stocks is that they are individually exhaustible at any moment although this is not contageous. They can be easily (?) restored, allowing an extensive aquaculture.

Hence, it all amounts to a stringent local management where a choice must be made between the exploitation and protection of the marine and submarine capital. Advisory agents are essential in this context.

Consequently, the monitoring should include:

- information towards fishermen,
- advisory agents,
- permanent diagnosis of stock by research authorities,
- strictly enforced regulations.

Finally, potential repopulation should be secured by research institutes. Marketing must be controlled in order to impose the trade balance.

This analysis of highly exploited coastal banks can be applied

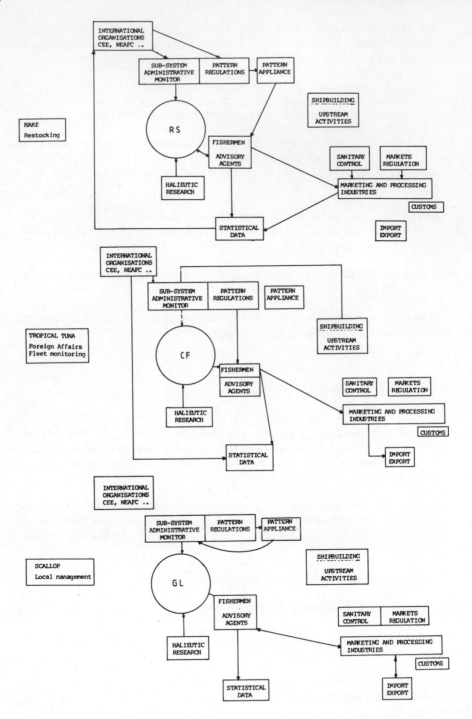

Figure 4. Graph 4: Decision Flows for Three Illustrative Fisheries

TABLE I

ANALYTICAL FRAMEWORK FOR A GIVEN FISHERY

FIELD	PROBLEM	COMMENT
Resource	Fishing statistics	Are existing statistics adequate If not, what is the problem ?
	Dynamics and biology	Are the dynamics and biology of the stock known ? If not, what kind of research should be developed ?
	Stability of the stock	Is the resource resistant to intensive exploitation ?
	Stock forecasts	Is it possible to foresee catches and yields. At what turn ?
	Restocking ?	Can the resource be restocked spontaneously in case of extinction ?
	Mixing of species	Is there a mixture species ? If so, is the problem analysed ?
	Resource management prospects	What is the state of the resource ? What can be done ?
	Technological improvements	What is to be gained due to a technological improvement ?
Legal aspects	Legal status of the resource	What is the legal status ? At the moment ? What could it be ?
	Law making	At what level ? Local ? National ? E.E.C. ?
	Problems of enforcement	Nature and cost of control. Legal problems ? Psychological problems ?
Production	Return on investment for the producer	Past, present and future. Return according to fishing simulations.

to a great number of invertebrate (urchins, abalones, clams, etc.) which have in common the fragility of the stock and a high demand leading to overexploitation.

Hake

This stock is not under threat but very ill-exploited. The overcapacity of the fleet is clear and its adaptation to present working conditions is difficult (energetic costs).

Moreover, hake is the typical example of the fresh fish French market. Therefore, the size, packaging and merchandising influence the marketing. Although these fish stocks are in the E.E.C. zone, only the French fleet exploits the stock.

A rational management of the stock thus look promising. Improved fishing gears and a good knowledge of the dynamics of the stock should result in a regular production of important catches of hake at optimal commercial size.

For these purposes, the monitor should foresee:

-     few advisory agents to collect statistical data and to train producers.

-     an important and technical research: management, selective gears, forecasts,

-     a technical and commercial research in order to innovate in the field of preservation, packaging and distribution.

CONCLUSION

The analysis of the two types of stocks through simple tools (analytical framework, commodity, - information - and decision flow charts) gives an idea of what the monitoring of a fishery system could be. This fragmentary approach is necessary as most fisheries have specific problems. These problems should be solved by the creation of a specific information planning and cooperation device regulating an adequate system.

However, a national synthesis under the direction of the monitor is necessary. Therefore, our last suggestion will be that of an "Inventory Conference" with a sufficient number of representations from each branch, regularly examining the halieutic research results in each fishery. The monitor could then elaborate the national fishing policy from these facts.

French fisheries have a future, the present constraints call for medium term solutions. The aim at the VARECH Group has been to suggest a few proposals in that direction.

This paper is the conclusion of the General Report of Research Group VARECH, entitled: "L'avenir des Peches Francaises (Reflexion sur une necessaire mutation)". This report was partially sponsored by the French Marine Fisheries Administration.

ON METHODOLOGIES WHICH BRIDGE THE GAP BETWEEN FISH POPULATION

MODELS AND FISHERY MANAGEMENT

C. C. Huang and A.R. Redlack

School of Business Administration
Memorial University of Newfoundland
St. John's, Newfoundland, Canada A1B 3X5

INTRODUCTION

To manage any fishing stock, one would like to know what is
the maximum sustainable yield and how can it be achieved. Although
under this challenge, considerable knowledge about fishing systems
has been learned, today this challenge still remains the centre
theme of most on-going fishery researches. However, in practice,
managing decisions must be made under whatever knowledge is
available. Thus, from the existing limited historical data of a
stock and the current available fish population models, one has to
estimate what is the maximum sustainable yield and decide how to
achieve it. To carry out this assignment seriously, generally
one has to face the following questions:

(a) If there are different models available, will different
    models give different results?
(b) How to deal with the situation that the system under
    study may be different from what is expected by the models?
(c) How sensitive are the estimates obtained to the model
    parameters and different statistical analysis?

These questions in fishery management have been less explored. In
the following sections, we examine the effects of the above questions on the ICNAF 3P redfish stock. A criterion based on predictability is then suggested for selection among the possible different
results obtained. After the maximum sustainable yield and other
estimates are established, the strategies on how it can be achieved
are discussed in the last section. The strategy which provides the
optimal catch is derived for any transition period taken to reach
the optimal equilibrium state.

ICNAF 3P REDFISH STOCK

ICNAF subarea 3P is located on south coast of Newfoundland, Canada. During 1955-71 exploitation of redfish in this area was almost exclusively by otter trawl. Newfoundland tonnage class 4 (N4) were the main participants until 1966. In 1967-68 non-member countries caught slightly more than 50%; during 1969-71 USSR vessels took approximately 65% of the total catch. The status of this stock has been evaluated by Parsons and Parsons[1] in 1975. Catches and standardized efforts during these years are copied in table 1A.

During 1955-68, the N4 hour was selected as the standard unit of effort and total efforts in standard unit were estimated by dividing the catch per unit effort of N4 into the total catch of all countries. For the period 1969-71, catches per hour for USSR tonnage class 4 trawlers were plotted against catches per hour for N4 for each month in which both fished at least 100 hours. A straight line drawn through the origin gave a conversion factor. Each year the combined effort and catch/effort of these two vessels were then calculated and the standard total efforts were estimated by dividing these values into the total catch of all countries.

DIFFERENT MODELS

In ICNAF div. 3P, there is a lack of adequate information on the length and age composition of redfish catches. For such a case, the Schaefer Model[2] is a popular method to examine catch/effort data and provides estimates of maximum sustainable yield. The Schaefer Model assumes

$$\Delta B_i + C_i = r\bar{B}_i (1 - \bar{B}_i/K) \tag{1}$$

with

$$\Delta B_i \text{ (change of biomass in year i)} = (\bar{B}_{i+1} - \bar{B}_{i-1})/2 \tag{2}$$

$$C_i \text{ (catch in year i)} = qE_i\bar{B}_i \tag{3}$$

where $\bar{B}_i$ is the average biomass in year i and E_i the total effort expended in year i. Using equations (2) and (3), equation (1) can be expressed solely in terms of the variables of catch and effort,

$$\{(\frac{C}{E})_{i+1} - (\frac{C}{E})_{i-1}\}/2 (\frac{C}{E})_i + qE_i = r - r(\frac{C}{E})_i /qK \tag{4}$$

From the data of catches and efforts, we can determine

Table 1. Nominal catch and standardized efforts of redfish from ICNAF Division 3P.

1A   1955-1971

Year	Catch (metric tons)	Standard catch/hr (kg)	Standardized effort (hrs fished)
1955	4,601	996	4,800
1956	3,275	1,020	3,211
1957	2,387	998	2,392
1958	3,510	911	3,853
1959	3,774	733	5,149
1960	9.225	573	16.099
1961	9.776	611	16.000
1962	13,439	512	26,248
1963	13,747	682	20,157
1964	13,807	692	19,952
1965	18,733	905	20,699
1966	20,868	874	23,876
1967	31,991	804	39,789
1968	13,884	780	17,800
1969	32,051	736	43,548
1970	37,370	707	52,716
1971	27,500	619	44,426

IB   1972-1976

Year	Catch (metric tons)	Standard catch/hr (kg)	Standardized effort (hrs fished)
1972	26,037	679	38,346
1973	18,368	612	30,013
1974	22,158	493	44,945
1975	28,250	507	55,720
1976	19,967	510	39,151

(a) q by maximizing the correlations between the term
on the left hand side of equation (4)   and $(\frac{C}{E})_i$,
then for this q value;

(b) r and r/K by regression analysis; and

(c) the maximum sustainable yield (MSY) by $y_s = rK/4$

As shown in the literature, the y_s derived from the Schaefer
Model sometimes has been taken as equivalent to the maximum
sustainable yield obtained by a simpler analysis suggested by
Gulland.   The Gulland Model[3] assumes

$$C_i/E_i = a - b(E_i + E_{i-1+}... +E_{i-n})/(n+1) \qquad (5)$$

Thus from the data of catches and efforts, we can determine

(a) a and b for a predetermined n value by regression
analysis; and

(b) the maximum sustainable yield by
$y_s = a^2/4b$

Although in the equilibrium situation, both equations (4) and (5)
result in the same form and their coefficients have the following
equivalent relationships

$$a = qK \quad and \quad b = q^2K/r,$$

the actual coefficients obtained from equations (4) and (5) by
fitting the nonstatic data are not necessarily going to satisfy the
above condition.   As shown in Table 2A, results obtained from
Schaefer Model are quite different from those obtained from Gulland
Model (no acceptable estimates can be derived from Gulland Model).

SOME BASIC ANALYSIS

The extremely low adjusted R^2 for both models and unacceptable
slope for Gulland Model indicate poor fitting of the data against
the models.   From the plots of $Y_E/\bar{B}$ against $\bar{B}$ and C/E against
$\bar{E}_{0-5}$ shown in Fig. 1, it is clear that neither the Ricker Model[4]
nor the Gulland-Fox Model[5] will provide a better answer.   Some
examination of the data is thus required.   Catches versus efforts
are displayed in Fig. 2.   The general pattern of the data seems to
agree pretty well with the assumption $C = q\bar{B}E$,   except for the
relatively low slope of the cluster of 59-64 data points. Fluctuation in recruitment is a possible explanation.   But we think this
may be just a general pattern of decreasing and then recovery of
catchability because of exploitation of the stock.   Taking 2/3 as

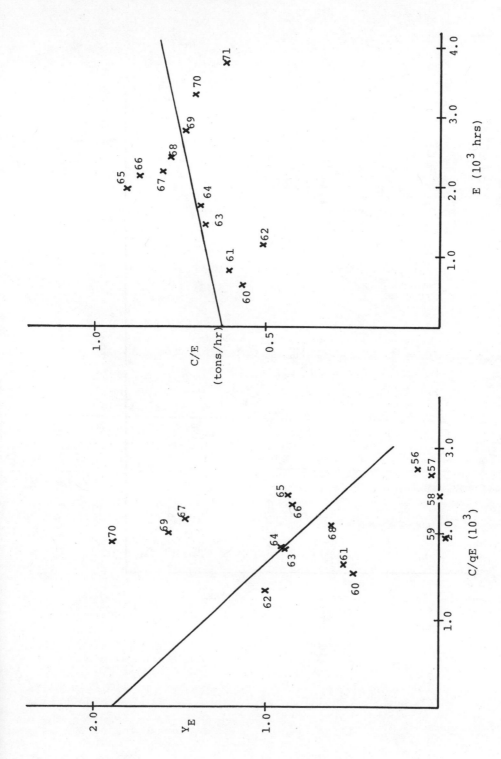

Figure 1. The Plot of the Data as Used in the Schaefer and Gulland Models

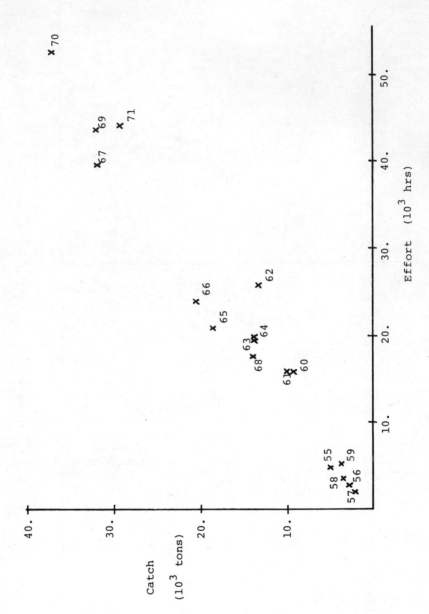

Figure 2. A Plot of the Catch Versus Fishing Effort

Table 2. Comparison of estimates obtained from different
models

2A Original Data

Models	MSY $(10^3$ tons)	Effort at MSY $(10^3$ hrs)	Catch/hr at MSY (tons)
Schaefer $Y_E/\bar{B}=1.89-5.54\text{x}10^{-5}\bar{B}$ $q=3.74\text{x}10^{-5}$ Adjusted $R^2=0.08$	16.1	25.3	0.64
Gulland (O-5) $C/E=0.618+4.41\text{x}10^{-6}\bar{E}_{O-5}$ Adjusted $R^2=0.05$			

2B Data with q adjustment

Schaefer $Y_E/E=3.52-11.07\text{x}10^{-5}\bar{B}$ $q=3.4\text{x}10^{-5}$ Adjusted $R^2=0.55$	28.0	51.9	0.54
Gulland (O-5) $C/E=1.022-1.01\text{x}10^{-5}\bar{E}_{O-5}$ Adjusted $R^2=0.65$	26.0	50.9	0.51

2C Data excluding 59-64

Schaefer $Y_E/\bar{B}=3-6.24\text{x}10^{-5}\bar{B}$ $q=2.12\text{x}10^{-5}$ Adjusted $R^2=0.89$	36.0	70.6	0.51
Gulland (P-P) $C/E=1.176-1.49\text{x}10^{-5}\bar{E}_{O-5}$ Adjusted $R^2=0.92$	23.2	39.5	0.59

the relative order of catchability for 59-64, the new estimates for
both models are then calculated and given in Table 2B. High adjusted R^2 shows a significant improvement of fitting after the simple
adjustment. Results obtained by discarding points 59-64 are given
in Table 2C for comparison. (The Gulland (P-P) model is the Parsons
and Parsons[1] calculation). Estimates obtained with q adjustment
from Schaefer Model are remarkably close to those obtained from
Gulland Model. However the estimates by discarding 59-64 data
points of MSY and effort at MSY by Schaefer Model are quite different
from those by Gulland Model. Criteria for selecting a reasonable
set of estimates will be discussed later.

DIFFERENT STATISTICAL APPROACHES

In calculating the estimates discussed previously, the
following questions arose:

(a) How sensitive are the estimates to the q in Schaefer
    Model?

(b) Can the n in Gulland Model be determined from the optimal
    correlations between C/E and $\bar{E}$, similar to the way q
    is determined in the Schaefer Model? And how sensitive
    are the estimates to the value of n?

(c) Are the basic assumptions made in general regression
    analysis for the dependent variable such as independence
    and normal distribution with constant variance valid for
    our analysis? If not, how will the estimates be affected?

To answer question (a) we plot the diagram of the correlation of
$Y_E/\bar{B}$ and C/E as a function of q. It shows that the correlation
is insensitive for a wide range of q values. This implies that
as far as the criteria of correlation is concerned any q in a
reasonable range will be acceptable. Comparisons of estimates for
$q = 3.4 \times 10^{-5}$, $q = 2.1 \times 10^{-5}$ and $q = 1.4 \times 10^{-5}$ are given in
Table 3. The difference between them is negligible. Thus in
practice, one may prefer to choose a q value in the acceptable
range which also conforms to the survival rate estimated from experiment, if one is available. To answer question (b), estimates for
different n values are calculated and displayed in Table 4. Comparison of estimates for different value of n indicates that it may
not be crucial but it is definitely important to obtain good intervals. Whether the n with best adjusted R^2 produces the best
estimates will be discussed later.

Once q or n is determined, the remaining task is the basic
regression analysis of obtaining estimators $\hat{\alpha}$ and $\hat{\beta}$ in the following regression equation:

Table 3.   Comparison of estimates obtained from Schaefer
           Model with different values of catchability

Models	MSY (10³ tons)	Effort at MSY (10³ hrs)	Catch/hr at MSY (tons/hr)	Adjusted R^2
$q = 3.4 \times 10^{-5}$	28.0	51.8	0.54	0.55
$(Y_E/\bar{B} = 3.521 - 11.07 \times 10^{-5}\bar{B})$				
$q = 2.1 \times 10^{-5}$	27.3	50.8	0.54	0.54
$(Y_E\bar{B} = 2.118 - 4.12 \times 10^{-5}\bar{B})$				
$q = 1.4 \times 10^{-5}$	26.3	48.7	0.54	0.51
$(Y_E/\bar{B} = 1.363 - 1.77 \times 10^{-5}\bar{B})$				

$$Y_i = \hat{\alpha} + \hat{\beta}x_i + e_i$$

Where $(y,x) = (Y_E/\bar{B}, \bar{B})$ in the Schaefer Model and $(y,x) = (C/E, E_{O-n})$ in the Gulland Model.   In order to make inferences from the observed sample about this population line, generally the following statistical assumptions are made:

1. For any fixed value of x, y is a random variable with a certain probability distribution (usually a normal distribution is assumed for convenience)
2. The y values are statistically independent of one another
3. The variance of y is the same for any x.

Although generally if the assumptions are not badly violated, the conclusion reached by a regression analysis will be reliable and accurate.   However if the assumptions are seriously violated then inaccurate results may be concluded.   For data collected over time, assumption 2 is rarely valid.   Therefore in fishery studies, it is necessary to examine the residual correlation.   If assumption 2 is not valid, the autocorrelation may be eliminated by considering the regression equation:

$$Y_t - r_1 Y_{t-1} = \hat{\alpha}(1-r_1) + \hat{\beta}(x_t - r_1 x_{t-1}) + \eta_t$$

where r_1 is the value of the autocorrelation parameter and η_t is taken to be normally distributed with zero mean and constant variance.

Checking assumption 2 in our study, we found that it is valid for the data with q adjustment or discarding 59-64 points, while it is violated for the original data. The results in table 5 show that substantial change in estimates may result by eliminating the autocorrelation in the study.

PREDICTABILITY

From previous sections, we learned that different models, different approaches of statistical analysis may give different estimates of maximum sustainable yield, fishing effort and catch per fishing effort at MSY. Naturally this brings up the question of which set of estimates should be adopted in setting the policy for fishing management. Judgemental decisions are usually the answer. In this section, we will suggest the predicability as a possible criterion.

The essential factor in fishery management is being able to know how the stock responds to the fishing efforts and natural environment. Thus it seems reasonable to assume that a model which can well describe the changes of the stock in future is better. Predictability of catches or catches/effort in future given the fishing efforts is therefore recommended as a criterion for selection among the different estimates (or different scenarios introduced to explain the existing data).

For this study, we have the new data available for 1972-1976. (Table 1B) Calculations of

$$H = \sum_{t=1972}^{1976} \{\text{observed } (\tfrac{C}{E})_t - \text{estimated } (\tfrac{C}{E})_t\}^2$$

for different models are given in Table 6. (For Schaefer Model, predictions are calculated using the solution of the Schaefer equation

$$\bar{B}_{i+1} = \frac{K}{r} \log\left[1 + (1 - e^{-rB_i/K})\ (r - qE_i)\ \frac{(e^{r-qE_{i+1}} - 1)}{(r - qE_{i+1})\ (1 - e^{-(r-qE_i)})}\right]$$

From Table 6, we learned that

(a) higher adjusted R^2 may not give better predicting power,

Table 4. Comparison of estimates obtained from Gulland
Model with different values of n

Models	MSY (10^3 tons)	Effort at MSY (10^3 hrs)	Catch/hr at MSY (tons/hr)	Adjusted R^2
n = 0	38.3	75.0	0.51	0.649
(C/E = 1.015 - 0.673x$10^{-5}\bar{E}_{0-0}$)				
n = 1	34.9	68.4	0.51	0.671
(C/E = 1.021 - 0.747x$10^{-5}\bar{E}_{0-1}$)				
n = 2	31.9	62.5	0.51	0.676
(C/E = 1.024 - 0.823x$10^{-5}\bar{E}_{0-2}$)				
n = 3	29.6	58.0	0.51	0.643
(C/E = 1.023 - 0.885x$10^{-5}\bar{E}_{0-3}$)				
n = 4	28.1	55.1	0.51	0.642
(C/E = 1.021 - 0.928x$10^{-5}\bar{E}_{0-4}$)				
n = 5	26.0	50.9	0.51	0.646
(C/E = 1.022 - 1.006x$10^{-5}\bar{E}_{0-5}$)				
n = 6	24.0	47.0	0.51	0.655
(C/E = 1.025 - 1.016x$10^{-5}\bar{E}_{0-6}$)				
n = 7	22.2	42.6	0.52	0.665
(C/E = 1.029 - 1.195x$10^{-5}\bar{E}_{0-7}$)				
n = 8	20.6	39.6	0.52	0.674
(C/E = 1.032 - 1.292x$10^{-5}\bar{E}_{0-8}$)				

Table 5.   Comparison of estimates obtained from models with
           different regression equations (with and without
           considering autocorrelation)

Models	MSY $(10^3$ tons)	Effort at MSY $(10^3$ hrs)	Catch/hr at MSY (tons/hr)
Schaefer			
$(Y_E/\bar{B} = 1.88 - 5.49 \times 10^{-5}\bar{B})$	16.1	25.2	0.64
$q = 3.72 \times 10^{-5}$			
$(Y_E/\bar{B} = 2.11 - 4.54 \times 10^{-5}\bar{B})$	24.6	28.6	0.86
$q = 3.72 \times 10^{-5}$			
$r_1 = 0.81$			
Gulland			
$(C/E = 0.62 + 4.41 \times 10^{-6}\bar{E}_{0-5})$			
$(C/E = 0.85 - 4.76 \times 10^{-6}\bar{E}_{0-5})$	38.0	89.3	0.43
$r_1 = 0.62$			

(b) Schaefer models have higher H values since the equation
    used for prediction is different from the equation used
    for regression, and

(c) estimates derived from models with better predicting
    power -- Schaefer ($q=1.4 \times 10^{-5}$), Gulland ($n=7$) and ($n=5$)
    p-p -- are very close.

Thus for ICNAF 3P redfish, we estimate that

1. MSY (10^3 tons)   : 22.2 $\sim$ 26.3
2. Effort at MSY (10^3 hrs)   : 39.5 $\sim$ 48.7
3. Catch/hr at MSY (tons/hr) : 0.52 $\sim$ 0.59

Table 6.   Comparison of H (predictability) for different
           Models

Models	Adjusted R^2	H(predictability) (1972 - 1976)
Schaefer		
$q = 3.4 \times 10^{-5}$	0.55	0.05
$(Y_E/\bar{B} = 3.521 - 11.07 \times 10^{-5}\bar{B})$		
$q = 2.1 \times 10^{-5}$	0.54	0.052
$(Y_E/\bar{B} = 2.118 - 4.12 \times 10^{-5}\bar{B})$		
$q = 1.4 \times 10^{-5}$	0.51	0.044
$(Y_E/\bar{B} = 1.363 - 1.77 \times 10^{-5}\bar{B})$		
Gulland		
$n = 2$	0.676	0.103
$(C/E = 1.024 - 0.823 \times 10^{-5}\bar{E}_{0-2})$		
$n = 5$	0.646	0.027
$(C/E = 1.022 - 1.006 \times 10^{-5}\bar{E}_{0-5})$		
$n = 7$	0.665	0.014
$(C/E = 1.029 - 1.195 \times 10^{-5}\bar{E}_{0-7})$		
$n = 5$ (P-P)	0.92	0.01
$(C/E = 1.176 - 1.49 \times 10^{-5}\bar{E}_{0-5})$		

STRATEGY IN QUOTA SETTING

After a set of estimates (or a model) is established, the next question in fishery management is what is the ideal quota to be set for the future years.   It is clear that the goal for quota setting is to maintain a maximum sustainable yield. However, the question of how to make the transition from current state into optimal equilibrium state still remains.

Suppose we adopt the Gulland (P-P) model, i.e.

$$(C/E)_i = a - b\bar{E}_{0-5} \qquad\qquad (5A)$$

with $a = 1.176$ and $b = 1.49 \times 10^{-5}$. Shall we set the quota for the future years as

(a) $C = MSY = 23.2 \times 10^3$ tons, or

(b) C_i determined by Eq (5A) with $E_i = $ Effort at MSY
$= 39.5 \times 10^3$ hrs, or

(c) something else.

Policy (a) may result in over fishing. Policy (b) will ensure the maximum sustainable yield after 5 years but this may not be the best transition strategy which gives optimal catch.   The best strategy for a policy taking n+5 years to transit into optimal equilibrium state can be stated as

Optimal total catch   =   max   (first year catch + optimal
with n+5 transition years   first year E
                                                total catch with (n-1)+5
                                                transition years)

Solving the above dynamic recursive equation, we obtain the best strategy for an $n \leq 6$ as

$$\text{first year } E(n+5) = \frac{\{(3 + \sum_{i=1}^{n} \frac{1}{2}i\,)a - b\sum_{i=1}^{5}E_{-i} - b\sum_{i=1}^{5}(i-1)E_{1-i}\}}{(n+1)b}$$

Optimal n can then be determined by comparing the average catch. Table 7 lists the 11 years average catches calculated for 5 to 11 years transition periods. It shows that the best strategy to have optimal catch is to take 10 years to make the transition from the current state into the optimal equilibrium state. The policy with 5 years transition period is the best strategy if one considers both optimal and minimum fluctuations of catches as the decision criteria. Of course the best strategy will be different for different current states. Similar arguments can be applied to Schaefer Model to obtain the best strategy in quota setting.

Table 7. Comparison of average catch for different
          strategies in quota setting

Number of years in transition period	Average catch per year over 11 years (tons)
5	22774
6	22773
7	22771
8	22775
9	22788
10	22827
11	22770

REFERENCES

1. L.S. Parsons and D.G. Parsons, An Evaluation of the Status of
    ICNAF Divisions 3P, 3O, and 3LN Redfish, Inter. Com. for
    the Northwest Atlantic Fisheries Research Bul.no.11 (1975)
    5-16.
2. M.B. Schaefer, Some aspects of the dynamics of populations
    important to the management of the commercial marine
    fisheries. Bull. Inter-Amer. Trop. Tuna Comm., 1 (1954) 25-56.
3. J.A. Gulland, A note on the population dynamics of the redfish,
    with special reference to the problem of age determination.
    Spec. Publ. Int. Comm. Northw. Atlant. Fish., 3 (1961)
    254-257.
4. W.E. Ricker, Computation and interpretation of biological
    statistics of fish populations, Bulletin 191, (Fisheries
    Research Board of Canada, 1975)

ACKNOWLEDGEMENTS

     This research is supported by Fisheries and Oceans, Government
of Canada. We would like to thank S.A. Akenhead and D. McKone of
the St. John's Biological Station for their encouragement and
fruitful discussions.

SYSTEMS DYNAMIC MODELS OF NEWFOUNDLAND FISHERIES

P.J. Amaria, A.H. Boone, A.D. Newbury and R.J. Whitaker

St. John's
Newfoundland, Canada

INTRODUCTION

For over three centuries, fisheries have been a very major
part of the life of Newfoundland people. The fishing operation up
until 25 years ago was mainly an inshore fishery, but in recent
years, fishing trawlers have been employed for the harvesting of
offshore marine renewable resources. The total annual Canadian
catch of fish during the past five years has ranged between 1 to
1.3 million metric tons. Newfoundland's share of the total
Canadian catch by weight has ranged between 0.3 to 0.5 million
metric tons. About 50% of the total Newfoundland catch is obtained
from distant waters, particularly from the area of the Grand Banks
in the North West Atlantic. The other 50% of the total catch
is obtained from inshore and coastal waters. The percentage of
employment in Newfoundland fish harvesting operation in relation
to inshore and offshore fishery is approximately ten to one. The
offshore fishery is highly capital intensive, whereas the inshore
fish harvesting operation is highly labour intensive.

The Newfoundland processing industry during the past 25 years
has also gone through major changes, particularly in relation to the
type of species and product forms produced, equipment, plant lay-
outs and facilities, however, very little change has been observed
in relation to the labour intensiveness in the fish processing
operation.

The supply of fish to the processing plants has been one of
the major factors in determining the production patterns. Normally
as inshore plant would buy fish from local fishermen mainly during
the months from April to November. These seasonal fish plants

therefore operate only part of the year, which makes it rather difficult to run at optimum efficiency throughout the year. The plant equipment and facilities are under utilized, manpower hiring, training and administrative activities, labour productivity suffers due to short term employment at the plant and the apportioned unit overhead costs are higher as compared to an all year plant operation.

Besides the problem of the supply of fish, there are other variables which the plant manager encounters, such as changes in prices of raw material, labour rates, equipment breakdowns, selling price fluctuations and market demands, etc.

Other requirements for an efficient plant operation are related to the managerial skills, mainly one of predicting or forecasting what actions need to be taken. These skills relate to the extent to which management is able to quickly and accurately analyse and establish the best course of action. Management should be able to forecast likely outcome of possible events that could occur during normal running of the plant. These changes in the variables could either be within the control of the management or they could be external variables upon which the management has no direct control, i.e., raw material prices, drop in the catching rate of inshore/ offshore fishery, etc. The management today continually ask questions such as:

- What is my profit margin per species?
- Do I have the necessary cash for the operation of the plant?
- What happens if market demand reduces?
- Should I process all fish or freeze part of it?
- What could be the projected cost of raw material, labour and overhead?
- What is the expected fish landings per species?

The above statements are only some of the questions which are continually being asked, for which adequate procedures have not been developed in the fish processing industry to provide quick and reliable answers. A manual calculating procedure could be established by the management which could take into account the time-variant and interactive nature of the variables of the processing operation. However, any such manual system is going to take a long time in calculating and establishing possible outcomes from a number of alternative courses of actions which management could take. In such an event, the delays in data processing and analysis may in many instances render the information useless since appropriate actions could not be taken at the right time. The efficiency of a management information system and analysis, is of paramount importance in the successful operation of the industry. It was therefore considered that a computer assisted systems dynamic models of fish harvesting and processing operations be developed so as to simulate on the computer various scenerios from which management

could observe the possible outcomes due to the changes of the
variables within the plant operation as well as observe the effects
of those variables which are outside the direct control of the
management, Amaria et al[1,2,3,4].

This study undertook to develop systems dynamic models of
Newfoundland inshore/offshore fishing industry.   The procedure of
systems dynamics was based on Forrester's works in industrial
dynamics, (Forrester 1961)[6,7].

INSHORE FISH HARVESTING MODEL

The model of an inshore fish harvesting operation was based on the
catching capability and environment factors.   No attempt was made
to relate environment factors with resource availability since
authors did not feel competent to consider this relationship in the
model.   The work done in this study should be considered as a
preliminary exercise.   The results of the study would need to be
verified in order to place confidence in the models.

The inshore fish harvesting model considered the following
parameters:
  A. Gear
      1.   Type of gear
      2.   Size of gear
      3.   Efficiency of gear-type
      4.   Number of gear usage
      5.   Lost or damaged gear
      6.   New gear purchased
      7.   Replacement for lost gear
      8.   Replacement for repaired gear
      9.   Species gear type weight coefficient

  B. Vessels
      1.   Type of vessels
      2.   Vessel-holding capacity
      3.   Vessels operating efficiency
      4.   Number of vessels

  C. Equipment
      1.   Type of equipment
      2.   Equipment sophistication.

  D. Environment
      1.   Number of personnel
      2.   Wind velocity
      3.   Sea-Ice condition and cover
      4.   Geographic location factor
      5.   Fishing competition

The above factors were considered for developing inshore fish harvesting models for cod, flat fish, (flounder and turbot), and pelagic fish, (herring, mackerel and caplin).   An inshore fishing area of Newfoundland was selected comprising of 25 communities in the districts of White Bay and Notre Dame Bay. The model coefficients for various parameters were established from an analysis of the actual data obtained for each of these communities from 1969 to 1976.   The historical data was available from FEDS*-2000 Data base system.   The inshore fish harvesting models were designed to establish estimates of fish catch per unit time. (lbs. per month per specie per gear type), based on various values of catching capability and environment factors. Computer simulation runs were carried out to observe the estimates of fish catch per unit time by selecting specific values for parameters of gear, vessels, equipment, personnel and environment conditions. The purpose of the simulation study was to observe sensitivity of the effect of each of the parameters of the inshore fish catching capability and environment factors on the estimated fish catch per unit time.

The concept used in this study of inshore fish harvesting sector is based on catching capability and environment factors. The catching capability include factors, (numbers, types, size, efficiency, etc.), such as for gear, vessels, equipment, and personnel. The environment factors include geographic location, fishing competition and the interaction of catching capability with ice conditions and cover, wind velocity and surrounding air temperature.

The interaction of carching capability with certain environment factors was based on the Bayesian probability theory.   The logic assumes that the probability of the catching capability being used for harvesting marine resource was dependent on certain states of the sea-ice condition and cover, wind velocity and air temperature. Bayesian theory was used to establish combination of the conditional probability of the catching capability being developed for harvesting of marine resource at any particular point in time. The conditional probability considered that the chance of catching capability being deployed was based on the combination of the sea state and atmospheric condition.   Historical data was obtained for each of the considered fishing locations from Ice Central Forecasting Centre and atmospheric weather stations, Ottawa.

The model logic considered that for marine resource to be harvested, three events should occur:

---

* Fisheries Experimental Data Base System developed by Department of Fisheries, Government of Canada, Ottawa.

1. Catching capability should be available for deployment.
2. Environment factors should be such that it would allow the catching capability to be deployed in the sea.
3. Availability of marine resource in the sea, for the time duration of the deployment of the catching capability.

The conditional probability coefficient of the environment component and the location factor, related to the availability of marine resource are combined with the catching capability function to establish the expected fish catch per unit time (Table 1)

Models were developed for cod, flounder, turbot, herring, mackerel and caplin. Simulation runs were performed on the computer to establish the expected fish catch per unit time for a certain number of changes to the model parameters. These changes related to gear, vessels, equipment, and personnel under certain environment conditions.

The model as developed for inshore fish harvesting operation is preliminary. Verification and validity of the model logic and procedure used is required. The validity could only be made by actually obtaining online data of a fishing community. This requires information of the amount of catching capability of number and types of gear, number and types of vessels used, etc., the actual environmental conditions of wind, air temperature and sea-ice conditions and cover, and comparing this information with the actual amount of fish landed onshore. Such a validity study could take at least 2 years. It is therefore stressed that the inshore fish harvesting model should not be used without validation.

OFFSHORE FISH HARVESTING AND PROCESSING OPERATION

The concept of computer application in the fishing industry was proposed in 1976 by Amaria et al during a workshop held under the auspices of the Canadian Institute of Food Science and Technology. Since then, a large fish processing company in Newfoundland has been involved in the application of a mini computer (P.D.P.-11), to develop forward operations plans for fish harvesting, processing and marketing. The computer planning system uses over 12000 variables in the establishment of operating strategies for 39 fishing trawlers and 8 fish processing plants.

The study undertook the feasibility of developing systems dynamic models for offshore harvesting operation involving 39 trawlers. The objective was to establish a desirable combination between trawler catching capacities, raw-material demand require- ments and production schedules for the processing of various species. The feasibility study established which species should be caught and how much increase in total catch was required per species for

Table 1.

Action	Wind Velocity (m.p.h.)	Air Temperature ($^{\circ}$F)	Sea-Ice Cover Condition	Probability of deployment of fishing gear
Definitely go out to deploy fishing gear	0-10	45°-over	0%-new ice	1.00
Possibly go out to deploy fishing gear	10-25	35°F-45°F	50%-new ice	0.50
Definitely not go out to deploy fishing gear	25-over	Less 35°F	100%-new ice or greater thickness	0.00

		Probability
Example:	If wind velocity = 7 mph	1.00
	Air temperature = 55°ᵣ	1.00
	Sea Ice Cover and	
	Condition = nil	1.00

The combined probability of fishermen going out to deploy their fishing gear will be: 1.00 x 1.00 x 1.00 = 1.00.

		Probability
Example:	If wind velocity = 15 mph	0.50
	Air Temperature = 40°F	0.50
	Sea Ice Cover and	
	Condition = nil	1.00

Then the combined probability of fishermen going out to deploy the fishing gear will be: 0.50 x 0.50 x 1.00 = 0.25.

		Probability
Example:	If wind velocity = 5 mph	1.00
	Air Temperature = 36°F	0.50
	Sea Ice Cover and	
	Condition = 100% multi year	0.00

Then the combined probability of fishermen going out to deploy the fishing gear will be: 1.00 x 0.50 x 0.00 = 0.00 (will not deploy).

each trawler for the desirable operating condition.   The combination
of species-mix caught per trawler was designed to increase the
overall efficiency of the offshore harvesting operation,Boone[5].

Further, the study also undertook the feasibility of developing systems dynamic models for eight fish processing operations
in Newfoundland.   The objective was to establish a combination
between raw material quality and quantity availability, plant
equipment and machinery capacities, market demands, labour
productivity, product yields and quality, species and pack-mix, raw
material price, labour rates, finished product price, inventory
holding and other operating conditions.   The feasibility study
established which species and pack should be processed by each of
the eight plants and how much increase in production was required for
the desirable operating condition.   This combination of species-pack
mix, labour production, equipment utilization was designed to increase
the overall efficiency of the eight combined plants.

The results of the study provided management with alternative
strategies for corporate planning for fish harvesting and processing
operations.   On the basis of these planning strategies, management
have since January 1979 been planning their operation of offshore
trawler fleets, operation of plants, equipment and machines, labour
force required and establishment of projected expenditures and
revenues.

The computer processed the data within a few hours, which if
compared to a manual operation would have taken weeks or months.
The accuracy of the computer output was dependent on the accuracy
of the input data, and sufficient care had to be taken in establishing reliable initial operating conditions.

The computer models related to the trawler harvesting, fillet
production, shrimp, breaded, other products and fishmeal operations
were developed over a period of months with close consultation with
the Systems Analyst, and other senior staff members of the company.
These models were built in a manner in which this company considered
the information of the variables which actually occurred within
their operation.

The trawler operation consisted of 39 trawlers operating from
five fish processing plants.   These trawlers operate all year round,
harvesting mainly ground fish such as cod, flounder, sole, turbot,
redfish and catfish, however, at particular times of the year some
of these trawlers are used for other types of fishing such as shrimp
or squid.

Forward planning for the trawler operation consisted primarily
of establishing the unit cost of trawler harvesting.   This unit
harvesting costs was established by dividing the total cost of

operating all 39 trawlers for a year by the total amount of ground
fish harvested. Using this constant unit harvesting cost, the new
profit or loss was then established for individual and groups of
trawlers assigned to a particular plant. Forward planning for
shrimp and squid follow the same procedure as ground fish.

Two basic computer models were developed for the trawler
operation: one for individual trawlers and another for groups of
trawlers which are similar in some respects. A sample of the input
data required for these models is given in Table 2.

Fillet production takes place in eight fish processing plants.
Five of these plants operate all year round, while the remaining
three are seasonal.

Forward planning for the fillet operation consists of: income
statements; unit costs of raw material, labour collection, polyphosphate packaging material, total direct, plant expenses and
total unit plant cost; total fillet production quantities and its
cost of production.

Two basic computer models were developed for fillet operation,
one for the fillet production and the other for fillet income statement. The fillet production model gives the unit cost of production
items, fillet production and its cost of production, while the
fillet income statement model gives sales value for species-packs
and the fillet income statement. A sample of the input data
required for the fillet operation are given in Table 3.

Forward planning for the precooked breaded operation consisted
of a possible of 24 finished products produced from each of the 7
fish species.

Forward planning for shrimp operation was divided into three
sections: processed at sea, offshore shrimp processed at plant and
inshore shrimp processed at plant.

Forward planning for processing of herring round, squid tubes,
herring fillets, squid round, offal, caplin, shredded, salmon,
mackerel and lumpfish roe were grouped together under the title
"Other Products." The computer model developed has the capacity to
handle up to ten different product forms. These products were
grouped by product name and products per plant.

The forward planning for fishmeal was designed for a combination
of eight fishmeal plants.

TABLE 2. TRAWLER INPUT DATA

COD (PERCENT)
DIEM RATE PER CREW MEMBER   PERCH (PERCENT)
CANADA PENSION PLAN PER PERIOD   YELLOWTAIL (PERCENT)
CREW INSURANCE PER PERIOD   FLOUNDER (PERCENT)
CREW STOCK - COD   GREYSOLE (PERCENT)
CREW STOCK - PERCH   CATFISH (PERCENT)
CREW STOCK - YELLOWTAIL   TURBOT (PERCENT)
CREW STOCK - FLOUNDER
CREW STOCK - GREYSOLE   NUMBER OF TRIPS/PERIOD
CREW STOCK - CATFISH   NO.OF CHARTERS/PERIOD
CREW STOCK - TURBOT   TYPE OF FISHING SELECTOR

CAPTAINS INCENTIVE OF LANDED VALUE
CAPTAINS COMMISSION OF LANDED VALUE
CREW GROSS STOCK OF LANDED VALUE
MATE & CHIEF COMMISSION OF CREW
BOSUM, SECOND & COOK COMMISSION OF CREW GROSS STOCK
ICERS COMMISSION OF CREW GROSS STOCK
WORKMEN'S COMPENSATION INC. OF TOTAL CREW SHARE
VACATION PAY OF TOTAL CREW SHARE
AVERAGE BREAK EVEN COST OF TRAWLER FISH
CHARTER COST PER TRIP
NUMBER OF SEA DAYS PER TRIP
TRAWLER LANDINGS PER TRIP
C.I.S. REFIT COST PER PERIOD
DEPRECIATION COST PER PERIOD
INTEREST COST PER PERIOD
LEASE COST PER PERIOD
INSURANCE COST PER PERIOD
ELECTRONIC RENTAL PER PERIOD
SHORE MANAGEMENT PER PERIOD

NUMBER OF CREW MEMBERS PER TRAWLER
TRANSPORTATION COST PER TRIP
GEAR AND EQUIPMENT COST PER TRIP
FISHING WARPS COST PER TRIP
FUEL COST PER TRIP
LUBE COST PER TRIP
MAINTENANCE COST PER TRIP
ICE AND ICE LABOUR COST PER TRIP
HOLD CLEANING COST PER TRIP
REFRIGERATION COST PER TRIP
MOVING TRAWLER COST PER TRIP
SOUNDER PAPER COST PER TRIP
SUNDRY COST PER TRIP

TABLE 3. PROCESSING INPUT DATA

SALES DATA

Selling Prices For All Species - Pack

Tariff Duties

Promotion
Commission
Frieght & Loading
Storage

SPECIES TYPES	PACK TYPES
Cod	Blocks
Flounder	Ones
Sole	Fives
Turbot	Layer 1-3 oz.
Catfish	Layer 3-5 oz.
Perch	Layer 5-8 oz.
Herring	Layer over 8 oz.
Mackerel	Layer skin-on
Squid	IQF 1-3 oz.
Shrimp	IQF 2-4 oz.
	IQF Over 4 oz.
	Twos
	Custom
	Whole and dressed

COST DATA

Unit cost of raw material for all species
Unit cost of packaging materials
Unit cost of polyphosphate
Hourly labour rates
Standard times for processing
Processing yields
Opening inventories
Plant expenses
Depreciation
Sundry expenses.

CONCLUSION

The work conducted so far is only a beginning in the understanding of the dynamic behaviour of the fishing industry.   Further
work is required in making the models as presented, more flexible.
The models could be refined so that they will accept multi-species
and multi-pack combinations with time varying and stochastic
characteristics of the species pack production rates, labour cost,
overheads, market demands, etc., thus making it suitable for
applications over a wide spectrum of the fish processing industry.

In building such large multi-variable stochastic models, a
word of caution at this stage would be appropriate.   The validity
and updating of the dynamic model is very important.   The validity
criteria for the model should not only include the correctness and
accuracy of the various values of the variables but also the truth
of the logical inter-relationships established for the various
parameters of the model with the real world situation, and these
iterative steps of validation and updating should be continuous.
It should be remembered that the level, sequence and time frame
work of the model objective and the inter-active variables are
also dynamic, since they also change with time, as new values and
understanding are realized by the management.   What it means is that
the design of the model should incorporate an adaptive component in
its logic.   Thus, the development and application of the dynamic
model should be truly dynamic.   This is the real challenge.

REFERENCES

1. P. Amaria, A. Boone, R. Whittaker & D. Newbury, System Dynamics
    of Newfoundland Fishery Operation, paper presented at the 4th.
    International Conference on Production Research, Tokyo (1977).
2. P. Amaria, A. Boone, R. Whittaker & D. Newbury, Working Paper
    on System Dynamics of Fishery Operation, unpublished paper
    Memorial University, St. John's (1977).
3. P. Amaria, R. Whittaker, D. Newbury & A. Boone, A System Dynamics
    as Applied to a Marine Renewable Resource unpublished paper
    presented to the "Teach-in-Session" at workship on System
    Dynamics of Newfoundland Fishery Operation Sponsored by CIFST
    (Newfoundland Section) St. John's October 1976.
4. P. Amaria, R. Whittaker, D. Newbury & A. Boone, A System Dynamics
    as Applied to a Fishery Operation unpublished paper presented
    at 21st. Atlantic Fisheries Technological Conference, Newport,
    Rhode Island, U.S.A. October 1976.
5. A.H. Boone, A Simulation Study of the Dynamic Behaviour of a New-
    foundland Seasonal Fish Processing Operation, unpublished
    M.Eng. Thesis, St. John's Memorial University of Newfoundland
    1978.
6. J.W. Forrester, Principles of Systems Wright-Allen Press 1968.

7. J.W. Forrester, <u>Industrial Dynamics</u>, Cambridge,Mass. M.I.T. Press (1961).
8. A.Douglas Newbury, <u>Materials Handling System to Increase Productivity of Inshore Fishery</u> unpublished M.Eng. Thesis St. John's, Memorial University of Newfoundland(1975)
9. T.B. Nickerson, <u>Systems Analysis in The Design of Operation of Fishing Systems</u>. Proceedings at conference on Automation and Mechanization in the fishing industry, Canadian Fisheries report no. 15, Montreal(1970)
10. A.L. Pugh III, <u>Dynamo II User's Manual</u>, Cambridge, Mass., The M.I.T. Press(1973).
11. B.J. Rothschild, <u>The System Approach in Fishery Management</u>, unpublished keynote address given at the "System Dynamics of Newfoundland Fishery Operation" workshop St. John's, Newfoundland (1976).

RESOURCE MANAGEMENT

EFFECTS OF ENVIRONMENTAL VARIABILITY AS THEY RELATE TO

FISHERIES MANAGEMENT

William J. Reed

Department of Mathematics
University of Victoria
P.O. Box 1700, Victoria, B.C., Canada V8W 2Y2

INTRODUCTION

Much of the theory on which fisheries management is based assumes that the dynamics of the exploited population are deterministic.   It is the purpose of this paper to investigate some of the consequences relevant to management of including a stochastic component, representing a small degree of environmental fluctuation in the basic dynamic model. Much of the discussion will be centred around the (deterministic) concept of maximum sustainable yield (MSY).

As starting point we take the discrete-time deterministic lumped-parameter model $x_{n+1} = f(x_n)$ and consider two stochastic generalizations of this model.   They are:

Model (A)

$$X_{n+1} = Z_n f(X_n) \qquad\qquad (1)$$

where the Z_n are independent, identically distributed (i.i.d.) random variables (r.vs.) with unit mean and with Z_n independent of X_n, X_{n-1}, ... etc.

Model (B)

$$X_{n+1} = f(X_n) + W_n X_n \qquad\qquad (2)$$

where the W_n are i.i.d. r.vs. with zero mean and with W_n independent of X_n, X_{n-1}...

In both models the stochastic process is Markovian and the deterministic model holds in terms of expected values, i.e.

$$E(X_{n+1}|X_n=x) = f(x) \tag{3}$$

It is assumed that the variances v_z and v_w of the r.vs. Z_n and W_n respectively are small relative to the average per capita growth rates of the population over the range of sizes considered.

SENSITIVITY OF A HARVESTED POPULATION TO ENVIRONMENTAL VARIABILITY

A number of authors have used either simulation models (Doubleday[4], Sissenwine[7] or analytic methods (Beddington and May[1], May et al.[5]) to study the effects of environmental stochasticity on population numbers and on yields. Summarizing the results of earlier papers May et al. state "... for any given level of environmental randomness, population numbers and yield exhibit greater fluctuation as harvest effort increases." Considering only the fluctuation in population numbers this statement is correct only in terms of relative fluctuation (coefficient of variation). For example in the model of Beddington and May[1] it follows directly (from their footnote 15) that the absolute fluctuation (variance) in population numbers decreases linearly with effort while the relative fluctuation increases.

Approximate expressions for the relative fluctuation(coefficient of variation of the steady-state distribution) of the population size N (immediately preceding the harvest), under constant effort harvesting for models (A) and (B) have been derived. For model (A)

$$CV_N \simeq \left(\frac{v_z}{1 - \left[\frac{\hat{x}f'(\hat{x})}{f(\hat{x})} \right]^2 (v_z+1)} \right)^{1/2} \tag{4}$$

and for model (B)

$$CV_N \simeq \left(\frac{v_w}{\left[\frac{f(\hat{x})}{\hat{x}} \right]^2 - \left[f'(\hat{x}) \right]^2 - v_w} \right)^{1/2} \tag{5}$$

where $\hat{x}$ is the deterministic equilibrium escapement level for the given rate of exploitation θ (i.e. $\hat{x}$ is the solution to $x = (1-\theta)f(x)$. The approximations are derived under the assumption that the variance σ_N^2 in population size is small so that powers of σ_N^2 and higher order moments can be ignored.

In the special case when the r.vs. Z_n in model (A) have a log. normal distribution and f is of the Cushing[3] form $f(x) = x^\beta$,* the exact steady-state distribution of N can be derived (Reed[6] p.298) and this provides a check on the approximation (4). It appears to be good for small values of β (large per capita growth rates) and small values of v_z. It can be seen from (4) and (5) that the way the relative fluctuation in population numbers responds to increased harvest rates depends on the precise form of the recruit-ment function f. This agrees with the findings of May et al.[5]. As a simple example consider the Beverton-Holt[2] form.

$$f(x) = \frac{(1+r)x}{1+rx} \qquad (6)$$

with r set at 1 and the "environmental" variance v_z and v_w set at 0.05. It can be seen in Fig. 1 that in model (B) the relative fluctuation CV_N decreases gradually with increases in the rate of exploitation θ up to the MSY rate ($\theta = .293$) and subsequently increases, while in model (A) CV_N shows a gradual increase over the whole range of harvest rates.

Approximate expressions for the absolute fluctuation (variance) σ_N^2 in a population size, are for model (A)

$$\sigma_N^2 \simeq \frac{v_z f^2(\hat{x})}{1 - \left[\dfrac{\hat{x}f'(\hat{x})}{f(\hat{x})}\right]^2 - \dfrac{\hat{x}^2}{f^2(\hat{x})}\left[f'(\hat{x})^2 + \dfrac{f^2(\hat{x})f''(\hat{x})}{f(\hat{x}) - \hat{x}f'(\hat{x})}\right]v_z} \qquad (7)$$

(7)

and for model (B)

$$\sigma_N^2 \simeq \frac{v_w f^2(\hat{x})}{\left[\dfrac{f(\hat{x})}{\hat{x}}\right]^2 - f'(\hat{x})^2 - v_w\left[1 + \dfrac{\hat{x}^2 f''(\hat{x})}{\{f(\hat{x}) - \hat{x}f'(\hat{x})\}}\right]}$$

---

* Here as in all other specific forms of the recruitment function used in this paper x is scaled so that the deterministic replace-ment stock size is x=1, i.e. so that f(1)=1.

Computations based on these formulae indicate that for most forms of f the absolute fluctuation in population numbers decreases as the harvest rate increases. Fig. 1 shows this for the example above.

STEADY-STATE YIELDS

The deterministic MSY is of size $f(x_1)-x_1$ where x_1 solves $f'(x)=1$. In a determimistic analysis policies of constant escapement at level x_1, constant effort with harvest rate $(f(x_1)-x_1)/f(x_1)$, and constant catch of size $f(x_1)-x_1$ all realize MSY. These policies differ only in the approach to equilibrium. On the other hand in a stochastic model the steady-states resulting from these policies in general differ. In evaluating the performance of a policy in a stochastic system two characteristics of importance are the long-run average yield and the variance in yield. We consider each of these in turn.

The following result concerning average yields is proved for model (A) in Reed[6] and the extension for model (B) is straight-forward.

> An upper bound on the long-run average yield of all policies in stochastic models (A) and (B) is the deterministic MSY. Furthermore this upper bound can be obtained only when the deterministic MSY escapement level x_1 is self-sustaining in the sense that
>
> $$\text{pr } \{X_{n+1} \geq x_1 \mid X_n = x_1 \} = 1$$
>
> and a policy of constant escapement at level x_1 is employed.

It follows that the deterministic constant-effort MSY policy will always result in an average yield less than MSY. This result could to some extent explain the consistent failure of many fisheries to produce yields as high as the predicted MSY, despite stringent management.

Average Yields of Deterministic MSY Policies

The extent to which the average yield of the deterministic MSY constant effort policy falls short of MSY depends on the precise form of the recruitment function of and the distribution of the r.vs. Z_n or W_n. Approximate formulae for the proportional drop Δ_y in yield are:

For model (A)

$$\Delta_y \simeq \frac{-x_1 f''(x_1)(1+\gamma)^2 v_z}{2\gamma^2(2+\gamma) - 2v_z\{\gamma+(1+\gamma)^2 f''(x_1)x_1\}} \tag{9}$$

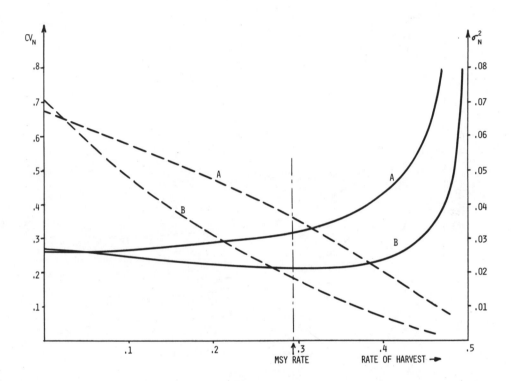

Fig. 1.   The solid curves marked A and B show the relative variation
          (coefficient of variation) of the population size as a
          function of harvest rate in models (A) and (B) for the
          recruitment function (6) with r=1 and $v_z=v_w=0.05$. The
          broken curves show the absolute variation (variance)
          in population size.

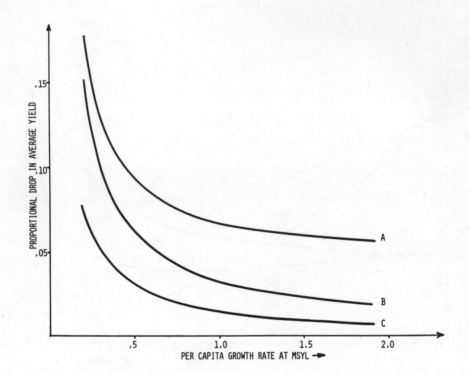

Fig. 2.   The proportional drop in average yield due to random
          variation of the constant-effort MSY policy in model (A)
          is shown as a function of the per capita growth rate γ
          at MSY, for the logistic form (A), the Beverton-Holt
          form (B) and the Cushing form (C) of the recruitment
          function. The environmental variance is v_z=0.05.

where $\gamma = (f(x_1) - x_1)/x_1$ is the per capita growth rate at the MSY level x_1.

For model (B)

$$\Delta_y \approx \frac{-x_1 f''(x_1) \ v_w}{2\gamma^2(2+\gamma) - 2v_w(\gamma + f''(x_1)x_1)} \qquad (10)$$

For the Cushing form $f(x) = x^{1/(1+\gamma)}$ in model (A) when the r.vs. Z_n have a log-normal distribution the exact formula (Reed[6] p.298) is $\Delta_y = 1 - (1+v_z)^{-(1+\gamma)/2\gamma(2+\gamma)}$ which agrees with the approximation (9) to the first order in v_z.

It can be seen from (9) and (10) that the proportional drop in average yield due to environmental fluctuation will increase with the degree of the fluctuation, at least when it is not too large. In Fig. 2 the approximate drop in average yield in model (A) is displayed for various values of   for the Beverton-Holt form (6), for the Cushing form and for the logistic form $f(x) =$ $(r+1)x(1-\frac{r}{r+1} x)$, with $v_z = 0.05$.   It can be seen that as the per capita growth rate γ increases so the effect of environmental variation diminishes.   Fig. 3 illustrates a similar trend for model (B) using the logistic form, the Pella-Tomlinson (see[5]) form $f(x) = (r+1)x(1-\frac{r}{r+1} x^m)$ with $m = 2$ and the Fox (see[5]) form $f(x) =$ $x(1-\gamma \log x)$. Also shown is the drop in average yield due to environmental fluctuation in Beddington and May's[1] continuous time logistic model with additive Gaussian white noise. From Figs. 2 and 3 it can be seen that the percentage drop in average yield of the constant-effort MSY policy is small when the environmental variance is small relative to the per capita growth rate at MSY.

For a policy of constant escapement at the MSY level x_1 when x_1 is not self-sustaining the amount by which the average yield falls short of MSY will depend in part on the probability, $pr\{X_{n+1} \leq x_1 | X_n = x_1\}$. For small values of this probability there will

be on average few years with no harvest and average yield will be close to MSY while the opposite is true for large values. For model (A) the above probability can be expressed $pr\{Z_n < \frac{1}{1+\gamma}\}$ and for model (B) as $pr\{W_n < -(1+\gamma)\}$. It follows that average yield will tend to decrease as environmental variability increases relative to the per capita growth rate at MSY.

<u>Variation in Yield of Constant-Effort Policies</u>

Since yield is proportional to population size for a constant-

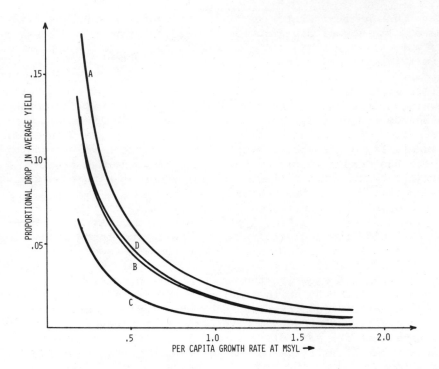

Fig. 3. This is similar to Fig. 2 for model (B) with the Pella-
         Tomlinson (m=2) (A), the logistic (B) and the Fox (C)
         forms of the recruitment function, v_z=0.05. Also shown
         (D) is the proportional drop in yield in the continuous
         time logistic model of Beddington and May[1].

effort policy the coefficient of variation of yield is the same as that of population numbers. Thus (4) and (5) are approximate formulae for the relative variation in yield.   Similarly approximate expressions for absolute variation (variance, σ_y^2) in yield can be obtained from (7) and (8)

$$\sigma_y^2 = \left[\frac{f(\hat{x}) - \hat{x}}{f(\hat{x})}\right]^2 \sigma_N^2 \tag{11}$$

Computations based on this formula seem to indicate that for most forms of f the absolute variation in yield increases with the harvest rate at least up to the MSY rate. The precise behaviour of the relative variation however depends on the form of f and the way noise is introduced into the model.   Fig. 4 shows the absolute and relative variation in yield for the Beverton-Holt form (6) with parameter values $r=1$, $v_z = v_w = 0.05$.   It can be seen that the relative variation in yield stays fairly constant at least up to the MSY rate even though the absolute variation increases. In this particular example harvesting at the deterministic MSY rate would neither cause excessive fluctuation in population numbers (Fig. 1) nor in yield. For lower growth rates or higher environmental variance this would not necessarily be true.

Variation in Yield of Constant MSY Policy

    For the constant escapement MSY policy when the MSY level x_1 is self-sustaining the yield variance in model (A) is $v_z f^2(x_1)$ and in model (B) is $v_w x_1^2$.   Using (11), the ratio in variances in yield of the constant effort and constant escapement MSY policies can be computed.   For example if v_w and v_z are small so that second and higher order terms can be ignored in (11), this ratio is approximately $\gamma/(2+\gamma)$.   This is less than 0.5 for $\gamma < 2$.   Thus for small degrees of environmental variance the constant effort MSY policy can result in significantly lower variance in yield than the constant escapement MSY policy. For example in model (A) with the Beverton-Holt form (6) with $r=1$ and $v_z=0.05$ the variance of constant effort is 21% of that of constant escapement (from (11)). The average yield of constant effort however is 94% of that of constant escapement. Or put another way the constant escapement MSY policy results in an average yield of 0.172 units with a standard deviation of 0.131 while the constant-effort MSY policy results in an average yield of 0.161 with a standard deviation of 0.060.

    The question of trade-offs between average yield and variation in yield is discussed in May et al.[5].   The analysis in this paper would indicate that in many cases the reduction in yield resulting from using a constant-effort MSY policy rather than the "optimal" constant-escapement MSY policy would be more than offset by the

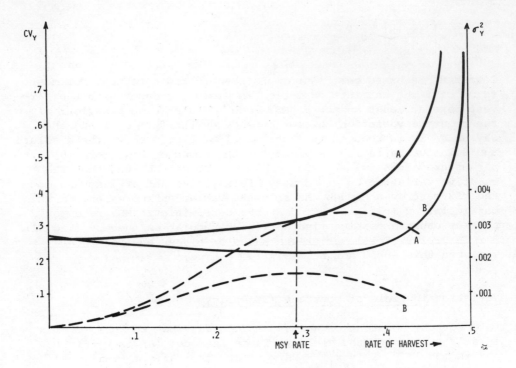

Fig. 4. The solid curves marked A and B represent the relative
variation in yield in models (A) and (B) for various
rates of exploitation for the recruitment function (6)
with r=1 and $v_z=v_w$=0.05. The broken curves show the absolute variation in yield.

reduction in variance and the elimination of zero yields.

CONCLUSIONS

Computations based on the approximate formulae used in this article indicate the following conclusions.

1. For most recruitment functions the absolute variation in population numbers decreases with increased harvest effort at least up to the MSY rate. However since the mean population size also decreases this is to be expected. The relative fluctuation gives a better indication of the population's response to increased harvesting. Whether this increases or decreases as the rate of harvesting approaches the MSY rate appears to depend on the form of the recruitment function and the way the random effects are introduced in the model. However for over-exploitation beyond the MSY rate the relative fluctuation eventually increases in all cases.

2. The average yield of constant-effort MSY policies is always less than MSY. The extent of this shortfall depends on the degree of environmental variation and the particularities of the recruitment function. In most cases it appears to be small especially when the environmental variance is small in comparison with the per capita growth rate at MSY.

3. When the deterministic MSY population level is self-sustaining long-run average yield is maximized by a policy of constant escapement at that level.

4. The variance in yield of the constant-effort MSY policy appears to be much less than that of the constant-escapement MSY policy, when the MSY level is self-sustaining. The drop in average yield that results from using constant-effort rather than constant escapement appears in most cases to be small compared with the corresponding drop in variance.

5. The variance in yield of constant-effort harvesting appears to increase with the rate of harvesting at least up to the MSY rate. The relative variation depends on the particularities of the model, but in most cases would appear to increase with harvest rate. This supports the conclusions of May et al.[5] in their continuous time analysis.

REFERENCES

1. J.R. Beddington and R.M. May. Harvesting natural populations
    in a randomly fluctuating environment. Science 197: 463-465
    (1977).
2. R.J Beverton and S.J. Holt. The dynamics of exploited fish
    populations. Fishery Invest. Ser. 2, London (1957).
3. P.H. Cushing. The dependence of recruitment on parent stock in
    different groups of fishes. J. Cons. Int. Explor. Mer. 33(3):
    340-362 (1971)
4. W.G. Doubleday. Environmental fluctuations and fisheries
    management. I.C.N.A.F. Sel. Pap. 1 (1976).
5. R.M. May, J.R. Beddington, J.W. Horwood and J.G. Shepherd.
    Exploiting natural populations in an uncertain world. Math.
    Biosc. 42: 219-252 (1978).
6. W.J. Reed. The steady-state of a stochastic harvesting model.
    Math. Biosc. 41: 273-307 (1978).
7. M.P. Sissenwine. The effects of environmental fluctuations on
    a hypothetical fishery. I.C.N.A.F. Sel. Pap. 2 (1977).

ABUNDANCE ESTIMATION IN A FEEDBACK CONTROL SYSTEM APPLIED TO

THE MANAGEMENT OF A COMMERCIAL SALMON FISHERY

P.R. Mundy and O.A. Mathisen

Fisheries Research Institute
College of Fisheries
University of Washington, Seattle, Washington 98195

INTRODUCTION

The economic benefits to the industry and the biological advantages to management from increasing the accuracy of yield estimates of commercial salmon fisheries were rigorously analyzed more than ten years ago (Mathews[1]).   However, the means to achieve the requisite accuracy and precision in forecasting and harvest control have not been forthcoming. It is possible that the lack of progress is due to the nature of the methods pursued and not to a lack of effort.

Conventional yield models (e.g. Schaefer, Beverton-Holt, Ricker) and refined derivations describe numbers of biomass of populations under equilibrium conditions. The residual variance obtained in applying such yield models to exploited fish populations is frequently prohibitively large (Sissenwine[2]).   In the case of sockeye salmon (Oncorhynchus nerka Walbaum) of Bristol Bay, Alaska, for example, the average absolute error in an estimation of annual adult abundance is 50% (1962-1975).   The lack of an accurate yield forecast has caused the spawning escapement to be excessive or deficient relative to fixed escapement goals by an average of 2.8 million individuals between 1959-1978.   At 1978 price levels, U.S. $1.50/kg, and at a long-term average weight of 2.4 kg/fish, an average spawning excess means a loss of U.S. $10.1 million in landed value, while an average spawning deficit means a loss of 4.8 million individuals in future returns at the average observed rate of 1.7 recruits/spawner.   While it is arguable whether or not the lack of accuracy in long-range yield forecasts is due to inadequate data, the fact of the poor predictive value of such applications presently remains.

As an alternative, commercial fisheries harvest management,
particularly contemporary salmon harvest management, is turning
to intraseason abundance estimation techniques in an attempt to
increase the accuracy of harvest control (e.g. Walters and
Buckingham[3]).   That timely estimates of daily and total seasonal
adult abundance are critical, but elusive, components of objective
solutions to salmon management problems was noted by two publicat-
ions which characterized salmon harvest management as problems in
linear programming (Rothschild and Balsiger[4]) and stochastic
dynamic programming (Lord[5]).   The intraseason abundance estimation
program developed for Bristol Bay sockeye salmon harvest control
is intended to support such objective solutions by estimating daily
and total adult abundance in an accurate and timely fashion.

The procedures to be described were developed during the
1976-1977 sockeye salmon fishing seasons in a cooperative effort
between the Commercial Fish Division of the Alaska Department of
Fish and Game (ADFG), the salmon industry and the Fisheries
Research Institute (FRI) of the University of Washington.   As an
average, Bristol Bay sockeye salmon spend one or two years in
nursery lakes (blackened areas, Fig. 1), two or three years in the
Gulf of Alaska and North Pacific Ocean and then return at the age
of 4-6 years to the tributaries of the nursery lakes to spawn and die.
An excellent fisheries data base which is accessible by computer,
as first developed at FRI (Knudsen, Poe and Mathisen[6]), makes this
fishery well suited to the study of the abundance estimation
processes.   The data base is now maintained by ADFG.

The relation of intraseason yield estimation to the whole
fishing system is shown in Fig. 2.   The interior loop defined by
boxes 1.3 - 1.6 is a process wherein the controlled variable is
the catch and the primary decision criterion is the abundance
estimate.   Yields labeled 1° - $(N-1)^\circ$ are functions of abundance
at various life history stages (e.g. parental stock-recruitment;
fry-recruitment...., marine immature-recruitment).   The
N° yield estimate is based on observations of adult abundance from
within and without the area of harvest of those same adults, so it
is a function of a "recruitment-recruitment" relation. Parenthetic-
ally it may be remarked that while the system developed here is
completely general and applicable to any life stage, the variance
involved increases with distance in time from prediction to event.

DESCRIPTION OF STUDY AREA

The first intraseason information on the abundance migration
is obtained from a gill net sampling program[1] which spans the
(1) All sample catch, commercial catch and estimates and counts of
    escapement are collected by programs of the Division of
    Commercial Fisheries, ADFG, Anchorage.

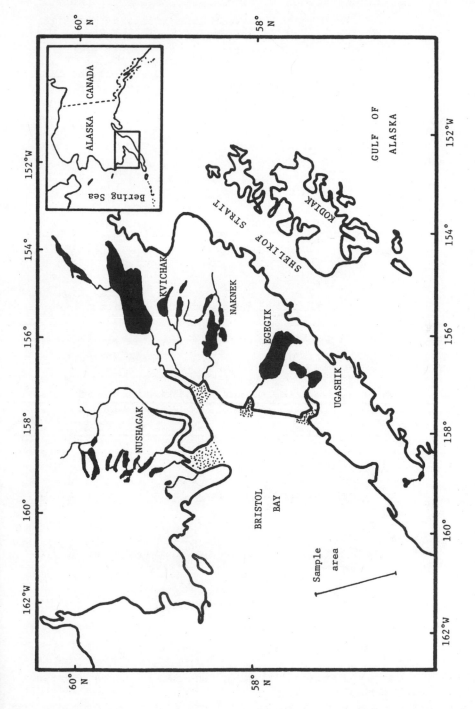

Figure 1. The Bristol Bay area, showing the sockeye salmon nursery lakes (blackened areas), commercial fishing districts (strippled areas) and sample transect line.

entire width of the migration off Port Miller, Alaska (Paulus [7]
and Straty[8]).   The second set of observations for the purposes
of this study is collected from the catches of the commercial gill
net fleet which is restricted to operate in the estuaries of the
major river system. (Fig. 1)   The numbers escaping the fishery
to spawn are estimated daily by gill net sampling inside the fishing
boundary lines in the rivers, by aerial surveys, or by both methods,
weather permitting.   Counts of escapement to all major watersheds
are eventually obtained by observers on towers who take advantage of
the clear water near the outlets of the nursery lakes.   The
evolution of this management system to its present form is given
by Mathisen.[9]

METHODS

     In the following list of symbols for key variables, the sub-
script t denotes "value assumed by the variable on day t", "BB"
refers to the fishing districts of Bristol Bay, and "S" refers to
the transect line off Port Moller.

$\hat{C}_t$ estimated total annual sample catch per unit effort
           (CPUE), S

c_t cumulative daily CPUE, S.

c_t' daily CPUE, S

f_t error sum of squares of estimated cumulative inshore
           abundance (BB) with respect to observed inshore
           returns.

g_t error sum of squares of estimated cumulative proportion
           of total sample catch with respect to expected cumulat-
           ive proportion of total sample catch

k upper bound in days of observed variability in the mean
           of the expected cumulative proportion of total sample
           catch

ℓ lower bound in days of observed variability in average
           transit time between S and BB

$\Delta\ell'$ the average transit time in days between S and BB in
           the current season

M_t total number of individuals in sample catch

$\hat{N}_t$ estimated total annual adult abundance

$\hat{n}_t$ estimated daily abundance, S

$\hat{q}_t$ estimated catchability as a function of inshore
           abundance

$\hat{q}_t'$ estimated catchability as a function of average weight
           of current migration

R_t total cumulative inshore returns (catch and escapement)

$\bar{t}$ mean of the distribution of expected proportion of
total sample catch

$\Delta t'$ the number of days by which the current migration is
early or late with respect to average migratory timing

u   upper bound in days of observed variability in average
transit time between S and BB

$\bar{w}_t$ average weight of inshore returns, BB

Y_t expected cumulative proportion of catch or total
abundance

Total abundance estimation

Total abundance estimates are made on each day of sampling in
two stages; total annual sample catch per unit effort (CPUE) in
the Port Moller area (called the same area hereafter) is estimated
and incorporated into a catch equation;

$$\hat{C}_t = \hat{q}_t \hat{N}_t \tag{1}$$

where $\hat{q}_t$ is estimated below and $\hat{C}_t$ is estimated by minimizing the
error function, g_t

$$g_t = \sum_{i=1}^{t+\Delta t} \left(\frac{c_{i+\Delta t}}{\hat{C}_t} - Y_i\right)^2 \quad \text{for } t > k \tag{2}$$

with respect to $\hat{C}_t$ while t assumes values on the open interval (–k,k)
in increments of unit time. Time shifts of ± 6 days, relative to
average migratory timing, have been observed in Bristol Bay sockeye salmon (1956-1975), hence k = 6 for unit time equal to one day.

At any value of Δt the minimum of g_t with respect to $\hat{C}_t$ is
found by setting the first derivative of g_t with respect to $\hat{C}_t$
equal to zero and solving for $\hat{C}_t$;

$$\hat{C}_t(\Delta t) = \sum_{i=1}^{t+\Delta t} c_{i+\Delta t}^2 \left/ \sum_{i=1}^{t+\Delta t} Y_i c_{i+\Delta t} \right. \quad \text{for } t > k \tag{3}$$

That value which minimizes g_t on (–k, k), $\Delta t'$, is the number of days
by which the present migration is early or late with respect to
average migratory timing. For the purposes of eqn. (1),

$$\hat{C}_t(\Delta t') = \hat{C}_t \quad \text{for } t > k.$$

For $t \leq k$, $\hat{C}_t(\Delta t=0) = \hat{C}_t$. The initial estimate, $\hat{C}_0$, is $\hat{q}_0 \hat{N}_0$

{from (1)} where q_O and N_O are defined below.

When C_t has been determined, the total annual adult abundance is found by solving (1) for $\hat{N}_t$;

$$\hat{N}_t \;=\; \hat{c}_t/\hat{q}_t \tag{4}$$

$\hat{N}_O = (N-1)^O$ yield (Fig. 2); the most recent estimate of annual adult abundance prior to the season.

Catchability as a function of total cumulative inshore returns, $\hat{q}_t$, is found by minimizing the error function, f_t, with respect to $\hat{q}_t$;

$$f_t \;=\; \sum_{i=1}^{t+\Delta l} \left(\frac{c_{i+\Delta l}}{\hat{q}_t} - R_i\right)^2 \quad \text{for } t > l \tag{5}$$

while Δl assumes values on the open interval (l,u) in increments of unit time.   The observed variation in average transit times, the difference between the mean of the migratory time density in the sample area and the mean of the migratory time density in the fishing districts (1968-1973; 1975) is 4-9 days; $\hat{l} = 4$; $u = 9$.   At any value of Δl the minimum of f_t with respect to $\hat{q}_t$ is found, after the procedure developed for (3), as

$$\hat{q}_t(\Delta l) \;=\; \sum_{i=1}^{t+\Delta l} c_{i+\Delta l}^2 \;/\; \sum_{i=1}^{t+\Delta l} R_i c_{i+\Delta l} \quad \text{for } t > l \tag{6}$$

That value which minimizes f_t on (l,u), $\Delta l'$, is the average transit time in days between the sample and inshore areas (Fig. 4). For the purposes of eqn. (1), $\hat{q}_t(\Delta l') = \hat{q}_t$ for $t > l$ and $\hat{q}_t = \hat{q}_O$ for $t \leq l$. The initial catchability, $\hat{q}_O$, is arbitrarily set at the observed catchability for the preceding year.

$\hat{q}_O$ = 1/18917 (1975) for 1976 season

$\hat{q}_O$ = 1/16500 (1976) for 1977 season.

The continuous cumulative probability density

$$Y(t) \;=\; 1/\{1 + \exp[-(a+bt)]\} \tag{7}$$

is used to approximate values of Y_t.   The parameters (a,b) were estimated from the cumulative daily CPUE as a proportion of total seasonal CPUE (dependent variable) of the 1975 sample data (Table 1).   For the 1976 season the estimation of parameters was

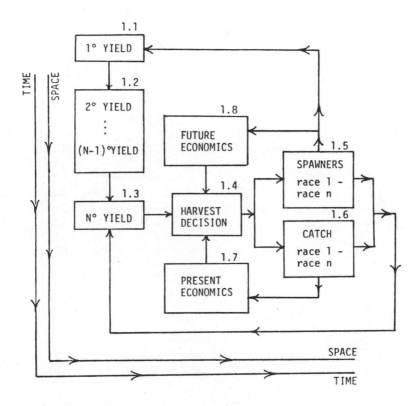

Figure 2. Schematic of the commercial fisheries management process

Table 1.   Port Moller sample catch data, 1975. (1) month,
            day; (2) daily number of sockeye caught/100
            fathoms/hour (CPUE); (3) cumulative daily CPUE;
            (4) cumulative daily CPUE as proportion of total
            yearly cpue.

(1)	(2)	(3)	(4)
613	1.0	1.0	.001
614	2.9	3.9	.003
615	4.7	8.6	.007
616	5.7	14.3	.011
617	22.5	36.8	.029
618	37.7	74.5	.058
619	11.2	85.7	.067
620	8.6	94.3	.074
621	17.6	111.9	.088
622	27.6	139.5	.110
623	37.6	177.1	.139
624	47.6	224.7	.176
625	57.6	282.3	.221
626	60.0	342.3	.269
627	261.7	604.0	.474
628	60.0*	664.0	.521
629	60.0*	724.0	.568
630	60.0*	784.0	.615
701	62.3	846.3	.664
702	60.5	906.8	.711
703	112.5	1019.3	.800
704	57.4	1076.7	.845
705	99.9	1176.6	.923
706	33.1	1209.7	.949
707	18.0	1227.7	.963
708	47.0	1274.7	1.000

* Interpolation, Alaska Department of Fish and
  Game. Missing CPUE = CPUE of 626

accomplished by linear regression (a = -5.6911; b = 0.37071) and,us-
ing the 1975 data again, by nonlinear least squares methods for
the 1977 season (a = -5.452; b = 0.3067).   The mean of the sample
area migratory time density, $\bar{t}$, is -(a/b). The mean of the inshore
migratory time density is $\bar{t} + \Delta \ell'$.

    In the event that total inshore returns cannot be tabulated
quickly enough to serve the estimation procedure of (5), an

alternative catchability function $\hat{q}'_t$ was constructed by describing the equation of a line connecting the data points (7.1, 7000) (5.5, 18917) which are the average weights of inshore returns and inverses of catchability in 1973 and 1975 respectively.

$$1/\hat{q}' \quad = \quad -3583.3(\bar{W}) + 38.982.6 \tag{8}$$

and

$$1/\hat{q}'_t \quad = \quad -3583.3(\bar{W}_t) + 38982.6 \tag{8A}$$

where $\bar{W}_t = \dfrac{1}{M_t} \displaystyle\sum_{i}^{M_t} w_i$ is the average weight of inshore returns as of day t.   Weights are in pounds (0.454 kg).

Daily abundance estimation

The estimation of daily abundance for inshore areas requires an estimate of abundance in the sample area to be projected ahead in time by $\Delta \ell$ days.

$$c'_t \quad = \quad \hat{q}_t \hat{n}_t \tag{9}$$

or

$$c'_t \quad = \quad \hat{q}'_t \hat{n}_t \tag{9A}$$

and

$$\hat{n}_t \quad = \quad \hat{r}'_{t+\Delta} \tag{10}$$

During the 1976 season, relation (9) was used exclusively while in 1977 it became necessary to use relation (9A) on occasion (see Results).

Summary of methods

Regardless of the form of equations (1) - (10), the nature of the process of daily and annual abundance estimation can be described in simple language. An estimate of abundance at time 1 is checked for accuracy against a count of abundance at time 2. If the estimate is found to be lacking, the procedure of estimation is corrected and the estimate is renewed until a desired level of accuracy is achieved. If the key variables of such a system (e.g. $\hat{q}_t$, $\Delta t'$, $\Delta \ell'$) can be accurately described by average performance, estimates of the key variables, based on observations at time 1, will be increasingly accurate. Note that the method could also be described as a Kalman filter.

DISCUSSION OF METHODS

The methods presented are based on the theory that expected proportion of total abundance by unit time of the migration in any year will share a common probability density with all other migrations, e.g. $N(t, \mu, \sigma)$, but with a differing mean and variance each year. In the anadromous salmonids so far critically studied, the interannual commonality of the form of the migratory time densities at a fixed spatial reference frame is indicated by the genetic heritability of migratory timing (Mundy[10]).

Cumulative abundance by date, divided by expected cumulative proportion by date, yields an estimate of total abundance (Walters and Buckingham[3]), but account must be taken of differences between the means and variance of the expected (eqn. 7) and observed $(c_t/\hat{C})$ probability densities.

To synchronize the observed mean, $\bar{t}$, to the expected $(-a/b)$, the minimization of eqn. 2 is necessary. That the variances may be unequal is recognized by relying upon $\hat{N}_{\bar{t}}$ as the best estimate of total abundance, since the common probability densities, $Y'(t)$, with the same mean but unequal variances will be unequal except at the inflection points. Hence, if the means of expected and observed densities are synchronized but the variances are unequal, the estimate $\hat{N}_t (t \neq \bar{t})$ will be less accurate than $\hat{N}_{\bar{t}}$.

The form of the probability density is arbitrary since it has no special biological significance. The logistic, $Y(t)$, was chosen since it is an appropriate model of the migratory timing of Bristol Bay sockeye (see Results) and since it had proven adequate for another salmonid population (Mathisen and Berg[11]).

RESULTS

The 1976 season was characterized by a clearly bimodal time density with an unusually late mean in the fishing districts. July 10 (average = July 5; $\Delta t' = 5$) (Tables 2 and 3). The $\hat{q}_0 = 1/18917$ was replaced by $\hat{q}_5 = 1/16500$ with a $\Delta \ell' = 8$. The "final" date estimate for management purposes, $\bar{t}$, was July 2. $\hat{N}_{\bar{t}} = 11.3$ million for an actual return of 11.4 million. The progress of $\hat{N}_t$ is plotted in Figure 3 (Table 3). Although the time density was bimodal, the fit of $c_t/\hat{C}_t$ to $\hat{Y}_t$ (parameters estimated by linear regression) was adequate to support retaining the model (Fig. 4). $\hat{q}_{\bar{t}} = \hat{q}_5$ was less than $\hat{q}_0$ because each fish was about 0.227 kg heavier than its 1975 counterpart, on the average. Even so, the change in catchability was fairly quickly detected and corrected so that the daily abundance estimates were reasonably accurate (Figure 5).

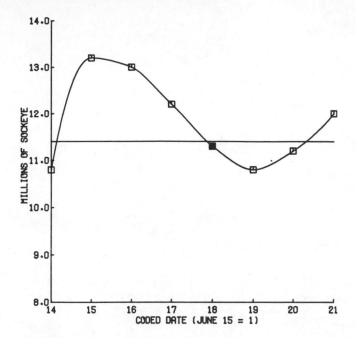

Figure 3. $\hat{N}_t$(blocks) in relation to actual total annual abundance (center line) for the 1976 season. Solid block locates $\bar{t}$.

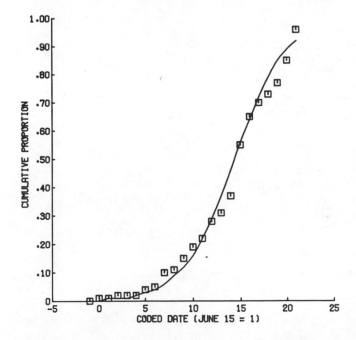

Figure 4. Fit of $c_t/\hat{C}_{\bar{t}}$, blocks, to Y(t), line, (1976) where t $\bar{t}$ = July 2, 1976.

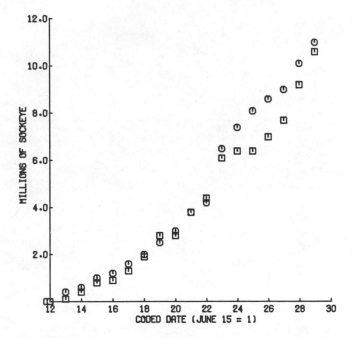

Figure 5. Fit of cumulative estimated daily abundance in the sample
          area moved ahead 8 days (octagons) to observed cumulative
          daily abundance (squares), 1976 season.

Table 2.   Port Moller sample catch data, 1976. (1) month,
            day; (2) daily number of sockeye caught/100
            fathoms/hour (cpue); (3) cumulative daily cpue;
            (4) cumulative cpue as proportion of total
            estimated cpue; (5) $Y(t) = 1/(1 + \exp(-5.6911 + 0.37071t)))$; (6) T.

(1)	(2)	(3)	(4)	(5)	(6)
613	1.9	1.9	.00323	.00336	1
614	2.8	4.7	.00799	.00486	2
615	3.0	7.7	.01309	.00704	3
616	1.5	9.2	.01564	.01016	4
617	0.0	9.2	.01564	.01465	5
618	1.5	10.7	.01819	.02109	6
619	16.3	27.0	.04591	.03027	7
620	3.9	30.9	.05255	.04326	8
621	26.7	57.6	.09795	.06148	9
622	10.0	67.6	.11496	.08669	10
623	21.4	89.0	.15136	.12089	11
624	23.4	112.4	.19115	.16612	12
625	18.8	131.2	.22312	.22398	13
626	33.1	164.3	.27942	.29486	14
627	19.2	183.5	.31207	.37726	15
628	31.7	215.2	.36598	.46742	16
629	111.6	326.8	.55578	.55977	17
630	57.5	384.3	.65357	.64815	18
701	28.1	412.4	.70136	.72743	19
702	18.6	431.0	.73299	.79451	20
703	21.5	452.5	.76955	.84852	21
704	48.9	501.4	.85280	.89029	22
705	62.5	563.9	.95913	.92161	23

The 1977 season exhibited an average Bristol Bay time density
with a mean of July 5 in the fishing districts ($\Delta t' = 0$), although
the observed time density was attenuated to the right of the mean.
The major abnormality was the size of the fish, 3.2 kg as a Bay-
wide average.   It was apparent soon after the start of sampling
that $q_o = 1/16,500$ (1976) was much too small.   A $q_t' = 1/8235$ was
used after sample day 3 (June 14) to estimate daily inshore
abundance.   Catchability as a function of inshore returns fluctuated
widely but $q_{15} = 1/10,100$ was used between June 26 – July 4 for total
abundance estimation with $\Delta \ell' = 7$.   The wide fluctuation in
catchability was noted in time to produce reasonably accurate
daily abundance estimates (Table 4). The sampling program was

Table 3.   1976 abundance estimation program, Bristol Bay
            sockeye salmon   (1) predicted total annual
            abundance (millions); (2) predicted cumulative
            daily abundance moved ahead 8 days; (3) observed
            cumulative daily abundance.

MO/DAY	(1)	(2)	(3)*
627		0.4x	0.1
628	10.8	0.6x	0.4
629	13.2	1.0x	0.8
630	13.0	1.2x	0.9
701	12.2	1.6	1.3
702	11.3	2.0	1.9
703	10.8	2.5	2.8
704	11.2	3.0	2.8
705	12.0	3.8	3.8
706		4.2	4.4
707		6.5	6.1
708		7.4	6.4
709		8.1	6.4
710		8.6	7.0
711		9.0	7.7
712		10.1	9.2
713		11.0	10.6

  * Preliminary figures, Alaska Department of
    Fish and Game, King Salmon, Alaska.

  x Back calculation, not prediction.

effectively eliminated from June 27 onward by poor weather so the
total abundance estimate did not change thereafter (Table 4) until
a last fit of projected inshore returns to observed yielded a
$\hat{q}$ = 1/8035.   The $N_{\bar{t}}$ = 10.5 million occurred on June 28 (July 5 -
$\Delta \ell '$) as compared to an actual return of 10.9 million.   The fit of
c_t/C to $Y(t)$ (Table 5) was quite good which reflected the average
nature of migratory timing in 1977.

DISCUSSION OF RESULTS

    The deviation from average migratory behaviour in 1976 and the
large size of the average fish in 1977 posed severe tests of the
estimation process. As implied by the form of the expected
cumulative migratory time density, $Y(t)$, the average migration has
a bell shaped time density.   In 1976 $\hat{N}_{\bar{t}}$ had an absolute error of
1.0 percent relative to N and at $\bar{t}$ (July 2) less than 30 percent of

Table 4. 1977 abundance estimation program, Bristol Bay
          sockeye salmon. All figures given in millions
          of fish. (1) predicted total annual abundance;
          (2) predicted cumulative daily abundance; (3)
          observed cumulative daily abundance.

MO/DAY	(1)	(2)	(3)*
620	9.1	0.1	0.1
621	9.4	0.2	0.1
622	9.9	0.2	0.3
623	9.6	0.3	0.5
624	8.7	0.5	0.7
625	9.2	0.7	0.9
626	10.5	1.0	1.5
627	10.5A	1.5	-
628	10.5A	1.7	-
629	10.5A	2.2	-
630	10.5A	2.6	-
701	10.5A	3.1	3.1
702	10.5A	3.5	-
703	10.5A	4.2	-
704	8.36	4.9	4.2
705		5.5	-
706		6.0	5.6
707		6.5	-
708		6.9	-
709		7.2	-
710		7.4	7.0

* Preliminary figure, Alaska Department of Fish
  and Game, King Salmon, Alaska.
- Figure not available during season.
A Sampling program not operational.

the migration had reached the fishing districts (see Fig. 5).   The
migration was late in 1976 by five days which is close to the
maximum observed at k = 6 days.   Thus, the estimation procedure
has been proven robust with respect to deviations in both the form
of the time density and its mean.

The average weight of fish in 1977 was about 40 percent greater
than the long term average. That $N_{\bar{t}}$ had an error of 4 percent
relative to N is an indication of·the critical nature of the scaling
factor, catchability, in determining the overall accuracy of
estimation.   At $\bar{t}$ (June 28) only about 20 percent of the migration
had reached the fishing districts.   The ability of the system to

Table 5.   Port Moller sample catch data, 1977.
(1) month, day;   (2) daily number of
sockeye caught/100 fathoms/hour (CPUE);
(3) cumulative daily CPUE; (4) cumulative
CPUE as proportion of total estimated
CPUE;   (5) Y(t) = 1/(1 + exp(-(-5.452 +
0.3067t))).

(1)	(2)	(3)	(4)	(5)
612	5.1	5.1	.001	.012
613	11.8	16.9	.017	.016
614	2.6*	19.5	.019	.021
615	6.4*	25.9	.025	.028
616	10.1	36.0	.035	.037
617	26.2	62.2	.061	.050
618	20.6	82.8	.082	.066
619	39.1	121.9	.120	.087
620	54.3	176.2	.174	.114
621	26.5	202.7	.200	.148
622	59.4	262.1	.258	.190
623	55.1	317.2	.312	.240
624	56.9	374.1	.368	.299
625	49.7	423.8	.417	.366
626	70.3	494.1	.487	.438
627	87.7A	581.8	.573	.512
628	71.4A	653.2	.643	.586
629	65.0A	718.2	.707	.656
630	57.0A	775.2	.764	.720
701	48.4A	823.6	.811	.776
702	39.8A	863.4	.851	.824
703	19.5	882.9	.870	

* Interpolation $t_n = (t_{n-1} + t_{n+2})/2$.

A Interpolation $t_n = Y(t_n) \times \hat{C}$.

calculate new values of critical variables on the basis of intraseason observations gives it a strong advantage over long range
forecasts.

    The experience gained in 1976 and 1977 was not to be tested
in 1978, since poor weather eliminated the sampling program, data
from the commercial fishery and escapement counts could be combined
with an inshore time density to produce timely total abundance
estimates (after Walters and Buckingham) if the mean and variance

of that time density could be estimated in advance of the fishery. Estimates of the median date of migration at the escapement counting tower (Nishiyama[12]) and in the fishing districts (Burgner[13]) as a function of gonadal maturity (Nishiyama) and ambient air temperatures (Burgner) in the Spring preceding the migration have given promise that characteristization of the migratory time density before the migration is possible. Such a forecast of the time density would also benefit the accuracy of abundance estimates from the sampling program.

REFERENCES

1. S.B. Mathews, Economic Evaluation of Forecasts of Sockeye Salmon Runs to Bristol Bay, Alaska. Ph.D. Dissertation, Univ. Washington (1967). Also, FAO, United Nations, Rome (1971).

2. M.P. Sissenwine, Is MSY an Adequate Foundation for Optimum Yield? Fisheries 3: 22-42 (1978).

3. C.J. Walters and S. Buckingham, A Control System for Intra-season Salmon Management, Proc. of a Workshop on Salmon Management, Intern Inst. Applied Systems Analysis, Schloss Laxenburg, Austria (1975).

4. B.J. Rothschild and J.W. Balsiger, A Linear Programming Solution to Salmon Management, Fish. Bull. 69: 117-140 (1971).

5. G.E. Lord, Characterization of the Optimum Data Acquisition and Management of a Salmon Fishery as a Stochastic Dynamic Program, Fish. Bull. 41: 1029-1037 (1973).

6. B. Knudsen, P.H. Poe and O.A. Mathisen, A Data File for the Bristol Bay Sockeye Salmon Fishery. Univ. Washington Fish. Res. Inst. Circ. 72-9 (1972).

7. R.D. Paulus, Bristol Bay Test Fishing Program, Anadromous Fish Act Proj. Completion Rep. AFC-31, Alaska Dept. Fish and Game, Div. Comm. Fish., Juneau, Alaska (1973).

8. R.R. Straty, The Migratory Pattern of Adult Sockeye Salmon (Oncorhynchus nerka) in Bristol Bay as Related to the Distribution of Their Home-river Waters, Ph.D. Dissertation, Oregon State Univ., Corvallis (1969).

9. O.A. Mathisen, Sockeye Salmon Management in Bristol Bay, Univ. of Alaska Sea Grant Publication (in press) (1979).

10. P.R. Mundy, A Quantitative Measure of Migratory Timing Illustrated by Application to the Management of Commercial Salmon Fisheries, Ph.D. Dissertation, Univ. Washington, Seattle (1979).

11. O.A. Mathisen and M. Berg, Growth Rates of the Char in the Vardnes River, Troms, Northern Norway, Report No. 48, Inst. Freshwater Res., Drottingholm (1968).

12. T. Nishiyama, Food Energy Requirements of Bristol Bay Sockeye Salmon (Oncorhynchus nerka Walbaum) During the Last Marine Life Stage, Spec. Vol.: 289-320, Res. Inst. No Pacific

          Fisheries, Faculty of Fisheries, Hikkaido Univ., Hakodate,
          Japan (1977).
13.   R.L. Burgner, Some Features of Ocean Migrations and Timing of
          Pacific Salmon, Contrib. No. 488, College of Fisheries,
          Univ. Washington (1978).

ACKNOWLEDGEMENTS

     The following persons contributed information or ideas;
Vincent Gallucci, Charles Meacham, Jr., Randall Peterman and Carl
Walters. Partial financial support of the research was provided by
the Alaskan salmon industry and the Division of Commercial
Fisheries, Alaska Department of Fish and Game.

Mr P.R. Mundy is currently at Department of Oceanography, Old
Dominion University, Norfolk, Virginia 23508.

A FISHERY-OILSPILL INTERACTION MODEL: SIMULATED CONSEQUENCES

OF A BLOWOUT

Mark Reed,   Malcolm L. Spaulding, Peter Cornillon

Department of Ocean Engineering
University of Rhode Island
Kingston, Rhode Island, 02881, U.S.A.

INTRODUCTION

Recent experience on both sides of the North Atlantic Ocean
has clearly demonstrated the degree to which coastal ecological
resources may be damaged by large oil spills.   In the case of the
Amoco Cadiz accident, for example, it appears that a majority of
the impacts can be assessed through direct observations.   In a
situation such as the Bravo blowout, however, the ensuing ecological
consequences are much less clear. This paper describes the present
state of an ongoing effort to construct a model system capable of
estimating the impact of petroleum spills on commercial fisheries
in offshore as well as coastal water. Ultimate system evolution
is aimed at a broader holistic concept, including a species-
specific ecosystem model to facilitate food web pollutant concentration studies, and a variety of impact sources and mechanisms operating
alone or in concert.   Possible future applications include, for
example, power plant entrainment, and effects relating to municipal
sewage outfalls, dredge waste disposal, heavy metals and other
water-borne toxic substances.   At present the system is composed
of a unit fish stock population model with spatial as well as
temporal dimensions, a model to simulate ocean transport processes,
and an oil spill behaviour and fates model.

MODEL SYSTEM DESCRIPTION

Development of the system has been facilitated by the selection
of a specific site and fish stock for preliminary applications.
Georges Bank, off the northeastern coast of the United States, was
chosen for its high productivity, its proximity to major oil supply

routes, and the presence of several areas within which offshore
oil exploration activities are projected (Fig. 1). The cod stock
on the Bank is simulated because, although small in international
terms, it is of considerable local economic significance. Furthermore, tagging experiments (Wise[1]) suggest little immigration or
emigration, whereas oceanographic observations as summarized by
Colton[2] support the belief that the stock is self-perpetuating.
These considerations permit a relatively valid application of unit
stock analysis (Cushing[3]). The presence of simpler fisheries models
of this particular stock (Silliman[4]; Walters[5]) and of more
sophisticated models of other cod fisheries (Hannesson[6]; Clayden[7];
Garrod[8]) supplies valuable sources for comparison of results and
methods. Whereas the ocean transport model is also site-specific,
the oil spill fates model is programmed more generally. As shown
in Fig. 2, the transport model supplies input to both the population
model and the oil spill model, and output from the latter is used
as input to the larval mortality sector of the fishery model.

Fishery Model

The population model cycles on a daily timestep when operating
in conjunction with other system components. The adult fishery
model is constructed along well established lines (e.g. Beverton
et al.,[9]; Walters[5]), whereas the egg and larval stages have been
modeled using techniques drawn from other areas of mathematical
ecology (Lotka[10]; Pielou[11]). The adults increase in length and
weight according to published growth equations (Pentilla et al.[12]
Heyerdahl et al.[13]) for the Georges Bank stock. The ith adult
age-class (i = 1,17) is subject to the usual mortality equation
following first order kinetics, and fishing mortality is assumed
linearly proportional to standardized fishing effort (Gulland[14]).

The fecundity of cod appears well correlated with weight
(Daan[15]; Bigelow et al.[16]; Nikolskii[17]; Gulland[18]). Temporal
distribution of spawning cod on Georges Bank remains something of
an enigma, it being uncertain whether cod spawning activity is
triggered by day length, temperature, or other environmental
factors (Wise[19]), and is modelled following subjective evaluations
by Bigelow et al[16], Colton et al.[20,21] and Walford[22].

The model takes on spatial dimensions through the simulation
of the spawning activity. On any given day of the year, a percentage
of the sexually mature females of each age class are assumed to
release their eggs. The total number of eggs available for
spawning on that day is divided into a set number of groups, or
sub-cohorts, each of which is deposited at random within the
modeled spawning area (Fig. 1). Thus each particle in the
transport model represents one sub-cohort of organisms in the
population model. As simulated time progresses, each particle is

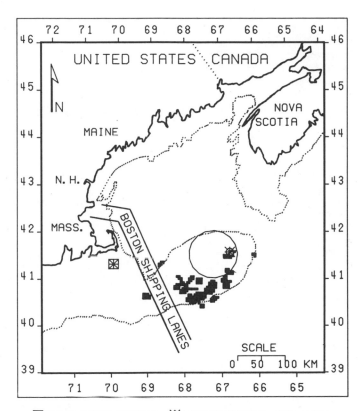

 WIND DATA SOURCE ✳ HYPOTHETICAL OIL WELL

Figure 1.   Map of the model area, showing spawning grounds
             (large circle), oil exploration tracts as of
             January, 1979 (dark rectangular regions), commercial
             shipping lanes, wind data source, and the location
             of the hypothetical blowout.

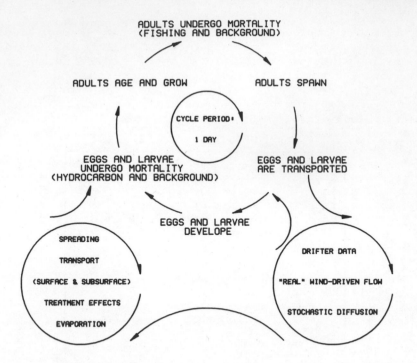

Figure 2.   Model system cycles, showing processes simulated
             in each subsector, and indicating interrelationships.

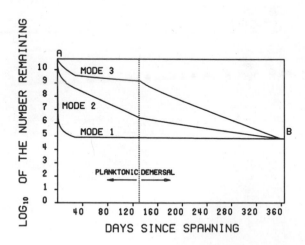

Figure 3.   Three possible piecewise continuous mortality curves
             for the first year of life.

transported, and represents a diminishing number of developing eggs, larvae, or post-larvae.   When a sub-cohort reaches the demersal stage, it loses its explicit representation in space, while of course retaining its presence in the population model.

There are five first year stages represented in the model. These are an egg stage, three planktonic larval and postlarval stages, and a demersal juvenile stage. The theoretical basis for the equations governing mortality during these developmental periods derives from Lotka[10], who employed the concept of a Taylor expansion about some hypothetical equilibrium value.   Density dependent mortality of an intragroup nature is included to represent competition among individuals of the group for limited resources (e.g. food), or mortality due to predation by some species external to the explicit model formulations (e.g. herring).   Density dependence between groups may result from competion for shared resources, or a parasitic (e.g. cannibalistic) relation.   There is considerable evidence of cannibalism among Gadoids (see for example Daan[15], or Bigelow et al.[16]), a theoretically plausible explanation for population control (Cushing et al.[23]).   Recent evidence reported by Ellertsen et al.[24] indicates that cannibalism amongst larval and postlarval Gadoids may contribute significantly to mortality, perhaps outweighing effects of food density.   This process is not yet explicitly incorporated here.   Egg and larval development rates follow information from Dannevig[31], Wise[19], Bigelow et al.[16] and Ellertsen et al.[32].

Sensitivity to the toxic action of hydrocarbons appears to result from a complex interaction among several variables (Malins[25]).   Along the same lines as suggested by Andersen et al.[26], efforts have been made to utilize published bioassay data (Kunhold[27,28]; Wilson[29]) to partition density independent mortality parameters of cod during planktonic stages (Reed et al.[30]). Although the model is designed to account for the variables relevant to toxicity, difficulties in interpretation of experimental results as well as uncertainty levels elsewhere in the present system (e.g. oil entrainment rates, details of ocean transport estimates) have prompted the adoption of a threshold assumption: any eggs or larvae entering an area in which the concentration exceeds their respective threshold values are deleted from the system.

The pre-recruit mortality rates used in the simulations reported here are the result of earlier efforts with steepest descent gradient search techniques using the sum of the squared differences between modeled and observed yield as the object function to be minimized (Reed et. al.[30]). Because of the complex topology of the surface, unique minima were not found, and three parameter sets were arbitrarily selected from several that were approximately equivalent. These mortality rates are graphically depicted in Fig. 3. Mode 1 corresponds to a case in which the demersal density dependent

parameters are essentially zero, suggesting a situation in which
the benthic food supply is non-limiting, and predation of older
cod on young-of-the-year is negligible. The United States National
Marine Fisheries Service stomach content data for the Georges Bank
cod does not reflect significant amounts of cannibalism at the
present time (Ray Bowman, personal communication), so that this
type of curve, with year-class strength completely determined during
the first 50 days of life, may be appropriate for this stock. Mode 2
contains significant mortality during both planktonic and benthic
stages.   The curve shows approximately 95% mortality before hatch-
ing, 6% per day during the larval and post-larval planktonic stages,
and 2% per day during the demersal fry period. These values correspond
reasonably well to estimates by Daan[34], Jones[35], and an average
estimate for the entire first year of 3% per day by Gulland[18].
Evidence reported by Daan[15] suggests that this type of regime, in
which cannibalistic mortality is a significant contributing factor
to population control, may better represent dynamics of North Sea
cod.   Mode 3, with 90% survival to the end of the planktonic stage,
is considered an unlikely upper limit to the family of curves
available with the model.   The significance of these mortality
modes in oil spill impacts modeling will be discussed below.

     Although the average error of the predicted yield is on the
order of 20% of the average catch, the model demonstrates a
reasonable ability to follow the trends in the record (r=0.81).
In addition, the simulated catch profiles agree quite well with
recent management analyses for this stock (Serchuck et al.[36]).
The model does not predict catch size with great accuracy, but the
impact estimates required here will be the differences between
predicted catches with and without a spill.   While the actual
predictions may be off by 20% of more, to the extent that the model
follows the yield trends, these differences may only be in error by
a few percent. The relative impact estimates themselves may be more
accurate than the absolute yield estimates from which they derive.

The Ocean Transport Model

     The marine transport model consists of three superimposed flow
processes, each representing transport energy spanning a different
portion of the energy spectrum.   Although such separation neglects
nonlinearities in the real system, this approach is adopted in
lieu of a three dimensional numerical modeling effort, and appears
to yield useful and credible approximations, as discussed below.

     The low frequency seasonal circulation patterns have been
deduced from oceanic drifter data gathered over approximately a
decade (Bumpus et al.[37]).   The resulting flow fields supply
estimates of net transports on temporal scales of weeks and months,
arising as the residual of tidal, meteorological, and river inflow

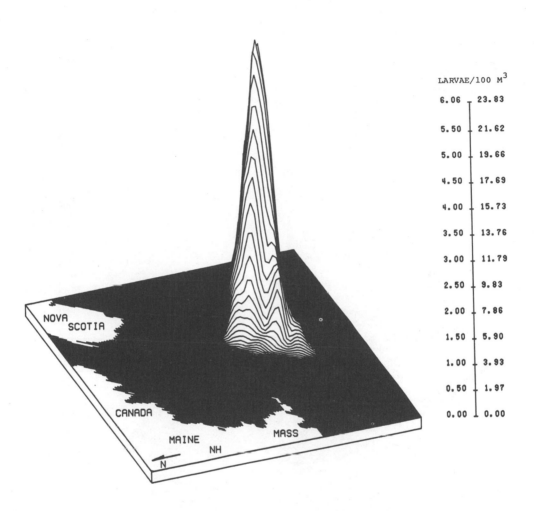

Figure 4. Observed larval code distribution for December, 1974.
(The larval concentrations south of Massachusetts
are associated with a separate stock, not modeled here).

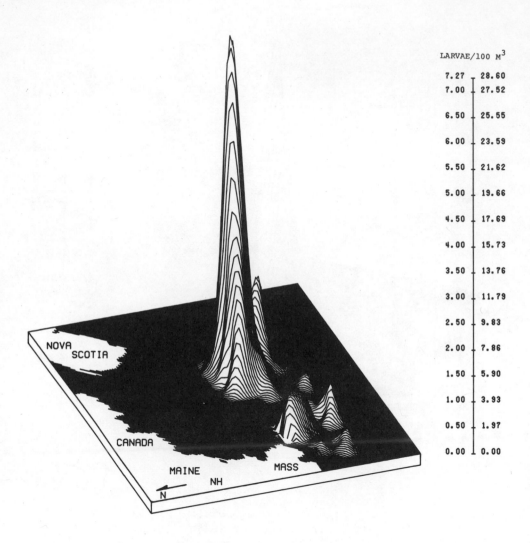

Figure 5.   Modeled larval cod distribution for 15 December.

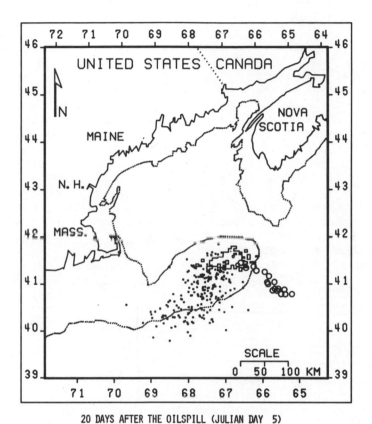

20 DAYS AFTER THE OILSPILL (JULIAN DAY 5)

Figure 6.   Simulated egg, larval, and oil distributions
              following the blowout. The points are clusters of
              developing eggs and larvae, the rectangular outline is
              the subsurface 50 part per billion (ppb) contour,
              and the circles are individual surface spillets
              emanating from the blowout.

events, as well as pressure forcing on the ocean boundaries of the
modeled area.

Intermediate frequencies of transport energy are input through
wind forcing on time scales of a few hours to a few days. Decoupling
of this input from the long-term effects it probably imparts to the
real system (Csanady[38]) is achieved by first passing a digitized
record of wind data for the time period of interest through a high
pass filter with a corner frequency of one cycle every ten days.
In this way only energy associated with trends occurring over time
periods less than this is retained in the record.
This filtered wind record is then used as input to a simple two-
dimensional model of the wind-driven surface layer (Reed[39]).

The third transport process in the model is dispersion.   This
is assumed to result from energy input at frequencies greater than
and equal to that of the tides, as well as energy cascaded into
this frequency range through dissipation processes in the real
system, and is modeled as a random walk process.

Preliminary validation of the combined codfish population-
ocean transport model has been performed using larval fish
distribution data reported by Colton et al.[20].   As shown in
Figs. 4 and 5, the model reproduces the observed patterns with
reasonable accuracy. The peak values are comparable, as are the
total numbers of ichthyoplankton.   (The smaller peaks south of
Massachusetts in the observed data sets represent a separate
spawning stock of cod, not modeled here.)

THE OIL SPILL FATES MODEL

The petroleum fates model has been structured in a modular
fashion to allow for easy implementation of subroutines dealing
with the various physical, chemical, and biological processes
determining the fate of a spilled oil .   The algorithms
used were generally selected for their simplicity, allowing for a
rapid identification of those areas in need of the most improvement.

Drifting of the surface slick is simulated by the vector
addition of mean ocean and wind induced currents. Spreading is
accomplished using the three region Fay[41] formalism with the
modification that interfacial tension as well as the volume of
oil in a spillet are allowed to change with time. Evaporation of
oil from the surface is performed by numerically solving the
evaporation equations given by Wang et al.[42]   Entrainment of the
spilled oil is performed using the algorithm presented by Audunson
et al.[43], modified by an exponentially decaying function of time
to simulate the effects of weathering. A three dimensional mass
transport equation solved by a Monte Carlo technique is used to

simulate the subsurface transport of the dispersed oil.   A detailed
presentation of this modeling procedure can be found in Spaulding[44].
Because of the rapidity with which changes occur in the early
stages of an oilspill, the model operates on a variable timestep,
and employs a double grid spatial system.   One grid is fixed in
space (an Eulerian reference frame), whereas the second expands
and translates to follow the spill (a Lagrangian frame), thereby
maintaining high resolution throughout the simulation.   The
interested reader is referred to Cornillon et al.[45,46] for a more
detailed discussion of these and other processes included in the
model.

RESULTS

The darkened rectangular regions of Fig. 1 locate about 4000
sq.km. of submarine land designated for lease by the United States
government to private oil companies for purposes of hydrocarbon
exploration as of January 1979.   A technical study (Offshore Oil
Task Group[47]) investigating the risks associated with actual oil
production in the area estimates that, for an oil find on the order
of one million barrels per day, something over a million gallons
per year will be spilled.   In the present paper some potential
implications of this type of activity are pursued.   Specifically,
a major oil well blowout is simulated, along with its impact on
the commercial cod fishery of the region, under several different
sets of assumptions.

This analysis assumes explicitly that the demersal O-group
stage and all adults will avoid areas contaminated by hydrocarbons,
that the stock will not suffer from bioaccumulation, and that the
adult feeding and spawning patterns will remain intact.   Thus only
the planktonic stages are modeled explicitly in space as well as
time, and it is through losses at these stages that losses in yield
are ultimately realized.

Using median values as estimated by Audunson et. al.[43] for the
North Sea, the blowout simulated here has a duration of 30 days,
and an output rate of 10 thousand metric tons per day. (This is
about twice the release rate but less than tenth the duration of the
presently ongoing Ixtoc I blowout in Mexico's Campeche Bay.)   Fig. 1
shows the location of the hypothetical oil well.   Fig. 6 depicts
a snapshot of the events subsequent to the blowout, showing both
surface and subsurface oil distributions.   The oil slick at the
surface dissappears from the biologically sensitive area shortly
after the end of the blowout on January 15, whereas subsurface
concentrations in excess of 50 parts per billion (ppb) continue to be
found more than two months later.   This phenomenon was also observed
following the wreck of the Argo Merchant oil tanker in December 1976.
Boehm et al.[48] reported dissolved hydrocarbons over a large area of
Georges Bank as averaging 44 ppb in February of 1977, versus only

TABLE I

Synopsis of Blowout Impact Estimates
(Millions of Kilograms Yield Reduction)

MORTALITY MODE*

		MODE 1	MODE 2	MODE 3
OIL TOXICITY THRESHOLD (PPB)	50	6.5	2.2	0.9
	100	5.2	1.9	0.7
	1000	1.6	0.5	0.1

*Reference Fig. 3 .

11 ppb in May, and less than 1 ppb the following winter.

Because of the uncertainties associated with the pre-recruit
mortality regime, and because the choice of a specific threshold
value for hydrocarbon-induced mortality is debatable, it was
decided to investigate the importance of these parameters to the
model results.   As shown in Table 1, mortality mode 1 (Fig. 3)
results in higher impacts at all thresholds.   That modes 2 and 3 are
associated with lower impact estimates reflects the fact that there
is virtually no density dependence in Mode 1 following the planktonic stages during which the eggs and larvae are susceptible to
the spill.

Because the simulated oil well is situated within the spawning
area, and because the majority of the biological impacts result
from eggs rising under contaminated water, the relative impacts of
the three toxicity thresholds do not scale with the total areas
within their respective contours. The integrated space-time
extents of the 50, 100, and 1000 ppb contours for this simulated
event scale as 1 : 0.7 : 0.04 respectively, whereas the areas over
the spawning grounds scale as 1 : 0.8 : 0.15, which more closely
parallels the ratios of the numbers in any given column of Table 1.

CONCLUSION

Responsible environmental impact analysis requires explicit
expression of the limitations of the approach used.   Although the
authors believe this sytem to be one of the more advanced single-
species fishery impact models extant, uncertainties introduced at
every level render the reliability of the results questionable.
Depending on the choice of toxicity threshold and mortality regime
alone, the estimated cost of this hypothetical event to the fishery
lies between $0.1 and $6 million U.S. (1979). Although this represents
the"best guess" presently available, it falls short of most legal
needs, and leaves something to be desired for even a fundamental
cost-benefit analysis.   Although some elements of this model
system, for example the ocean transport section, can be greatly
improved using existing knowledge and techniques, certain other
aspects, notably the pre-recruit mortality formulations, are and will
remain limiting until a sufficient data base exists for their proper
validation.   The advent of larger and faster computers will provide
the capability for incorporating explicit feeding behaviour models
at the multi-trophic, inter-species level of sophistication with
advanced hydrodynamic transport models. The real usefulness of
these endeavours will be limited by the data base behind them.   In
the case at hand specifically, this implies that any serious national
or international programme for advanced marine fisheries resource
management must include a concerted ichthyoplankton data acquisition
programme, systematic in both space and time. Although expensive, it

appears clear that basic progress in this field will be largely
dependent on the planning and implementation of such programmes.

REFERENCES

1. J.P. Wise, Cod Groups in the New England Area, Fish.Bull.
      63(1): 189-203 (1963).
2. J.B. Colton, Jr., History of Oceanography in the Offshore
      Waters of the Gulf of Maine, U.S. Fish. and Wildlife Serv.
      Spec. Sci. Rep., Fisheries 496 (1964).
3. D.H. Cushing, Marine Ecology and Fisheries Cambridge U. Press,
      London, (1975).
4. R. Silliman, Analog Computer Models of Fish Populations, Fish.
      Bull. Bus. Comm. Fish. 66(1): 31-46 (1966).
5. C.J. Walters, A Generalized Computer Simulation Model for Fish
      Population Studies, Trans. Amer. Fish. Soc. 98: 505-512 (1969).
6. R. Hannesson, Fishery Dynamics: A North Atlantic Cod Fishery,
      Con. J. Econ. 8: 151-173 (1975).
7. A.D. Clayden, Simulation of the Changes in Abundance of the Cod
      (Gadus morhua L.) and the Distribution of Fishing in the North
      Atlantic, Min. Agric., Fish., and Food (G.B.), Fish. Invest.
      Series II, Vol.27, No.1 (1972).
8. D.J. Garrod, Population Dynamics of the Arcto-Norwegian Cod,
      J. Fish. Res. Bd. Can. 24(1): 145-190 (1967).
9. R.E. Beverton and S.J. Holt, On the Dynamics of Exploited Fish
      Populations, Ministry Agric., Fish., and Food (G.B.), in
      Fish. Invest. Ser. II, Vol. 19 (1957).
10. A.J. Lotka, Elements of Physical Biology, Williams and Wilkins
      Co., Baltimore, (1925).
11. E.C. Pielou, An Introduction to Mathematical Ecology, Wiley
      Interscience, New York (1969).
12. J.A. Pentilla and V.M. Gifford, Growth and Mortality Rates for
      Cod from the Georges Bank and Gulf of Maine Areas, ICNAF,
      Res. Bull. 12: 29-36 (1976).
13. E. Heyerdahl and P. Wood, Data Summaries Prepared for Sub-area
      5 Cod Assessment, ICNAF Working Paper 76/IV/70 (1976).
14. D.J. Garrod, The North Atlantic Cod, pp. 216-239 in Gulland,J.A.,
      ed., Fish Population Dynamics, John Wiley and Sons, New York
      (1977).
15. N. Daan, Results of Dutch Cod Egg Surveys in the North Sea in
      1973, ICES CM 1973/F:37 (1973).
16. H.B. Bigelow and W.C. Schroeder, Fishes of the Gulf of Maine,
      Fish. Bull. Fish. Wild. Serv., U.S. Government Printing
      Office, Washington, D.C., (1953).
17. G.V. Nilolskii, Theory of Fish Population Dynamics, Oliver and
      Boyd, Edinburgh (1965).
18. J.A. Gulland, Survival of the Youngest Stages of Fish, and Its
      Relation to Year Class Strength, ICNAF Spec. Pub. 6: 363-372
      (1965).

19. J.P. Wise, A Synopsis of Biological Data on Cod, Gadus morhua, Fish. Div., Bio Branch, FAO, Rome (1961).

20. J.B. Colton, Jr. and J.M. St. Onge, Distribution of Fish Eggs and Larvae in Continental Shelf Waters, Nova Scotia to Long Island, Serial Atlas of the Marine Environment, Folio 23, American Geographical Society, N.Y., (1974).

21. J.B. Colton, Jr. and R.R. Byron, Gulf of Maine - Georges Bank Icthyoplankton Collected on ICNAE Larval Herring Surveys, September 1971 - February 1975, U.S. Department of Comm., NOAA Tech. Rep. NMFS-SSRF-717 (1977).

22. L.A. Walford, Effect of Currents on Distribution and Survival of the Eggs and Larvae of the Haddock (Melanogrammus aegle- finus) on Georges Bank, U.S. Dept. Comm. Bus. Fish. Bull. 29 (1978).

23. D.H. Cushing and J.W. Horwood, Development of a Model of Stock and Recruitment, in J.H. Steel, ed., Fisheries Mathematics, Academic Press, N.Y., (1977).

24. B. Ellertsen, E. Miksness, P. Solemdal, S. Tileth, T. Westgaard and V. Oisted, Growth and Survival of Three Larval Populat- ions of Cod (Gadus morhua L.) in an Enclosure, presented at ICES Early Life History of Fish Symposium at Woods Hole, Massachusetts, USA, April 2-6 (1979a).

25. D.C. Malins, Effects of Petroleum on Arctic and Subarctic Marine Environments and Organisms, 2 vols., Academic Press, N.Y., (1977).

26. K.P. Andersen and E. Ursin, The Partitioning of Natural Mortality in a Multispecies Model of Fishing and Predation, in J.H. Steele, ed., Fisheries Mathematics, Academic Press, N.Y., (1977).

27. W.W. Kunhold, The Influence of Crude Oils on Fish Fry, pp.318- 322, in M. Ruivo, ed., Marine Pollution and Sea Life Fishing News (Books) Ltd., Surrey, (1972)

28. W.W. Kunhold, Effects of the Water Soluble Fraction of a Venesuelan Heavy Fuel Oil (No. 6) on Cod Eggs and Larvae, ICES C.M. Fisheries Imp. Comm. (mimeo) (1977).

29. K.W. Wilson, Acute Toxicity of Oil Dispersants to Marine Fish Larvae, Mar. Biol. 40: 65-74 (1977).

30. M. Reed and M.L. Spaulding, An Oil Spill-Fishery Interaction Model, Part X in Environmental Assessment of Treated Versus Untreated Oil Spills: Second Interim Progress Report, U.S. Dept. of Energy Contract No. E(11-1) 4047 (1978).

31. H. Dannevig, Aarsberetning om Flodevigens Udklackningsanstalt Virkosomhedi Femaaret 1883-1888 (in Norwegian, 1899).

32. B. Ellertsen, E. Miksness, P. Solemdal, S. Tileth, T. Westgaard and V. Oisted, Feeding and Vertical Distribution of Cod Larvae in Relation to Availability of Prey Organisms, Int. Counc. Exp. Sea/Early Life History Symposium (1979).

33. M. Reed and M.L. Spaulding, An Oil Spill-Fishery Interaction Model: Comparison of Treated and Untreated Spill Impacts,

33. (cont.) Proceedings of 1979 Oil Spill Conference, pp. 63-73
     (March, 1979).

34. N. Daan, Comparison of Estimates of Egg Production of the
     Southern Bight Cod Stock from Plankton Surveys and Market
     Statistics, Int. Cons. Exp. Sea/Early Life History Symposium
     (mimeo), 28 pp. (1979).

35. R. Jones, Simulation Studies of the Larval Stage and Observat-
     ions on the First Two Years of Life, with Particular
     Reference to the Haddock, Int. Cons. Exp. Sea/Early Life
     History Symposium (mimeo) 26 pp. (1979).

36. F.M. Serchuk, P.W. Wood and B.E. Brown, Atlantic Cod (Gadus
     morhua): Assessment and Status of the Georges Bank and Gulf
     of Maine Stocks, NMFS, NEFC, Woods Hole, Lab. Ref. 78-03
     (1978).

37. D.F. Bumpus and L.M. Lauzier, Surface Circulation on the
     Continental Shelf off Eastern North America Between Newfound-
     land and Florida, Am. Geographical Soc., Serial Atlas of the
     Marine Environment, Folio 7 (1965).

38. G.T. Csanady, Mean Circulation in Shallow Seas, J. Geophysical
     Research, 81(30): 5389-5399 (1976).

39. M. Reed, Development and Applications of an Open Ocean Model
     for Oil Spill-Fishery Impact Assessment, Unpub. Ph.D. Thesis,
     Univ. Rhode Island (in progress).

40. A. Okubo, Some Speculations on Oceanic Diffusion Diagrams,
     Rapp. Proc. Reud. Cods. Int. Exp. Mer. 167: 77-85 (1974).

41. T.A. Fay, The Spread of Oil Slicks on a Calm Sea, in D.P. Hoult,
     ed., Oil on the Sea, Plenum Press (1969).

42. M. Wang, W.C. Yang and C.P. Huang, Modeling of Oil Evaporation
     in an Aqueous Environment, Ocean Engineering Report 7, Depart-
     ment of Civil Engineering, Univ. of Delaware, 38 pp. (1976).

43. T. Audunson, P. Steinbakke and F. Krosh, Fate of Oil Spills on
     the Norwegian Continental Shelf, presented at the 1979 Oil
     Spill Conference, Los Angeles, USA (mimeo) 37 pp. (1979).

44. M.L. Spaulding, Numerical Modeling of Pollutant Transport Using
     a Lagrangian Marker Particle Technique, NASA Tech. Memorandum
     TM X-73970 (1976).

45. P. Cornillon and M. Spaulding, An Oil Spill Fates Model, U.S.
     Dept. Energy, Division of Environmental Control Technology,
     Contract No. E(11-1) 4047 (1978).

46. P. Cornillon and M. Spaulding, Oil Spill Treatment Strategy
     Modeling for Georges Bank, Proceedings of the 1979 Oil Spill
     Conference, pp. 685-692, Los Angeles, March (1979).

47. Offshore Oil Task Group, The Georges Bank Petroleum Study Report
     MITSG73-5, Mass. Inst.of Tech.,Cambridge, Mass., 3 vols (1973)

48. P.D. Boehm, W.G. Steinhauer, D.L. Fiest, N. Mosesman, J.E. Barak
     and G.H. Perry, A Chemical Assessment of the Present Levels
     and Sources of Hydrocarbon Pollutants in the Georges Bank
     Region, in: Proc. 1979 Oil Spill Conference, Los Angeles
     (American Petroleum Institute, 1979).

A MULTIPLE SPECIES FISHERY MODEL: AN INPUT-OUTPUT APPROACH

Frank C. Hoppensteadt[1] and Ira Sohn[2]

1. Professor of Mathematics, Courant Institute of
   Mathematics, New York University & the University
   of Utah.
2. Associate Research Scientist, Institute for Economic
   Analysis, New York University

INTRODUCTION

One indirect consequence of the 35 year absence of conflict
among the world's developed nations has been the transformation of
the world's fish resources from a renewable resource like land
and forecasts to near non-renewable ones, such as the metallic and
non-metallic, fuel and non-fuel minerals. This process can be
directly attributed to the unparalled and unprecedented rise in the
standard of living in these countries which has resulted in part
from various world-wide institutional agreements on trade and
commerce, along with the increased level of the world's population,
and from the introduction of modern technology to exploit these
major fisheries. If fish could vote they undoubtedly would favour
war among the developed nations.

The imperiled state of many of the world's leading fisheries
has prompted "host" governments (under pressure from domestic
fishing interests at least as a short run policy measure) to place
restrictions on area, duration, type of gear and absolute tonnage
of fish caught, first on foreign vessels and later on their domestic
fleets. The first policy, not unexpectedly, resulted in a transfer
of income, from foreign to domestic fishermen. Although politically expedient, these policies fell short of facing up to the real
problem of checking overexploitation of the fishery.

Ecologists and marine biologists, claimed that the use of
"back-of-the-envelope" estimates of permissible landings both for
foreign and domestic fleets were not acceptable to maintain the

115

required fish stocks and they argued for broader and longer term
policy measures by government management experts.   In the United
States, these considerations led to the establishment of research
teams in fishery centers which bring together biologists, economists and applied mathematicians to develop long-run management
schemes based on optimization analysis of models introduced over
the past forty years.

Theoretical advances in fishery modeling have resulted in the
development of internationally recognized definitions, such as
maximum sustainable yields, optimal yield, etc., and they identified
many of the parameters used in world-wide discussions on fisheries.
In spite of this, practical policy measures still prove to be
difficult to formulate within the analytical framework of existing
models.   This limitation on policy modeling resulted not so much
from a lack of awareness on the part of fishery researchers but
rather from their methodological framework or, more precisely, their
lack of one.

THE METHODOLOGICAL PROBLEM OF INTERDEPENDENCIES

The formulation of policy measures derived from an analytical
framework which is based on partial analysis such as predator-prey
models is bound to have serious limitations and, consequently, to
be of limited usefulness for policy makers. This is so because in
large, interdependent economic or ecological systems it is often the
case that the direct effect of one sector or level of the system on
another is dwarfed by the larger, indirect effects of a third or
fourth sector of the system.   The inability to incorporate the
indirect effects - the spillovers of the system - can and often
does preclude the promulgation of internally consistent policies.
For example, natural resource legislation currently being enforced
by the U.S. Environmental Protection Agency requires that domestic
copper petroleum and steel producers contain their sulfur oxide
emissions and this has had and will continue to have very serious
consequences for the domestic sulfur industry. As a matter of fact,
according to experts at the U.S. Bureau of Mines, if current abatement policies are maintained, the amount of by-product sulfuric
acid produced by the domestic copper, steel and petroleum
industries should be sufficient to satisfy annual U.S. demand for
sulfuric acid by the year 1985, thus completely eliminating the
need for a domestic elemental sulfur industry. Whether this is
desirable from a social, economic or political standpoint is an
entirely separate question. Hence there is a need to describe a
system which captures the mutual inter-dependency of its parts or,
in other words, of formulating the system within the context of a
general equilibrium framework.

While economists as far back as Adam Smith and David Ricardo

recognized the existence of indirect effects on an economic system,
it was not until 1884 that Leon Walras, a Swiss economist,
formulated an internally consistent, simultaneously determined,
general equilibrium framework incorporating the factors of production, and demand and supply patterns. From its theoretical formulation until its empirical implementation another 60 years elapsed
until Leontief, in the 1930's, developed Input-Output Analysis,
which described in quantitative terms a general equilibrium framework of the American economy. For the first time the structure of
industry outputs could be derived in an internally consistent way
from exogenously given levels of consumer expenditures, government
spending, business investment and foreign trade activity. This incorporated the direct and indirect sectoral requirements in addition
to providing quantitative values of the derived demand for the
primary factors of production, i.e., labour and capital.

The input-output approach to economic problems is based on a
structural description of the component parts of the system. It
is a framework which is founded on facts, that is, each producing
sector of the economy can be described in terms of its inputs --
intermediate materials and primary factors (labour and capital)
required per unit of its respective output. The approach is
designed to answer what the industry output levels must be in order
to satisfy a given, exogenous vector of goods and services delivered
to final demand: that is, expenditures made by consumers, government,
business firms for the replacement and additions to productive
capacity, and foreign trade. The tool is remarkably flexible for
it permits the introduction of changes in the level and/or composition of final demand, changes in technology, i.e., the intermediate material requirements per unit of output. These changes
can be favourable, such as less material requirements per unit of
output, or unfavourable, for example, in the case of natural
resource extraction, additional energy, labour, and capital are
needed per ton of extracted metal because of deteriorating ore
grades. Consequently, its general equilibrium format and its
flexibility in being able to incorporate changes in various parameters, make input-output analysis particularly suitable as a tool
for describing and analyzing the economic implications of
alternative policy measures. This was the impetus that initiated
the work on the World Input-Output Model, recently developed for
the United Nations, in which a whole array of alternative sets of
assumptions regarding population growth, availability of natural
resources, pollution abatement levels, balance of payments
restrictions, labour force constraints, growth rates of Gross
Domestic Product, and alternative export and import policies were
introduced as scenarios -- with each scenario depicting a future
development course that the world economy would assume to the year
2000.

As for a more practical, short-term policy tool, is here in

Norway, at the Central Bureau of Statistics and at the Ministry of Finance in Oslo, that input-output analysis which forms the backbone of the successive generations of MODIS models of the Norwegian economy, has reached its most refined form in providing short and medium term management and co-ordination of a modern developed economy.

In contrast to most of the other papers being presented at this symposium, we are reporting on a research program in progress. This began about six months ago and has as its purpose the development of a methodological framework, designed along the lines of Input-Output Analysis, for a multiple species fishery which can be used for the rational management of that fishery. The model combines biological and economic data with optimization techniques. Our remarks are limited to a presentation of the biological component of the model.   A detailed description of the model is to be presented at a future date.

A STRUCTURAL DESCRIPTION OF A FISHERY

At a recent Congress[1], Richard C. Hennemuth, Director of the Northeast Fisheries Center at Woods Hole reported:
        "Successful management, the fulfillment of expect-
        ations, will depend to a large extent on adequate
        advice based on good models.

            The simpler models include only one effect.
        There are no interactions and multiple effects
        are ignored. Regulations of fishing mortality
        on a single-species stock assumes no interactions
        with any other components of the system."

Continuing on this same theme, although somewhat more critic-ally, Earl E. Werner of the Kellogg Biological Station and the Department of Zoology at Michigan State University recently wrote in a book review of Ecology of Freshwater Fish Production, edited by S.D. Gerking (Wiley, N.Y., 1978), which appeared in Science (11 May 1979):
        "Fisheries biologists have traditionally been
        concerned, understandably, with production and
        yield of a single-species populations and have
        paid relatively little attention to how a species
        interacts with the abiotic environment or with
        other species. The latter concerns, of course,
        have been the abiding preoccupation of ecologists...

_____

1. Man As a Predator in the Marine Ecosystem," presented at the
   Second International Ecological Congress, Jerusalem, Israel,
   September, 1978.

> th{is} book provides no evidence of any integration
> of recent ecological theory into the thinking of
> fisheries biologists..." the collection as a whole
> reflects as readily what is lacking in modern
> fisheries biology."

One implication from the above citations is that fishery biologists, and in particular those mandated to prescribe management policies governing fish landings, are in search of a methodological framework that in an analytical and internally consistent way incorporates the interspecific activity among fish to which their management plans can be aligned.

We are currently developing a model of this type based on the Input-Output technique. This structural approach is founded on facts, observations and the technical relationships which govern the way in which fish species interact with one another.

On the basis of existing data which describes energy flow among several species of groundfish and two species of pelagic fish, we are developing a "technology" matrix for various species of the fishery. After proper mathematical manipulation, this will generate estimates of the required stocks of fish that must be made available for a given vector of fish landings - either by foreign or domestic fleets. Whether or not these prescribed stocks are available for the desired level of landings can be estimated by classical effort-yield curves or by the periodic stock samples made by the various management agencies. In the event that the required stocks for the desired level of fish landings -- species-by-species -- are insufficient, policy measures can be directed at the species level e.g., by restricting catch by the usual policy instruments available to the management agencies.

THE FORMAL STRUCTURE OF THE MODEL (BIOLOGICAL COMPONENT)

The first consideration in constructing such a framework is to specify which species will be taken into account and at what degree of aggregation; e.g., by broad class of ground-fish by individual species such as cod, haddock, hake, pollock, etc., or by a sharper description of the individual species' age distribution. In most investigations the degree of detail desired depends on the kinds of data available. In fisheries if two species have similar diets, are in turn preyed upon by a third species and one is a significant by-catch of the second, it may be desirable to disregard the distinction between these two species and instead focus on the age-structure of the fish. For example, juvenile and adult groundfish will have significantly different diets.

At the present time, the model we are constructing considers these eight species: cod, yellowtail flounder, haddock, silver hake, herring, mackerel, pollock and perch. In addition, requirements of "other finfish," "other flatfish" and "other invertebrates" are also determined, although their "input requirements" are not des- cribed in our study.

The internal consistency of the system, given the predetermined system of classification, leads to a set of balance equations -- the axioms of the system. For example, the existing stock of haddock must be sufficient to satisfy the requirements of, for example, cod and pollock, which are based on food-intake levels, the mortality and recruitment rates, as well as the amount of haddock landed by both domestic and foreign fleets.

Our basic framework is the one developed by Leontief in 1970[4], While our study is concerned with the development of a general equilibrium framework for modeling a fishery, the structure of the model presented below is admittedly "primitive" in its assumptions. However, we believe that the model can be refined to incorporate the "peculiarities" of the structural profile of fisheries. The modifications of the model are discussed in Section V.

Let $x_{i,t}$ denote the mass (or caloric content) of standing stock i (say, i = 1 corresponds to Cod, i = 2, Pollock, etc., up to i = n) in the tth year. Then we denote by X_t the vector of standing stocks:

$$X_t = \begin{pmatrix} x_{i,t} \\ \vdots \\ x_{n,t} \end{pmatrix}.$$

A_t denotes the ecology matrix which describes internal consumption at time t, B_{t+1} is the renewal matrix for the next year, and D_t is the vector of exogenous demand.

Ecology Matrix

The components $a_{ij,t}$ of the ecology matrix measure the units (mass or calories) of species i needed to sustain one unit of species j.
We write

$$A_t = (a_{ij,t}),$$

and observe that the i-th component of the vector is the number of units of species i required by the internal needs of the fishery in year t.

Renewal Matrix

The components $b_{ij,t+1}$ of the renewal matrix measures the units of species i needed per unit increment of species j in the next year. We write

$$B_{t+1} = (b_{ij,t+1}),$$

and observe that the i-th component of the vector

$$B_{t+1}(x_{t+1} - x_t)$$

is the number of units of species i required by the fishery to cause increments in the standing stock of sizes.

$$x_{t+1} - x_t.$$

B_{t+1} will be diagonal when age structure is not considered since we suppose that the presence or absence of species i does not affect the birth rate of species i, $(i \neq j)$, and the element $b_{ii,t+1}$ represents the biotic potential of the i-th species.

Exogenous Demand Vector

The vector D_t, is the vector of fish landings (by classification) in time period t.

It should be emphasized that since the internal consumption matrix, A, and the stock recruitment matrix, B, are dated, their coefficients can change in successive time periods, thereby allowing substitution in the energy flow and changing recruitment and mortality rates.

These ingredients are brought together to account for distribution of the standing stock among internal consumption, renewal and external demand:

(Balance Equation) $X_t = A_t x_t + B_{t+1}(x_{t+1} - x_t) + D_t$

The biological parameters of this model are as follows:

   A. The food-intake coefficients $(a_{ij}'s)$

   B. The mortality rates $(a_{ii}'s)$

   C. The stock recruitment rates $(b_{ii}'s)$

   D. The fish landings $(d_i's)$

E.   The time lags

The flexibility and comprehensiveness of the Input-Output
framework permits the development of alternative scenarios based
on different sets of assumptions regarding changes in the above
parameters.   Quantitative estimates of the fish stocks required to
sustain the system based on the particular values of parameters that
were selected can then be determined.   For example, consider an
exogenous restriction of hake landings.   Because of internal
consumption, mackerel, herring and hake (juveniles and young adults)
will be unfavourably affected, resulting in depleted adult stocks
of these species a few years hence.   Recruitment rates of hake will
also increase, again affecting the internal consumption matrix,
a_{ij}'s, and the stock recruitment coefficients, b_{ii}'s.   These and
other effects resulting from changes in the above parameters can be
incorporated into this internally consistent general equilibrium
framework.

At the present time, we are assembling data on internal consumption and fish landings for the selected fish species.   Data
and fish landings are readily available from Fishery Statistics of
the United States (issued by the National Marine Fisheries Service)..
The internal consumption coefficients are more difficult to obtain
although some data on energy flow do exist.   However, significant
efforts are currently being directed to describe fish diets and
their effects on fish stocks {Grosslein, Lagton and Sissewine,[2]
Steele[9]}.

Modeling and empirical investigation can take two divergent
paths: One which makes use of existing data and, consequently,
requires the structural representation to comply with the data.
The other seeks to describe analytically the structural relationships
and then forces the investigators to collect the relevant data.
Research on fish diets, the flow of energy in the food chain and the
structure of fish stocks has been progressing.   Input-Output
Analysis will enable this data as well as successive generations
of more accurate and refined information to be used in a systematic
and comprehensive way to model a multiple species fishery.

REFINEMENTS AND MODIFICATIONS OF THE INPUT-OUTPUT FRAMEWORK

The inherent flexibility of the input-output technique permits
the model to be structured around the "facts," instead of vice versa.
This is a particularly important characteristic due to the many
"peculiarities" that are found in fisheries.

A. Non-Linearities:

Recruitment (hence renewal) and internal demand are

almost definitely not linear functions of standing stocks.
Nonlinear terms can be introduced to account for this, but
this results in models containing more parameters than
current data can determine. We are currently pursuing this
in a theoretical study.

B. By-Catch

As Hennemuth reports in[3], incidental species can be quite
significant. Estimates claim that in the North-west Atlantic
38% of total fishery mortality is generated as by-catch, which
heretofore has not been introduced into modeling efforts.
"Facts", such·as these, can be systematically incorporated
into the input-output framework.

C. Time Lags

Because of the dynamic nature of fisheries, species interactions are often unrecognized or neglected due to the passage
of time between successive generations of adult classes of
fish. Fish stocks, not unlike capital must be maintained,
replaced and even augmented through time due to increased
future demand. With a sharper classification scheme, fish
can be tracked, year by year using the input-output technique,
incorporating the structural changes that are age-dependent
such as diet, recruitment rates, mortality rates, etc.

D. The Economic Component

The fishing sector, like all other sectors in our
economy, has introduced modern technology to economize on
labour. Current technology, while not species specific is
unique, at least in the short-term, for groups of species.
For example, otter trawls and longlines are used for landing
most species of groundfish while purse seiners and floating
traps are the most common gear for pelagic fishing.

The input-output framework is particularly suitable for
integrating the structural profile of the fishing industry
into the economy at large -- both in terms of the intermediate
and primary inputs required, directly and indirectly, and in
terms of the distribution of its output to the various sectors
that make up our economy.

CONCLUSION

The approach to a multiple species fishery model presented
here is an attempt to describe the effects of alternative management by means of a comprehensive, internally consistent and detailed

description of a fishery. This type of modeling requires considerable investment -- both human and financial -- to secure the necessary data. However, we have already seen the effects on the fishery of holding to the short-term view of attempting to manage individual species rather than looking at the broader, more comprehensive framework. Describing the component parts of the fishery within the context of a general equilibrium framework, like input-output models attempt to do, will require continued collaborative work by biologists, applied mathematicians, economists and ecologists if we are to develop a realistic and accurate profile of multiple species fisheries.

BIBLIOGRAPHY

1. E. Cohen, M. Grosslein, M. Sissenwine, and F. Steimle, "Status Report on the Production Studies at the North-east Fisheries Center," National Marine Fisheries Service Unpublished Manuscript, February, 1978.

2. M.D. Grosslein, R,W. Langton and M.P. Sissenwine, "Recent Fluctuations in Pelagic Fish Stocks of the Northwest Atlantic, Georges Bank Region, in Relationship to Species Interactions," National Marine Fisheries Service - Unpublished Manuscript, 1978.

3. Richard C. Hennemuth, "Man As a Predator in the Marine Ecosystem," presented at the Second International Ecological Congress, Jerusalem, Israel, September, 1978.

4. W. Leontief, "The Dynamic Inverse," Contributions to Input-Output Analysis, A.P. Carter and A. Brody (eds.), 1970. North-Holland Publishing Company, Proceedings of the Fourth International Conference on Input-Output Techniques Geneva, 8-12 January, 1968, Vol. 1, pp. 17-46.

5. W. Leontief, with A. Carter and P. Petri, The Future of the World Economy, A United Nations Study, New York, Oxford University Press, 1977.

6. J.H. Steele, The Structure of Marine Ecosystems, Harvard University Press, Cambridge, MA., 1974.

ACKNOWLEDGEMENTS

The authors wish to acknowledge the financial support of this project by the Sea Grant Institute, State University of New York.

AN OVERVIEW OF APPLICATIONS OF OPERATIONS

RESEARCH IN FISHERY MANAGEMENT

J.A. Gulland

Fisheries Department
FAO
Rome, Italy

INTRODUCTION

The title of the present note differs slightly from that
suggested by the organizers "A review....... in resources management",
and included in the programme. This change calls for some explanation, which can also serve as an introduction to the theme of the
paper.

Firstly, a review suggests a careful and detailed examination
of the state of the art, with due acknowledgement to all of the
significant contributions.   This has not proved possible; emphasis
is given to a broad look at the subject.   Perhaps this may give a
better general picture of the wood (or jungle) that is fishery
management, but at the cost of failing to mention some of the
important trees that make up the forest. Certainly this note does
not attempt a serious review of the literature though the papers
mentioned should provide a useful entry to this literature.

Secondly, the change from resource management to fishery management emphasizes the point that the manager (in the sense used here)
is not just concerned in conserving the stocks of herring or cod,
but in seeing that these resources are used to produce benefits for
society.   Thus he is not concerned with the resources alone. He has
to look forward to what happens to the fish after it is caught and
the impact on processing and marketing, as well as backwards to the
environment in which the fish lives.   Fishery management, if all the
aspects are taken into account, is a highly complicated subject and
therefore a priori likely to be one for which an operations research
approach would be valuable. Certainly this approach has been
successfully used for several parts of the whole management system,

125

especially that part of the biological system dealing with the impact
of fishing on fish stocks. The present symposium provides many
examples, from biological through the operations of fishing fleets
to the question of processing and distribution.   However, the
individual systems described are usually treated independently. For
example, the biological models treat the determination of a figure
of Total Allowable Catch under certain constraints (e.g., maintaining
the spawning stock at a given size) as an end in itself, while models
of the shore side of the industry tend to take the catch quota as a
pre-determined and uncontrollable quantity.   A discussion of the
application of operations research should therefore start with a
description of what is involved in fishery management.

THE NATURE OF FISHERY MANAGEMENT

     A common view of fishery management in the past is that it is
something imposed on fishermen by scientists - and here scientists
means biologists rather than the scientific community in general.
As a result the inputs or controls thought of in connection with
management are usually limited to restrictive measures such as size
limits, closed seasons or catch quotas, and the outputs to such
quantities as the magnitude of the total catch.

     These views are much too restrictive and out of date.   There
is a growing awareness that fishery management (as opposed to
resources management) should consider the total benefits that
fisheries can contribute to society, should take account of all
decisions that can affect the way fisheries operate, and therefore
must involve anyone investing or influencing investment in fisheries,
for example tax authorities as well as the more obvious fishery
authorities.   This aspect, and others, was examined by a working
party of ACMRR.   Figures 1 and 2, taken from the report of its
first meeting in December 1978[4] shows some of the interests
involved in fishery management, and the flow of information and
influence between them.

     In terms of using the methods of operations research to approach
the problem of fishery management, an explicit identification of the
inputs and outputs of the system is important. The traditional
inputs are the restrictive regulations of the resource manager (in
the narrow sense). These have been discussed and reviewed many
times[6] and lists of the better known methods are contained in the
constitutions of several international fishery commissions. They
include controls on the sizes or conditions of fish that can be
landed, or the types of gear that can be used; closed areas or
seasons; and direct controls on the fishing effort, e.g., by
different types of limited entry.   For the present purposes these
fall into three main groups - those which affect the sizes caught
(particularly the size at first capture); those which reduce the

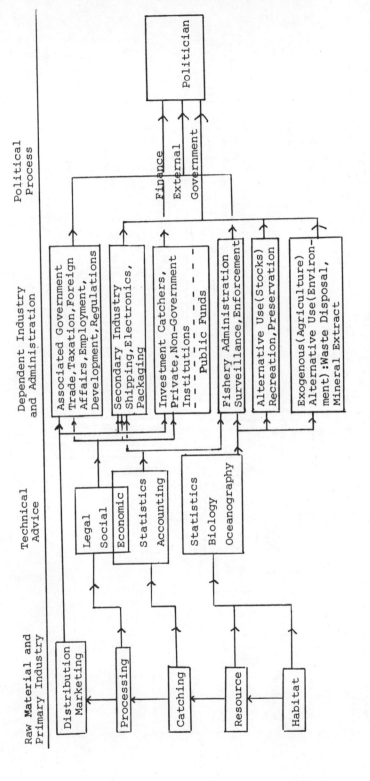

Figure 1. Information Flow

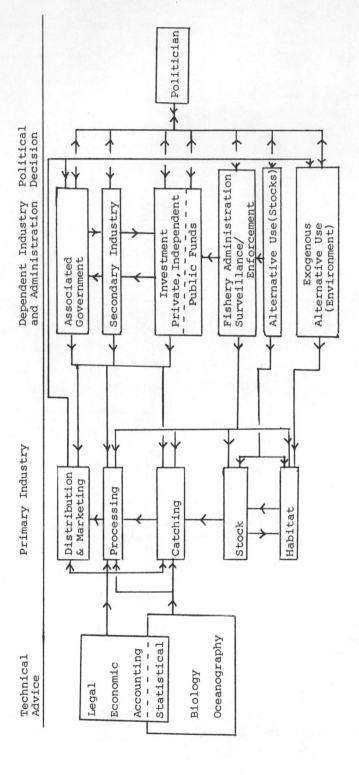

Figure 2. Management Influence

fishing mortality, and at the same time allow the costs to be reduced more or less in proportion; and those that reduce the fishing mortality but do not facilitate a reduction in costs. The distinction between the last two categories is often vital. If, as in many stocks, the biological yield curve is flat-topped, the benefits from management come from reducing costs; one step in doing this is to reduce fishing mortality, but the point of the operation is lost if the regulations impose some form of inefficiency, or additional costs. Most do.

In addition all the other decisions that affect the way fisheries operate and their success or failure should be taken into account in considering the fishery management system. Fiscal incentives are often given to fisheries in the form of low duty for fuel, tax exemptions on profits re-invested in fishing, etc. These are fine when the resource is under-exploited, but these measures can easily be continued long after this stage, and even after another section of the government is considering or applying measures to limit the amount of fishing. A particularly frustrating situation to those who would like to see fisheries treated as a single system is the payment of subsidies to alleviate the economic troubles brought about by excess fishing capacity, thus ensuring that this capacity remains, and the economic troubles continue.

The most important input is investment, in ships and shore facilities. The decisions on investment have almost always been taken entirely independently of management decisions of the traditional type, with mutually frustrating results. Since decisions on investment in vessels is normally based on the gross catches expected to be taken by the additional vessels rather than the net additional to the total catch (the marginal yield)[5] which will be much smaller or even negative in heavily fished stocks, investment continues long after it has become much too high from the viewpoint of the fishery manager. Conversely, management regulations can easily upset investment decisions. The location and size processing plants will be determined partly by the presence of shore facilities, and by access to markets, but also by the location of recent catches, and their pattern of variation within years, and from year to year. These latter can easily be changed by management regulations, e.g., a closed season. It seems unlikely that the potential benefits from a well-managed fishery can be achieved unless there is some coordination of inputs, i.e., that the management regulations are chosen to permit the most rational pattern of investment and that this pattern is based on the situation that will arise under that management system, rather than some extrapolation of current conditions.

As regards outputs, the traditional measure is the size of the total catch, as typified by use of the Maximum Sustainable Yield as the Holy Grail of fishery management. This particular concept has

come under heavy fire especially from economists and from theoretical
ecologists[10] though sometimes for the wrong reasons, e.g., because
decisions are taken on erroneous and over-optimistic estimates of
MSY.   Attempts to replace MSY by another single concept, e.g.,
Optimum Yield[14] are likely to meet much the same difficulties. The
outputs from a fishery that need to be taken into account in
managing a fishery will be as diverse and as difficult to reconcile
as any other set of national policy objectives.   Among these
maximizing the gross yield for a particular stock of fish is of
very minor importance.   Increasing fish supplies, or protein supply
in general, may well be an important objective, but it is doubtful
whether this is best achieved by heavy investment with low marginal
returns in one fishery while other fisheries - or aquaculture or
chicken rearing - offer much better net returns.   Economic objectives
- maximizing net economic returns, net earnings of foreign exchange
(a key objective in the establishment of many fisheries, e.g. for
shrimp in developing countries) etc. - can be more important than
getting the greatest catch.   Also important are social objectives
such as improving the lot of the poorer section of the community,
maintaining employment in isolated areas, etc.   It is hardly
possible to combine all these objectives to provide a single measure
of how well the management system is performing, or to identify a
certain system as being the optimum, even if it were not for the
fact that different people are interested in different outputs.   The
latter difficulty may be less where the government has control of
many of the inputs (e.g., investment, tax patterns and resource
management controls) but management decisions will normally have to
be based on a balance between a number of different outputs, and this
balance may change from year to year or at least from decade to
decade.

SYSTEM MODELS

    Few general descriptions of the complete management system
have been given.   It is therefore not surprising that models of
the complete system do not exist. Rather there are a number of
models describing individual parts of the system; these can be
grouped into biological models (describing the fish stocks, and
the environment in which they live); bio-economic models (describing
the large-scale interactions between fish stocks and the fisheries);
and industrial models (describing the actual operations of the
individual elements of the fishing industry ashore and at sea).
The following brief overview of these models will concentrate on
the weaknesses and strengths of these models, and the degree to
which the lack of a model of the complete system, or of adequate
links between models of different sub-systems, can lead to poor
decisions.

    Biological models lie at the heart of fishery management.

Without an understanding of how the fish stock behaves and reacts
to exploitation it is impossible to arrive at a sensible exploitation policy.   Descriptions of biological models can be found in
any text book dealing with fish population dynamics[7],[13] and need
not be reviewed in detail again here.   For the present purposes
the important questions are how adequate the models are as descriptions of the biological processes relevant to management, and how
easy it is to integrate them into a general description of fisheries.
Two classes of model have been used - the production model, which
treats the population as a single entity; and the analytic model,
which considers changes in the population as being the net effect
of recruitment and growth and mortality of individual fish.

   As descriptions of how actual fish population behave the two
types of model have been about equally successful. Production
models make less demands on data and have therefore been applied
more widely than analytic models. Analytic models, even in the
simple form commonly used (with the parameters of recruitment,
growth and natural mortality constant, and independent of stock
abundance) do provide a better insight of why fish population
should behave as they do in response to exploitation. This
theoretical background gives added confidence in applying even
simple production models to fisheries where direct observations on
how the stock is behaving are very scarce.

   These models, as generally applied, give only incomplete
descriptions of fish stocks; usually providing a set of equilibrium
conditions corresponding to the effects of different patterns of
steady fishing over a period in which the other factors that affect
the stock have also been constant. They do not take account of the
variations in the environment (including man-made effects such as
pollution or land-reclamation) that can cause short-to-long term
fluctuations in catches, or stock abundance or of the biological
interactions between species and hence between fisheries on these
species, nor do they describe the interim situations that occur
as the exploitation pattern changes from one steady state to another.
As management guidelines the usual output of these models, e.g.,
total catch, describes only a part of the factors that should be
taken into accounting in reaching management decisions.

   At this stage there is a distinction in the ways the two types
of model are used. Analytic models, with an explicit statement of
the numbers of fish of each age, provide a good tool for studying
transitional effects, some environmental effects, such as strong
or weak year-classes, and some interactions between species. On
the other hand, the simpler production models are more convenient
for combining with economic and other data.

   Development of biological models to provide good advice on
the management of fisheries on several species faces severe

difficulties.   The simpler models[11] tend to produce what is
intuitively obvious, or directly follows from the assumptions made
and are little help in specific management situations.   The more
realistic and complex models, e.g., of the North Sea[1] or of the
Bering Sea[9] at present make such demands on data, and the estimation
of numerous parameters that they only provide some guidelines on the
qualitative effects that might be expected - and in particular
some warnings that the effects may be quite different from those
expected from simple single species models.   They cannot as yet be
used to give quantitative predictions.   A small change in some of
the parameters, e.g., of the predation rate of one species on
another, which is well within the range of values consistent with
available observations, can make a big difference to the expected
results.   Nevertheless, the careful use of available multi-species
models can provide useful guidance in managing these fisheries, and
result in better decisions than those based on a single-species
approach[3].

    The point on which weaknesses in the models cause greatest
practical problems concerns recruitment, and especially the
interaction between environmental factors and adult stock abundance
in determining recruitment.   This is a weakness of our biological
understanding rather than in the modelling process itself,[12] but
the result is the same.   For most stocks no reliable prediction
can be made of the effect of reducing the adult stock on either the
average recruitment, or on the likelihood of a particularly strong
(or weak) year-class.

    The interest in and use of bio-economic models, which combine
some at least of the salient features of the population dynamics
of fish stocks, and the economics of fisheries has increased greatly.
This has partly been due to dissatisfaction with the purely
biological models and concepts as guidelines for management, and
partly due to the increased possibilities for practical implementation of management.   Establishment of 200 mile exclusive economic
zones has also given an impetus to providing a respectable basis
for rational fisheries management.   Some of these models have used
considerable mathematical rigour in determining optimum policies[2]
but the price of obtaining workable models has tended to be a
simplification of the biological picture (a simple production
model is most often used) and a reduction in the scope of factors
considered - sometimes to the economics in the catching side of the
industry, without reference to the shore side.   The marketing side
may also need more attention.   When the catch from a given stock
is only a small part of the total supply (e.g., to the market for
fish blocks) questions of demand can probably be ignored in
determining the best harvesting policy.   In other cases the
elasticity of demand can be important in determining the desirable
level of fishing effort and total catch.

These models tend to be generalized and academic, rather than specific and practical. That is they deal with the pattern of interactions, and how, for example, the optimum fishing effort changes with changes in fish prices or discount rate, or what the general effect of a high licence fee would be. They have in the past been less concerned with determining the optimum fishing effort for a particular stock. This is probably neither surprising nor disappointing. Specific measures are usually arrived at as a balance between competing interests. In this the calculation of some theoretical optimum for that fishery will have little weight (see for example[8] and[15]), though an understanding of general principles, e.g., that the economic outcome of applying a quota as the only control will be an influx of extra effort, can be estremely useful.

In contrast, another set of models, those used in making industrial decisions, tend to be specific and practical. They have been used - usually at the request of the institution funding the development - to decide, for example, where a new fish meal plant should be located. Even models with somewhat general outputs, e.g., on the optimum size of trawlers, tend to be set within a specific fishery, rather than aimed at a solution giving general guidance on the optimum size of vessels.

The most striking feature of these two classes of model is the degree to which they have been developed in isolation. Parameters, e.g., the catch rates or the length of season which are the variable output from the bioeconomic models are treated as fixed inputs to industry models. This is bound to lead to sub-optimization. It is particularly dangerous, and likely to lead to poor decisions, because fishery management - in the somewhat narrow sense of the application of restrictive regulations - tends to become a contest between the manager on the one hand and the fishermen or the fishery enterprise on the other. Restrictive measures become necessary because decisions taken by the latter group which individually are logical and aimed at maximizing individual benefits, when taken together lead to collective nonsense - overfishing, economic losses and possibly a fall in catches. Once a regulation has been introduced the same situation can arise. Industrial decisions are taken to maximize short-run individual welfare within the constraints set by the regulations, which together reduce the benefits to the fishery as a whole, and hence, ultimately, to each individual. There are numerous examples of these, such as the scramble for whaling or halibut quotas, or perversions of good vessel design so as to fit within controls on length or tonnage. Regulations need to be based on some prediction (e.g., from a model) of how the fishermen will change their practices, not on their present practices. Equally, investment and similar decisions taken by the industry should be based on the situation as it will be in the future under regulations, not on conditions in the recent past. Indeed the two classes of decision should be taken in a coordinated manner; if they are done on the basis of a single integrated model so much the better.

DISCUSSION

Fishery management – that is, the rational use of fishery resources to achieve the greatest net social, economic or other benefits from these resources – is a complex process.   It involves a large variety of interests and activities, and its achievement will require inputs from several diverse disciplines. An approach using the methods of operations research would be useful. It would be difficult at the present to construct a model describing in an integrated manner all the elements concerned, but useful models exist and are being steadily improved, for many individual elements. In some cases, e.g., detailed modelling of the interaction between fish stocks and the natural environment, the absence of interactive links with other elements may not be serious. In others it can lead to a serious degree of sub-optimization. A particular need is to understand the interactions between decisions concerning the regulatory aspect of management, and those concerning investment, marketing and similar industrial activities.

REFERENCES

1.   K.P. Andersen and E. Ursin, A multispecies extension to the Beverton and Holt theory of fishing, with accounts of phosphorus circulation and primary production. Meddr.Dan.m. Fiskeri-og-Havunders. NS 7: 319-435 (1977).
2.   C.W. Clark, Mathematical Bioeconomics; the optimal management of renewable resources. John Wiley, New York. 352 p. (1976).
3.   FAO, Expert Consultation on Management of Multispecies Fisheries. Rome, Italy, 20-23 September 1977. FAO Fish.Tech.Pap., (181). 42 p. (1978).
4.   FAO/ACMRR, Working Party on the Scientific Basis of Determining Management Measures, Interim report of the ACMRR Working Party on the Scientific Basis of Determining Management Measures.   Rome, 6-13 December 1978. FAO Fish.Circ., (718): 112 p. (1979).
5.   J.A. Gulland, The concept of marginal yield from exploited fish populations. J. Cons.Int.Explor.Mer., 32(2):256-261. (1968).
6.   J.A. Gulland, The management of marine fisheries. Scientechnica, Bristol. 198 p. (1974).
7.   J.A. Gulland, The analysis of data and development of models, in Fish Population Dynamics, J.A. Gulland, ed., John Wiley and Sons, London, pp. 67-95. (1977).

8. R. Hilborn, Some failures and successes at applying systems
    analysis to ecological management problems. University of
    British Columbia, Vancouver, B.C. 7 p. (1978).

9. T. Laevastu and F. Favorite, Evaluation of standing stocks of
    marine resources in the eastern Bering Sea. Northwest and
    Alaska Fisheries Center Processed Report, 35 p. (1976).

10. P.A. Larkin, An epitaph for the concept of maximum sustainable
    yield. Trans.Am.Fish.Soc., 106(1):1-11. (1977).

11. R.M. May, J.R. Beddington, C.W. Clark, S.J. Holt and R.M. Laws,
    Management of multispecies fisheries. Science. (205) No.4403,
    pp.267-277. (1979).

12. B.B. Parrish, ed., Fish stocks and recruitment: Proceedings of
    a symposium held in Aarhus, 7-10 July 1970. Rapp.Proc.v.
    Cons.Int.Explor.Mer., 164. (1973).

13. W.E. Ricker, Handbook of computations for biological statistics
    of fish populations. J. Fish.Res.Board Can., (119): 300 p.
    (1958).

14. P. Roedel, Optimum sustainable yield as a concept in fisheries
    management. Spec.Publ.Am.Fish.Soc., (9):89 p. (1975).

15. K.E.F. Watt, Why won't anyone believe us? Simulation, 28,pp.1-3.
    (1977).

MATHEMATICAL AND NUMERICAL MODELING OF PHYSICAL AND BIOLOGICAL

PROCESSES IN THE BARENTS SEA

Professor Jens G. Balchen

Division of Engineering Cybernetics,
The Norwegian Institute of Technology
University of Trondheim, N-7034 Trondheim, Norway

SUMMARY

A short review is given of a long range program which has
been started in Norway to develop mathematical and numerical models
of the physical, chemical and biological processes in an ocean in
order to be able to estimate past, present and future states of
the system. Included in the term state are the least number of
quantities which give an adequate description of the total system
behaviour in terms of growth, migration and distribution of both
plankton and fish. An important aspect of the program is to
develop techniques for recursive updating of the states and parameters of the models against data received from measurements taken
in the real system. The final goal is an operational set of models
in a dedicated computer system which can be used in planning the
utilization of the resources, investigating consequences of
different management strategies and in planning fisheries operations.
The program started in 1975 and the first system is planned to
operate in 1984.

Published in Environmental Biomonitoring, Assessment, Prediction,
and Management-Certain Case Studies and Related Quantitative Issues,
Edited by J. Cairns, G.P. Patil and W.E. Waters. Statistical
Ecology Volume S11. International Co-operative Publishing House,
Fairland, Maryland, U.S.A. 1979.

FISHING OPERATIONS

APPLICATIONS OF OPERATIONS RESEARCH IN FISHERIES

Cyrus Hamlin, N.A.

Ocean Research Corporation

Kennebunk, Maine, USA

INTRODUCTION

The future of civilization needs OR. In the fisheries context, the expanding scope and complexity of the fisheries, and the increasing political and nutritional pressures on the fisheries, will cause the rational and orderly techniques of OR to become of crucial importance in solving the difficult problems lying ahead in the best possible way.

This paper will present case histories of three projects in which OR played a major part. These cases range in scope from a review in broad functional and geographic terms of regional fishery potential (P.R.), to a detailed study of harvesting in a small local fishery (New England Shrimp fishery). Actually there are no bounds to the scope in which OR can be applied - from projects of global dimensions to the improvement of one small facet of a single fishery.

OR must always be considered a tool; a most powerful and useful tool, thanks to the computer, but a tool nonetheless. It can do only what we ask of it, using only the information and instructions we build into it. This tool can help us make sounder decisions, but it cannot make the decisions for us. Like any powerful tool, OR can do serious damage to the user if he is not careful. Accordingly, the users of OR and its implements - models, flow charts, computers, etc. - must constantly interact with the technique to monitor the computations and avoid mistakes.

OPERATIONS RESEARCH AT OCEAN RESEARCH CORPORATION

At ORC we have been working formally with OR, or as we call it "Systems Engineering", for about eleven years. In the course of our experience in a score or more of projects of a wide range of magnitude, we have developed a set of rules which we find useful as a check list for any decision-making task, no matter how small. To avoid errors, backtracking, and other frustrations, these rules should be followed, from number 1 to the end. The more difficult it is to follow one of the rules, the more important it may be to do so.

Rule 1. Clearly state the question which the projected engineering study is to answer, i.e. the purposes for which it is being carried out. (This is the most important, and often the most difficult, step to take.)

Rule 2. Define the limits of the system to be studied. (Locate as precisely as possible the interfaces between the system under study and the rest of the universe.)

Rule 3. Establish the criterion which will be used in evaluating the output of the study. (Usually this is an economic criterion, such as Return on Investment (ROI), but it can be non-financial, such as number of people employed, or amount of protein made available.)

Rule 4. List the constraints and assumptions under which the study will be conducted. (Many simplifying assumptions and constraints can be reasonably made before getting into the project, such as that the demand for a particular fish species will remain high despite the amount landed. This does not mean that these constraints and assumptions cannot be altered, removed, or later added.)

(NOTE: Rules 1 through 4 are preparatory in nature. Their function is to simplify and improve the work of the systems engineer by eliminating all extraneous steps and generally tidying up the project.)

Rule 5. Construct a model of the system. (This obviously does not refer to a hardware model, but to creating a model on paper to give form to the system, by identifying its elements and their interrelationships. Block diagrams and flow charts are examples of models.)

Rule 6. Quantify the model. (Put numbers wherever they are required to describe the model. For example, to establish the size of refrigerated sea water (RSW) tanks requires entering the amount of fish which can be held in a unit volume of RSW, such as "45 lbs. per cubic foot".)

Rule 7. Operate the model, carrying out all mathematical and logical operations. (Until a few years ago, this required the services of computer specialists. The hand-held calculator now makes it possible to carry out complex calculations quickly and accurately in the field. Mini-computers at prices of $US1,000 to $US3,000 are becoming available which will have ample capacity for most computations required by fisheries system engineering.)

Rule 8. Analyze the results, or output, of the calculations, checking for personal and computer errors. (It is essential to be vigilant continuously for errors in the mathematical statements used in the model, for errors of input, and even for errors made by the calculator or computer. Techniques which can be used for checking include visual examinationm hand-made mathematical checks, graphing of data, and sketches. For every formal program there should be a well-verified set of input and output data to check the program.)

Rule 9. Prepare a report, recording all the data and other concepts used in building and operating the model, the method of operating the model, and the input and output values. (Even if the study is done for in-house use only, there should be a full record kept so that someone using the output data a year hence can determine exactly the method of arriving at the results.)

CASE HISTORY NO. 1 - "PUERTO RICO: Distant Water Fishing Study"

Background

P.R., a U.S. island in the Caribbean of 3421 square miles, had a population in 1977 of 3,337,000. Annual income per capita is $2472 (1977) and unemployment is 20.1% (1976/1977).

The U.S. Department of Agriculture instituted a study to improve the general economy of the island, more particularly to provide stable employment opportunities. I contracted to do that portion of the study relating to the fishing industry.

Purpose of Study

Identify those fisheries which might be engaged in by P.R., evaluate potential economic and employment returns of each, and recommend for further study those with acceptably high benefits to P.R.

System Boundaries

Geographical: All waters south and west of latitude 40°N and longitude 40°W to the continental land mass.   P.R. lies close to the centre of the area and about 1800 miles from its farthest boundaries, a distance arbitrarily considered as the range which could be expected from a country whose population was not considered traditionally of fishermen.   Functional: All elements of the fishing system were assumed to be within the P.R. purview, up to and including the sale of the finished product in P.R. markets or its shipment to external markets. In other words, the value added by P.R. capital and labour would be maximized. (By contrast, the U.S. tuna industry used P.R. only as a transhipment base for supplies to the vessels and unprocessed catch from the vessels, with very small net benefit to P.R.)   Resources: Any economically viable fishery resource.

System Restraints

Resources which were fully or over-exploited would be rejected; P.R. would not gain at the expense of another fishing interest.

Criteria of Successs

The primary criterion was employment; how many persons could be stably employed at an acceptable income level. The secondary criterion was the input to the financial health of the country, i.e. ROI. Because these criteria were not precisely quantified at this stage, the results were presented covering ranges of values from which the planners could select fisheries with the optimum combination of characteristics.

Assumptions

The following assumptions were used in making the study:
1.   The P.R. labour pool would provide enough persons to man any fishing industry selected for promotion, who would also be willing to be trained for the industry and to alter their tradition- ally land-bound way of life.
2.   Financing would be available to carry out the public (harbours, etc.) and private (vessels, processing plants, etc.) investments required by any selected fishery.
3.   Adequate markets would exist for any selected species.

Methodology

The project was carried out in the following steps:

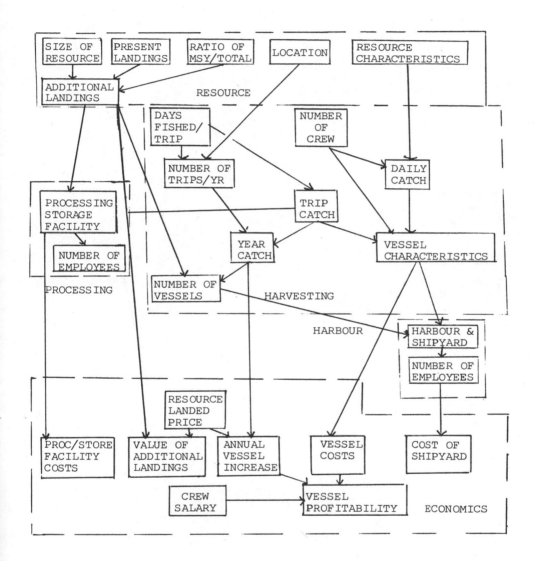

Figure 1. Block Diagram: Potential Puerto Rico Fisheries

FIGURE 2

PARTIAL COMPUTER PRINTOUT OF RESULTS

RUN

RATIO OF MSY TO TOTAL, K1= .5
RATIO OF MSY TO POTENTIAL LANDINGS, K2= 1
IS PRINTOUT OF CONSTANTS DESIRED? IF YES SET H=1; IF NO,
SET H=-1? -1
ACTUAL RESOURCE POPULATION, LBS=T1
PRESENT CATCH, LBS=T2
VALUES TO BE USED FOR T1, T2? 1.12E8,2.24E6
MSY,LBS/YR= 5.600000E+7
UNCOUGHT POTENTIAL MSY LBS/YR= 5.376000E+7
ADDITIONAL POTENTIAL LANDINGS, LBS/YR= 5.376000E+7
CATCH RATE, LBS/FISHERMAN/DAY FISHED=V2
DENSITY OF CATCH, LBS?CU FT=V3
FOR TUNA LONGLINING, V2=616 AND V3=34.3
FOR TROUT AND CROAKER TRAWLING USING A NO.45-A NETO3.0 KN,
V2=377 0, V3=30.0
VALUES TO BE USED FOR V2,V3? 616,34.3
FOR TUNA, SET M=1; FOR TROUT AND CROAKER, SET M=2;
VALUE FOR M=? 1
SIZE OF CREW=? 10
ACCOMMODATION AREA, SQ FT= 698

DAYS FISHED/TRIP,B6=? 10
TRIP CATCH, LBS= 30800
HOLD AREA SQ FT= 174.2562
PARTIAL AREA SQ FT= 872.2562
CUBIC NUMBER OF VESSELS= 16131.15
SHAFT HORSEPOWER OF VESSEL= 393.9544
TOTAL HORSEPOWER OF VESSEL= 448.9513
TOTAL VESSEL COST, DOLLARS= 162685.6
LENGTH OF TIME UNDERWAY, DAYS/TRIP= 18.33333
NUMBER OF TRIPS?YR= 16.36364
ANNUAL CATCH/VESSEL, LBS= 504000
NUMBER OF VESSELS= 106.6667
LENGHT OF VESSEL,LOA,FT= 88.14965
GROSS INCOME/VESSEL AT $ .15 /LB=$ 75600
NET INCOME/VESSEL= $-6495.162

```
IF FISHERMAN'S ANNUAL SALARY= $3000
OWNER'S NET SHARE= $-36459.16
OWNER'S ROI= -22.43294 %


IF FISHERMAN'S ANNUAL SALARY= $5000
OWNER'S NET SHARE= $-56495.16
OWNER'S ROI= -34.72659 %
IF FISHERMAN'S ANNUAL SALARY= $7000
OWNER'S NET SHARE= $-76495.16
OWNER'S ROI= 47.02023 %


IF FISHERMAN'S ANNUAL SALARY= $9000
OWNER'S NET SHARE= $-96495
OWNER'S ROI= -59.31388 %


GROSS INCOME/VESSEL AT $ .18 ?LB=$ 90720
NET INCOME/VESSEL=$ 8020.037


IF FISHERMAN'S ANNUAL SALARY= $ 3000
OWNER'S NET SHARE= $-21979.96
OWNER'S ROI= -13.5107 $
```

1. A simple block diagram (Figure 1) was constructed to assure that all sectors of any fishery covered by the project were considered. Note the emphasis on employment and economics.

2. A detailed block diagram was prepared to include every element and relationship required to give the results which would fulfill the purpose of the study.

3. The detailed block diagram was translated into a flow chart which when quantified, programmed and operated would provide the desired outputs.

4. The flow chart was quantified, using constants wherever possible and variables as required by the various fisheries. The best available data were used throughout, ranging from precise hard data to good guesses;   it should be pointed out that since the study was intended to provide comparisons among a number of fisheries and that further detailed studies would be made, it was sufficient for inputs to be consistent, and absolute values were not of great importance at this stage.

5. The flow chart was programmed for computer operation in BASIC language.

6. Based upon an extensive literature search, as well as contacts in person and by letter with many authorities, likely fish resources were selected for computer operation, and input was prepared for each.   The species and the fishing methods used in the study were:

Resource	Fishing Method
Tuna	Long-lining
Croaker*	Bottom Trawling
Sea Trout*	Bottom Trawling
Shark	Long-lining, and hand-lining w/powered reels.
Snapper	Hand-lining w/powered reels
Red Crab	Pots
Groundfish (Grand Banks*)	Bottom Trawling

Presentation of Results

The computer output (Figure 2) was analyzed and presented in both graphical and narrative form. A graph for the sea trout and croaker trawl fishery presented the ROI plotted against annual fishermen's earnings, for three values of landed price per pound and four values of days per trip spent fishing. The narrative discussion of the output follows.

---

* The possibility of croaker and sea trout trawlers also fishing for cod, etc., on the Grand Banks off Newfoundland was investigated briefly to see if that northern fishery would provide a viable option.

"Croaker/Sea Trout Fishery. The croaker/sea trout fishery
was tested for variations in days fished/trip, crew salary, and
landed price/lb. For a crew salary of $4000/year, the highest ROI
was achieved fishing for 20 days each trip; the actual value of ROI
varied from +58.0% to +98.0% for unit price of $.15 and $.21/lb.
respectively. The associated fishery system included vessels
of 130' LOA with 720 shaft horsepower, each having trip catches of
301,600 lbs. of trout and croaker, and costing $462,900 (a maximum
of 118 vessels could be expected to be in port simultaneously), a
shipyard 1070' long and 195' deep having a capital cost of
$3,828,000 and employing 256 yard workers, and a storage and
processing facility occupying 165,000 square feet, having a capital
cost of $31,874,000 and employing 2440 processors."

Use of Results

The project provided the planners with a hard core of the
results presented in a useful form. In addition, there was a
description of the method of deriving the results, and discussions
of each. Armed with this material, the planners could make
sensible decisions as to the next step to take, i.e. which fisheries
warranted further study. Although no major fishery has yet been
established in P.R., it is thought that this study has contributed
significantly to the thinking in P.R. about fisheries.

CASE HISTORY NO.2 - "JAMAICA: Identification of Fishery Development
                     Project"

Background

Jamaica, also an island in the Caribbean, is a member of the
British Commonwealth. With an area of 4400 square miles and a
population of slightly over 2,000,000, it has an economy based
largely on agriculture and minerals. Per capita income (1976) was
$US1300, and unemployment was estimated at about 25%.

Fishing is an important food industry, with fishermen
constituting approximately 1% of the total labour force, and perhaps
an equal number in the fishery distribution, processing, governmental,
etc., sectors. Landings amount to an estimated 9500 metric tons
annually, or about 10 lbs per capita. The importance of fish as
an item of diet is illustrated by the ex-vessel cost in 1976 of
36cUS to 63cUS per lb., high for a developing country. Also,
not only is virtually the entire catch consumed locally, but in
1976 fish were imported to the value of $US17,300,000, contributing
to a staggering negative balance of payments.

75% of all fish landed in Jamaica are from 3500 beach-based

dugout canoes, half of which were fitted with outboard motors in
1976.   Of the balance, most are landed by carrier vessels serving
fleets of canoes based on small offshore islands.   There were in
1976 only 8 fishing vessels larger than canoes, and two of these
were FAO-operated training vessels.   All fish are landed in
excellent condition and distributed and marketed rapidly.

In 1976, at the request of the Jamaican government, a FAO/IDB[1]
Cooperative Programme mission was sent to Jamaica to analyze the
fisheries sector and recommend a program of expansion and development.   The mission included an economist, processing-marketing
specialist, naval architect-fleet specialist (myself), fresh-water
biologist, marine biologist, civil (harbour) engineer, and an
institutional specialist, all under the direction of an FAO/IDB
staff member.

A major factor in the Jamaica study not present in the P.R.
survey was the 200-mile fishery management zone.   Although not all
countries in the area had adopted it, prudence required that, in
seeking new fishing areas, the 200-mile zone be considered a fait
accompli for the entire area.

Purpose of Mission

Identify all fishery resources potentially available to
Jamaican vessels, and select those, if any, suitable for funding
by a loan from the IDB.

System Boundaries

Geographic: In general, the Caribbean Sea.   However, the Gulf
of Mexico shrimping grounds, the Bahama Banks, and the continental
shelf off Guyana were also examined.   Major emphasis was placed on
resources within Jamaica's 200-mile zone and in international
waters. Functional: From the fleet standpoint, the study would
consider the resource from the bottom of the sea to its landing at
Jamaica.   It would include, but not be restricted to, traditional
Jamaican craft and gear.   Resource: Since the landed fish would be
consumed in Jamaica, only those species acceptable to the domestic
market were considered.

System Restraints

Only those resources were considered which were under or non-
exploited.   All countries in the area were assumed to have adopted
the 200-mile fishery management zone.

Criteria of Success

The final test of any proposed fishing system was whether it
would land fish at a price competitive in the open Jamaica market.
Less precise tests were whether the proposed area was politically
accessible to Jamaica vessels and whether there were Jamaicans
willing and trainable to man the vessels.

Assumptions

1. There was ample room to introduce an additional 2000 tons
annually into the Jamaica market without affecting prices materially.
2. Fishermen would be available to man whatever vessels were
recommended, but not without an intensive training programme.
3. Harbour and support facilities would be provided for whatever fleet was recommended.
4. Tropical fish, well iced, can be kept up to 10 days before
landing, with ample shelf life remaining for processing and marketing.

Methodology

The following steps were taken leading to a proposed fishery
project:
1. Identify every possible fishery which might be available
to Jamaica. Initially these were restricted to Jamaica waters and
the banks of SW of Jamaica, but selected additional areas were
included, specifically the Bahama Banks, the Campeche and West Florida
shrimp grounds (with the shrimper discards in mind), and the
continental shelf of Guyana; only the last one survived early
screening.
2. Eliminate those fisheries which on first examination do not
meet the various criteria and restraints.
3. Prepare for each of the rest one or more operational models,
including practical fishing methods and such variables as trip
length, crew size, etc.
4. Collect and analyze available data and reduce to usable
terms. Where data were lacking or were insufficient, make
educated guesses.
5. Quantify the models and operate them through the desired
range of variables. In this study the final output would be the
projected landed cost per pound. (Note: All necessary calculations
were carried out in the field with a pocket calculator, working
to a format made up for the purpose.

Presentation of Results

The useful data from calculations were entered onto a uniform

FIGURE 3

CONDENSED DATA - REVIEW OF POTENTIAL JAMAICA FISHERIES

FISHERY DESCRIPTION

HARVESTING SYSTEM
CHARACTERISTICS

1   **VESSELS**
2   LENGTH OF OVERALL - LOA   FT
3   LENGHT OF WATER LINE - LWL   FT
4   BEAM   FT
5   DRAFT MAN   FT
6   DEPTH   FT
7   DISPLACEMENT, TO LWL   LONG/TONS
8   CUBIC NUMBER (LWL x B x D)   CU FT
9   SHAFT HORSEPOWER, CONTINUOUS
10 STEAMING SPEED   KNOTS
11 CREW SIZE
12
13 **OPERATIONS**
14 MILES TO GROUNDS   NAUT MILES
15 LENGHT OF TRIP (DAYS AT SEA)   DAYS
16 TIME SPENT FISHING PER TRIP   DAYS
17 TIME BETWEEN TRIPS   DAYS
18 NUMBER OF TRIPS PER YEAR
19 CATCH PER DAY, MARKETABLE FISH   LBS
20 CATCH PER TRIP   LBS
21 CATCH PER YEAR   LBS/TONS
22 POTENTIAL RESOURCE   TONS
23 NUMBER OF VESSELS REQUIRED
24
25
26 **ECONOMICS**
27 COST OF VESSEL & GEAR, TVC   J$
28 TOTAL ANNUAL COSTS   J$
29   .VARIABLE COSTS, LESS FUEL & CREW   J$
30   FUEL & LUBE COST   J$
31   CREW SHARE   J$
32   .FIXED COSTS   J$
33 TOTAL ANNUAL COSTS LESS INT. 4 DEP   J$
34 LANDED COST PER LB. OF FISH, AVG.   J$
35

format (Figure 3).   There were a total of 17 variations in 5
different fishing areas. For each of these, the data presented
included vessel descriptions, operations, and economic data.

Use of Results

From the data presented, the optimum operation was selected,
taking into account not only the landed price criterion, but other
less precise factors.   For instance, the Pedro Bank area, selected
for the project, is within Jamaica's 200-mile zone and the resource
is thus assured; also, the vessels and the fishing gear and methods
are all familiar to Jamaican fishermen.   This is an excellent
example of how the rational and orderly procedures of systems
engineering provided a sound basis for making important decisions
regarding highly complex conditions.   Although one fishery, based
on the Pedro Banks, was obviously the most suitable for exploitation,
the study also provided usable models of other fisheries which may
be ripe for exploitation in the future.

CASE HISTORY NO. 3 "DESIGN STUDY: An Optimum Fishing Vessel for
                   The New England Shrimp Fishery".

Background

In 1968 my company, ORC, contracted to the Division of Economic
Research, U.S. Bureau of Commercial Fisheries (now the National
Marine Fisheries Service) (NMFS) to carry out a design study aimed
at establishing the optimum characteristics of a fishing vessel
for the Georges Bank groundfish fishery.   To do this a model of
the fishery was constructed and operated over ranges of variables
to give as output the characteristics of 128 vessels. The vessel
with the highest ROI was selected, and was described fully in the
report and in conceptual plans.   (This report is available in print.)

In 1969, it appeared that there would be a continuing and
steady fishery in the Gulf of Maine based on the northern shrimp,
Pandalus borealis.   This species had begun to show in commercial
quantities in 1959 and a fishery rapidly developed. Approximately
50% of the catch went to north European markets, which demanded
a high quality product.   This profitable fishery has since declined
(it is not certain whether through water temperature increase or
over-fishing, or both) but during its peak years there began to be
built for it an increasing number of excellent small fishing craft.
In 1969, ORC was contracted to the NMFS to develop the optimum
characteristics of a vessel suitable for the New England shrimp
fishery.

This work was carried out as a development of the systems

approach utilized for the Georges Bank design study. Unfortunately,
just about the time the study was completed, the shrimp resources
began to disappear, so the results of the study have not yet been
put into practice.

The major purpose of both these studies was not to produce
designs from which vessels would be built for the respective
fisheries, but to develop a rigorous and consistent method based
on OR of selecting fishing vessel characteristics. In our
experience this purpose has been achieved; we rarely approach any
problem relating to fishing vessels without resorting to one of
these models, in whole or in part.

Purpose of Study

Identify the optimum characteristics of a vessel for the New
England (Gulf of Maine) shrimp fishery.

System Boundaries

Geographical: The U.S. northern shrimp fishery is carried out
within 30 miles of the north and west shores of the Gulf of Maine,
and this was the zone in which the vessels would operate.
Functional: Shrimp are caught with bottom trawls in waters to about
100 fathoms deep. However, the concept of the multi-fishery
combination vessel was considered desirable, if not necessary, for
sound economics, and this concept was incorporated in the study.
Resource: Primarily the northern shrimp, P.borealis. Secondarily
any available commercial marine resource within 30 miles of the
coast.

System Restraints

1.   Vessels would be built of steel (although a comparison
among other normal building materials was also carried out).
2.   Otter trawling will be the fishing method.
3.   The hold was sized for groundfish; a fish hold adequate
for a trip-load of shrimp would be too small for groundfishing.

Criteria of Success

The sole criterion of success was the highest ROI in the
shrimp fishery.

7.4.1 "BLOCK DATA" Subroutine (Constants and independent variables)

Symbol		Quantification	Used in Subroutine	Commentary
OC	Opening coefficient of trawl net	0.055	7.4.4	7.4.4.3
RETRAT	net retrieval time coefficient	10.0	7.4.4	7.4.4.4
QC	Catching coefficient of trawl net	0.021	7.4.4	7.4.4.5
UCDW	Hydrodynamic drag coeff of warps	0.455	7.4.4	7.4.4.7
UCDD	Hydrodynamic drag coeff of doors	0.8	7.4.4	7.4.4.9
UF	Coefficient of frictional drag (mid) on bottom doors (bottom)	0.0 0.82	7.4.4	7.4.4.9
UCDN	Hydrodynamic for resistance coeff for net (av. shrimp trawls)	0.237	7.4.4	7.4.4.11
UCFN	Hydrodynamic friction resistance coeff of net	0.0262	7.4.4	7.4.4.11
HPC	Horsepower constant	550.	7.4.5	
KTFPS	Knots of (ft/sec) Multiplier	1.689	7.4.5	
SHPCOF	Shaft HP Coeff	14.5	7.4.5	7.4.5.2
SHPEXP	Shaft HP exponent	0.8	7.4.5	7.4.5.2
TRCOF	Trip Coefficient	0.75	7.4.5	7.4.5.3

Figure 4. Portion of Flow Chart

PHYSICAL	FT	HL								
TRAWL HEADLINE	FT	HL	40	40	40	40	50	50	50	50
TOWING SPEED	Knots	VT	1 5	2 0	2 5	3 0	1 5	2 0	2 5	3 0
Lgth WATERLINE	FT	LWL	37 15	40 00	43 53	45 39	39 17	41 75	46 87	50 99
Lgth OVERALL	FT	LOA	40 12	43 20	47 01	49 02	42 30	45 09	50 62	55 07
BEAM	FT	BEAM	12 53	13 35	14 35	14 87	13 11	13 85	15 28	16 41
DEPTH	FT	DEPTH	5 08	5 47	5 96	6 22	5 36	5 71	6 43	7 01
CUBIC NUMBER		CUBE	363 63	2921 81	3724 13	4199 75	2751 97	3305 11	4603 99	5864 69
DISPLACEMENT	Lg.Tn	DISP	19 08	23 81	30 70	34 80	22 37	27 10	38 31	49 35
SHAFT HP		SHFHP	53 83	125 42	151 29	225 53	72 25	127 70	204 87	306 01
WINCH HP		WINHP	31 36	31 36	31 36	0 00	46 44	46 44	46 44	0 00
GENERATOR HP		GENHP	0 78	2 01	4 45	8 72	1 28	3 34	7 42	14 59
TOTAL HP		TOTHP	85 97	158 79	187 09	234 25	119 96	177 47	258 72	320 61
HOLD SIZE	Cuft	HOLS	196 30	261 74	327 17	392 61	302 41	403 21	504 00	604 81
No of CREW		NUMCR	2	2	3	3	2	2	3	4

OPERATIONAL										
SPEED W/SHFHP	knots	VSX	7 67	9 35	9 45	10 38	8 11	9 20	9 93	10 68
MIN ALLOW SPD	knots	VST	8 22	8 41	8 62	8 73	8 35	8 51	8 82	9 04
CATCH / MIN	LB	Q	2 7	3 70	4 62	5 54	4 33	5 77	7 22	8 66
CATCH / CYCLE	LB	CATCY	498 96	665 28	831 60	997 92	779 62	1039 50	1299 37	1559 25
CATCH / YEAR		catyr	206 73	275 64	344 54	413 46	318 46	424 62	529 44	636 93

ECONOMIC										
TOTAL VESSEL COST	$x1000		47 00	66 12	83 56	97 78	57 88	75 28	108 39	138 16
OWNER'S INVEST	$x1000		same	54 72	68 40	82 07	63 22	84 30	105 37	126 44
ANNUAL INCOME	$x1000		24 62	32 83	41 04	49 24	37 93	50 58	63 22	75 86
CREW SHARE	$x1000		2 59	4 13	5 42	7 12	3 06	4 32	6 70	9 70
CREW EARN EACH	$x1000		11 02	14 35	11 87	14 04	17 44	23 12	18 84	16 54
OWNER'S GROSS SHARE	1000		16 41	21 88	27 36	32 83	25 29	33 72	42 14	50 58
OWNER'S NET SHARE	$x1000		8 68	11 33	13 98	17 20	15 95	21 81	25 62	28 60
RETURN ON INVEST	%		18 47	17 14	16 93	17 60	27 56	28 48	24 66	20 70

Figure 5. Characteristics of Optimum Vessel for Each Net and Towing Speed, 131 Days Fishing per Year

Assumptions

1.   The resource would be exploitable the year round.

2.   Catch rates recorded at the time of the study would continue into the future.

3.   The demand and prices for the landed shrimp would remain constant.

4.   There would be no construction subsidy available for these vessels. (Shortly before, the U.S. had had a construction subsidy program which paid the owner up to 50% of the cost of the vessel and its gear in the form of a grant from the government. This subsidy was no longer in effect.)

5.   Shrimp are best caught towing at $1\frac{1}{2}$ to 2 knots.

Methodology

1.   A block diagram was constructed of the New England shrimp industry consistent with the conditions given above.

2.   A linear block flow chart of the shrimp program was constructed.

3.   This chart was expanded into a flow chart (Figure 4) including all elements of the system and their inter-relationships. These were categorized into blocks to simplify revisions, alterations, and general computer operations.

4.   The expanded flow chart was programmed for computer operation in FORTRAN IV language.

5.   The program was quantified, based upon the best available data.
(Note: the beauty of the computer is that the effect of changing a single input (based upon more recent data, for instance) can be done in minutes, even though it might affect every calculation.)

6.   The program was punched on cards and operated on an IBM 360 computer utilizing as inputs constants and ranges of variables.

7.   The output was utilized to determine the "Optimum" vessel characteristics.

Presentation of Results

Large masses of data are best presented in tabular and graphical form, covered by explanatory notes. For this study, two major sets of tables were prepared, 1) Trawl net characteristics and 2) vessel characteristics (Figure 5). Several graphs were also presented; typical of these is a plot of ROI versus net size for various crew sizes and towing speeds. Presented in this fashion the calculated data can be used in similar fisheries elsewhere. Conceptual drawings were prepared showing a vessel of the selected characteristics.

Use of Results

As mentioned above, we at ORC make frequent use of the results
of these two studies. We find it expedient and simple to rework
them to suit different conditions and problems found in other
fisheries.   There has also been a certain amount of informal use
of these design studies by researchers, shipyards, and naval
architects.

COMMENTARY

After eleven years of working with OR, largely in real world
situations, some of our thoughts and conclusions regarding its use
in the fisheries may be helpful:

1. OR is necessary for fishery development.   No one disagrees
that fisheries benefit from, even need, vertical integration, in
planning and organization. Yet the complexities at each level -
resources, harvesting, processing, marketing, etc. - are so
bewildering and indeterminate that it is not possible to comprehend
them all even at one level, let alone for the entire system.   Only
through a discipline such as OR can rational planning and decision-
making be carried out.

2. There must be recognition of the "generalist". This is
particularly true of the fishing industry. There are fishery
specialists at every turn, and most of them are very necessary.
However, the proliferation of specialists has only increased the
fragmentation which so seriously plagues fisheries.   Our work with
OR has demonstrated the need for generalists, individuals who can
work intelligently with any of the fishing industry sectors, yet
are dedicated to maintaining an objective, even, overall viewpoint.
Classified as a naval architect, I am involved directly with fishing
vessels, gear, and oceanography, and less intimately but importantly
with marine biology, harbours, and processing; working in all these
sectors, I inevitably find myself functioning as a generalist.
There should be a formally designated discipline of generalism,
whose practitioners are trained to view all sectors of the fishing
industry with equal knowledge and objectivity.

3. High sophistication must be avoided: OR should become a
standard and mundane factor in the fisheries. To do so, though,
requires that it be couched in terms fully understandable to the
user.   The very name "Operations Research" can cause awe-inspiring
visions of "Star War" laboratories manned by academics and
bureaucrats.   It is for this reason I prefer the term "System
Engineering" as better describing the practical down-to-earth work
we do.

4. <u>We must spread the word</u>. We must sell OR to the fishing
industry from the artisanal fisherman to the giant corporation.
This involves explaining what it is and what it is not, in
understandable and graphic terms at every opportunity - at business
luncheons, to industry organizations, to individuals.

In this paper I have endeavoured to give some useful
impressions of that branch of OR which I call systems engineering -
the application of OR to practical real world problems. Above all,
I hope I have kindled in others my own conviction of the essential
place OR must have in the increasingly complex world of tomorrow.

BIBLIOGRAPHY

1. C. Hamlin, J.R. Ordway, "The Application of Systems Engineering
        to Practical Marine Situations", Marine Technology Society,
        Washington, D.C. (1972) English.
2. C. Hamlin, "Puerto Rico Distant Water Fishing Study for
        Robert E. O'Brian, Inc., (1971) English.
3. Mission Report, Hamlin et al., "Jamaica: Identification of
        Fishery Development Project" FAO/IDB Cooperative Programme.
        Washington, D.C., (1977) English.
4. C. Hamlin, "Design Study: An Optimum Trawler for the Georges
        Bank Groundfish Fishery", National Technical Information
        Service, No. COM-72-11298. Springfield, Virginia (1971)
        English.
5. C. Hamlin, "Design Study: An Optimum Fishing Vessel for New
        England Shrimp Fishery", National Marine Fisheries Service,
        Washington, D.C. (unpublished) English.

REFERENCES

1. FAO - Food and Agriculture Organization of the United Nations
    IDB - Interamerican Development Bank.

A MATHEMATICAL MODEL USED FOR PRE-FEASIBILITY

STUDIES OF FISHING OPERATIONS

C.T.W. Curr

White Fish Authority
Industrial Development Unit
St. Andrew's Dock, Hull HU3 4QE, England

INTRODUCTION

When deprived of traditional sources of fish, for whatever
reason, various sectors of a fleet may be forced to consider
previously unexploited fisheries as an alternative. In such cases
it is difficult to predict the economic performance of any given
class of vessel, because of the scarcity of catch rate data. The
problem is all the more severe when possible catch rates are high
in relation to vessel size. An example of this arose in
considering the likely performance of pair trawlers catching blue
whiting for human consumption, where the hold capacity is very
much reduced when compared to bulk stowage for fish meal.

In this situation, a single haul, or the last haul of a short
sequence may far outstrip the remaining hold capacity. The vessel
can easily catch much more than it can stow, so that simple
calculations based on average catch rates will produce seriously
optimistic predictions of vessel performance, by including fish
that can be caught but not landed. Any economic analysis,
therefore, must take full account of the large random haul by haul
variation in catch rate. Although this problem has previously
been attacked using simulation models, that method can be cumbersome, tedious and expensive. In examining the economic prospects
of the blue whiting fishery, we developed an efficient alternative
statistical approach, which used to the full the limited catching
data available at the time.

This paper deals only with fishing operations which are
essentially unimpeded by on-board restrictions in fish handling.
The approach is intended solely for none processing vessels, which

can stow their catch more or less as rapidly as it can be brought abroad.   The method described proved to be very useful in select- ing from alternative methods of exploiting blue whiting at a pre- feasibility stage.

APPROACH TO THE PROBLEM

We consider two outcomes resulting from a defined amount of fishing effort: a given sequence of hauls may or may not generate a total catch in excess of hold capacity.   In the long term, the average catch will be a combined effect of repeated occurrences of the two outcomes:-

Case I: a given number of hauls generates a catch greater than hold capacity. The long-term contribution of fish per trip is given by:-

p (catch exceeds hold cap.) x (hold cap.)

where p denotes probability

Case II: the catch from the same number of hauls is insufficient to fill the hold, but the vessel has run out of fishing time. The long-term contribution of fish per trip arising in this way is:-

p (catch less than hold cap.) x catch, summed over all possible catches from zero to hold capacity.

Adding the two contributions, we obtain the long-term average catch per trip, and can deduce the average time on fishing grounds.   All that is required is a method of predicting the various probabilities. This can be done by using a probability function chosen to represent and extend the available catch data. In the case of blue whiting, the only catch data obtained from "commercial" fishing, and available to us in 1976, came from the Sea Fisheries Institute at Gdynia, Poland. Although this consisted of a mere 155 hauls from one trip, it was sufficient to formulate a mathematical model of the catch distribution, in the shape of the Erland function.   The average haul weight was 13.3 tonnes, with a maximum of 6 hauls per day and an average of 3.6 hauls/day.

Table 1 gives frequency distributions of the catch from one haul, and from independent groups of 2 or more consecutive hauls, up to 8.   The nature of these distributions is the main topic of this paper.

Table 1. Frequency Distributions of Catch Weight from
Consecutive Hauls, taken in Independent Groups
(Polish data)

Catch from
group of con-
secutive hauls.

tonnes	1	2	3	4	5	6	7	8
0 - 9	60	12	5	1	1			
10 - 19	51	17	4	3	0	2	1	
20 - 29	29	18	8	5	5	0	0	1
30 - 39	10	15	10	2	1	2	0	0
40 - 49	5	8	6	5	1	1	2	1
50 - 59		4	9	5	1	1	0	1
60 - 69		2	5	5	8	4	4	0
70 - 79		1	2	6	3	3	2	1
80 - 89			3	1	3	3	2	2
90 - 99				2	2	2	2	2
100 - 109				3	3	0	2	1
110 - 119					1	4	1	2
120 - 129					2	2	1	2
130 - 139						1	3	1
140 - 149						0	1	1
150 - 159						1	0	1
160 - 169							0	2
170 - 179							0	0
180 - 189							1	1
No. in sample	155	77	52	38	31	26	22	20
Mean, tonnes	13.3	27.2	40.8	54.3	67.9	81.6	93.3	108.8
S.D., tonnes	10.4	15.8	20.4	26.0	31.1	36.7	38.7	43.6

FORMULATING THE MODEL

It had been found by Coverdale[1] that the herring catches of
Scottish inshore vessels follow a pattern which can be satisfactorily represented by the negative exponential distribution. That
is, individual hauls of x tonnes can be regarded as samples from
a distribution whose probability density function is given by:-

$$f(x) = (1/m) \exp(-x/m), \quad \text{where } m = \text{mean catch}$$

Since this type of function is easy to handle, it had been
useful, e.g. in predicting the proportion of trips when a catching
unit would catch more fish in a single haul than it could stow.
In the context of herring, it was used to predict the effective
utilisation of chilled sea water containers on board fishing
vessels, and hence to get a much better picture of the economics of
the operation than could be gained by using simple average catch
rates.

In order to assess the likely performance of non-freezer
vessels catching blue whiting, it was important to try a similar
method, which if successful, would properly account for the enormous
variation in catch rate, and produce quick answers appropriate for
a pre-feasibility study.

The negative exponential distribution is a special case of a
family known as Erlang distributions, whose probability density
functions take the general form:-

$$f(x) = ((k/m)^k/(k-1)!) x^{k-1} \exp(-kx/m)$$

where k is a positive integer: k = 1 gives the negative
exponential distribution
x is an individual sample (catch)
and   m is the mean catch.

There is some theoretical basis for proposing this type of
approach, as the Erlang distribution can be regarded as describing
an activity which takes place in stages, when each stage behaves
according to an exponential distribution.   Single-species
fishing operations might be expected to fit into such a scheme,
where a single haul represents a stage.   In this case, we should
attempt to represent the result of a sequence of hauls as a sample
from some Erlang distribution.

We start by finding the appropriate value of the parameter k.
In an Erlang distribution, the coefficient of variation (i.e. the
standard deviation divided by the mean) is given by the reciprocal
of the square root of its k value. The best estimate of k for
an actual distribution is therefore given by:-

$$k_{est} = (mean/S.D.)^2$$

It is interesting to examine how the estimate of Erlang k varies with the number of hauls. A regression analysis on the data, weighting each observation by the sample size, gave the relationship:-

$$k_{est} = 2.07 \ln(n) + 1.58, \text{ where } n = \text{no. of hauls.}$$

Fig. 1 shows that this equation gives a good fit to the data. All of the observations lie close enough to the line for the nearest integer estimate of k to be unchanged by the regression analysis.   Intermediate estimates of k can be read from the graph.

Having found estimates of Erlang k parameters, it remains to compare the mathematical distributions with actual data, to see how well the model fits reality.   We first construct Erlang frequency distributions, and then use the chi-square test to see how well they fit the data.   To find the Erlang frequencies, we need the integral of the probability density function, which is given by:-

$$\int f(x) = -\exp(-kx/m) \sum_{i=o}^{k-1} (kx/m)^i / i!$$

The process is therefore straightforward, if somewhat repetitive, as the number of terms in the summation increases with k. In Table 2, the 1-,2-,3- and 4- haul distributions from Table 1 are compared with the appropriate Erlang approximation.   In each case, the Erlang function provides a good fit to the actual data, as none of the chi-square values is statistically significant. Fig. 5 gives a better impression of the validity of the Erlang model for the 1-haul and 3-haul distributions, whilst Fig. 4 shows some of the family of Erlang distributions derived from the analysis.

CALCULATING THE EXPECTED CATCH PER TRIP

If f(x) is the probability density function of catch from N hauls and H is the hold capacity, then:-

$$p(x>H) = \int_{H}^{\infty} f(x) \, dx$$

is the probability of catching more than the vessel can hold. In this case:-

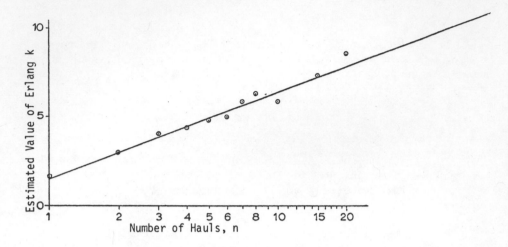

Figure 1. Estimate of Erland K-value v No of Consecutive Hauls, n

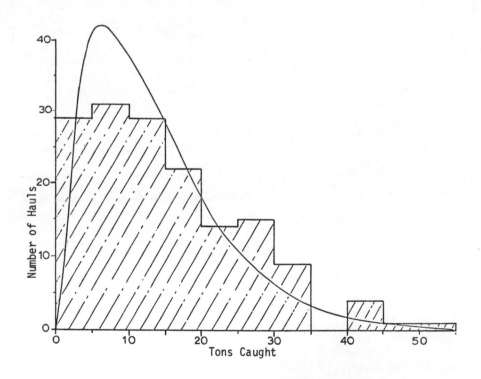

Figure 2. Catch Rate Distribution for 1 Haul of
Blue Whiting, Spring 1976

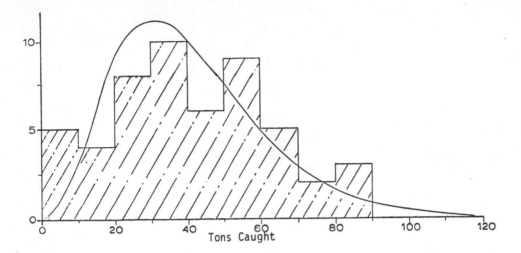

Figure 3. Catch Rate Distribution for 3 Consecutive Hauls of
Blue Whiting, Spring 1976

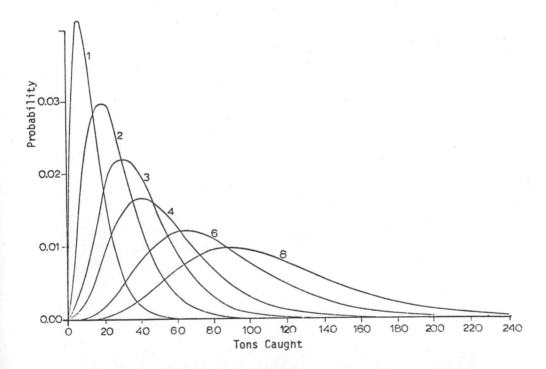

Figure 4. Probability Distribution for Total Catch of
Blue Whiting from Sequence of Hauls

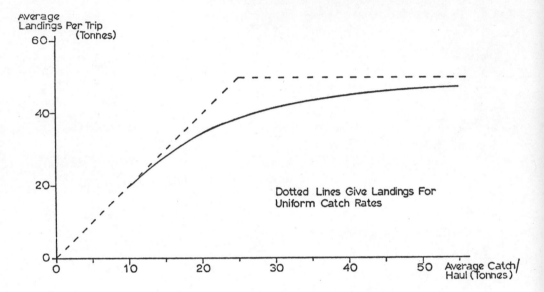

Figure 5. Weight Landed vs Catch Rate for a
50 tonne Capacity Vessel

Table 2   Comparison between actual Polish 1976 catch rate distributions and those expected after fitting the appropriate Erlang function to the data

Catch Tonnes	-- 1 Haul -- Actual	Erlang	-- 2 Hauls -- Actual	Erlang
0 - 9	60	68.7	12	7.8
10 - 19	51	55.6	17	21.7
20 - 29	29	21.3	18	20.3
30 - 39	10	6.7	15	13.3
40 - 49	5	1.9	8	7.3
50 - 59		0.8 (over	4	3.6
60 - 69		50)	2	1.7
70 - 79			1	0.7
80 - 89				0.6 (over 80)
Total	155	155.0	77	77.0

Chi-square (d.f.)   7.6 (3)   3.8 (5)

Catch Tonnes	-- 3 Hauls -- Actual	Erlang	-- 4 Hauls -- Actual	Erlang
0 - 9	5	0.9	1	0.3
10 - 19	4	6.1	3	2.1
20 - 29	8	10.6	5	4.6
30 - 39	10	11.0	2	6.1
40 - 49	6	8.8	5	6.2
50 - 59	9	6.1	5	5.4
60 - 69	5	3.8	5	4.2
70 - 79	2	2.2	6	3.1
80 - 89	3	1.2	1	2.2
90 - 99		1.3 (over	2	1.4
100 - 109		90)	3	0.9
110 - 119				1.5 (over 110)
Total	52	52.0	38	38.0

Chi-square (d.f.)   3.9 (5)   3.7 (3)

None of the Chi-square values is significant

$$\int_{H}^{\infty} f(x) \, dx = \exp(-kH/m) \sum_{i=o}^{k-1} (kH/m)^{i}/i! = A$$

as $f(x)$ denotes an Erlang distribution with parameter k, and mean m.

The average catch when N hauls fail to fill the hold is given by:-

$$\int_{o}^{H} xf(x) \, dx = m(1-\exp(-kH/m)) \sum_{i=o}^{k} (kH/m)^{i}/i = B$$

The long-term average weight landed per trip is therefore $AH+B(1-A)$, which can be worked out with ease on a pocket calculator, although a computer is more convenient for repeated calculations. A computer program has been written, which will tabulate average values of weight landed per trip, no. of hauls per trip, trip duration, no. of trips and weight landed during the catching season, for any values of hold capacity, catch rate, max. no. of hauls/trip, steaming times, etc.

USE AND VALIDATION OF THE MODEL

The average Polish catch rate was fairly low, at around 13 tonnes per haul, but the pre-feasibility study required vessel performances predictions for much higher catch rates. In order to do this, the mathematical model was simply used with bigger numbers, on the assumption that higher catch rates were distributed in the same pattern as in the Polish data. That this extrapolation was valid was shown by demonstrating that some Swedish blue whiting data obtained a year later, with a mean catch rate of 62.5 tonnes per haul, fits the same model in the same way[2] and Table 3. For a given number of consecutive hauls, the shape of the catch distribution can be regarded as fixed, whatever the average catch. (See Fig. 4).

We have noted that the Erlang function, with integral values of k, the "shape-controlling" parameter, is very convenient to handle mathematically. It is therefore no problem to calculate the probabilities required for predicting vessel performance in terms of average landings per trip. Fig. 5 shows an example of results for a 50 tonne capacity boat taking up to 2 hauls per trip over a wide range of average catch rate. It will be seen that the short-fall between the landings predicted by the Erlang analysis and those based on assumptions of uniform catch rate, is greatest when the expected total catch is equal to hold capacity. In this case simple calculations could overestimate landings per trip by up to 37%.

Table 3   Comparison between actual Swedish 1977 catch rate
distributions and those expected after fitting the
appropriate Erlang function to the data

Catch	-- 1 Haul --		-- 2 Hauls --	
Tonnes	Actual	Erlang	Actual	Erlang
0 - 19	16	12.4	1	0.6
20 - 39	17	21.2	1	2.8
40 - 59	13	18.9	6	4.7
60 - 79	16	14.1	3	5.8
80 - 99	7	9.6	9	5.9
100 - 119	9	6.2	6	5.5
120 - 139	9	3.6	3	4.7
140 - 159	2	2.4	3	3.9
160 - 179	2	1.4	4	3.1
180 - 199	0	0.8	2	2.4
200 - 219	1	1.4 (over 200)	3	1.8
220 - 239			1	1.4
240 - 259			2	1.0
260 - 279			1	2.4 (over 260)
Total	92	92.0	46	46.0
Mean	62.51		125	
S.D.	43.10		72.5	
k-value (est)	(2.1)	2	(2.97)	3
Chi-square (d.f.)		8.00 (6)		3.01 (4)

Catch	-- 3 Hauls --		-- 3 Hauls --	
Tonnes	Actual	Erlang (k=4)	Actual	Erlang (k=5)
0 - 39	0	0.3	0	0.1
40 - 79	2	2.5	2	1.8
80 - 119	1	4.8	1	4.6
120 - 159	7	5.7	7	6.1
160 - 199	6	5.2	6	5.8
200 - 239	6	4.0	6	4.5
240 - 279	1	2.8	1	3.0
280 - 319	3	1.9	3	1.8
320 - 359	2	1.2	2	1.0
360 - 399	2	0.8	2	0.6
400 - 439	0	0.8 (over 400)	0	0.7 (over 400)
Total	30	30.0	30	30.0
Mean	187		187	
S.D.	88		88	
k-value (est)	(4.5)	4	(4.5)	5
Chi-square (d.f.)		3.88 (3)		2.59 (3)

None of the chi-square values is significant

CONCLUSIONS

This work demonstrates strikingly how a straightforward analytical model, built from very little data, can sometimes be used well beyond the numerical bounds of the original information. More recent data confirmed the validity of this extension of the model.   It appears sufficiently robust to apply to any fishery exploited by simple vessels, provided that the catch rate distributions can be represented by Erlang functions.

Economic analyses based on this method formed a pre-feasibility study which indicated how the White Fish Authority's research and development programme should proceed towards the commercial exploitation of blue whiting.

REFERENCES

1.   Coverdale, I.L., The Application of a Chilled Sea-Water
        Container System to the Scottish Inshore Fishing Industry.
        M.Sc Thesis, University of Hull (1972).
2.   Hagberg, A., Palmen, L-E., Silverfjall, K-A, Swedish Blue
        Whiting Fishery Trial in 1977, National Board of Fisheries,
        Sweden (1977).

SIMPLE COMPUTATION MODELS FOR CALCULATING

PROFITABILITY OF FISHING VESSELS

Torbjørn Digernes

Vessel Division,
Institute of Fishery Technology Research
Marine Technology Centre, Trondheim, Norway

INTRODUCTION AND BACKGROUND

The Norwegian fishing fleet is above all characterized by diversity. This applies to vessel sizes, fishing gear types, fishing grounds and patterns of operation for the vessels (combination of seasonal fisheries and gear type in the yearly operation cycle).

The economics of fishing vessel operation is therefore an equally diverse matter.

Another typical fact about our fishing fleet is that a large majority of it is owned and operated by active fishermen. The best background knowledge for establishing the necessary data for performing profitability computations lies with the fishermen themselves.

Our experience, however, is that thorough precalculations of operational economics of the projects are being done only to a very small degree. This situation called for the development of simple models for economics of fishing vessel operation. The models would have to be made available to the regional fishery advisory system, which is the body that assists the fishermen in planning vessel projects. A requirement for the models therefore was that they should be simple enough to be operated by people with no training in use of quantitative models. In an investment situation the fishermen seldom consider alternative investment opportunities to fishing vessels. Their problem therefore mostly is stated as, "I wish to invest in this vessel; will I be able to pay interests and instalments on the loans I need?"

Most difficult in this respect, has up to now been the first
year of operation.   If you are able to pull through this first
year, the inflation rate normally helps you to manage your
responsibilities later on.

Therefore, both vessel owners and financing institutions have
considered a first year plan of operation and first year economic
result as being an adequate base for decision.   Consequently,
it was decided to make the models produce a one year economic
result.

THE MODELS

Two models have been developed, KALKYLE and MINIKALKYLE.   These
are designed to supplement each other in the following way,

- KALKYLE gives a thorough, easy to read documentation of the
  planned operation (on a seasonal basis), the underlying assumptions, and the economic results. The report is suitable for
  presenting the project to a third party, e.g. a financing
  institution. The report contains consequence calculations in
  the form of parameter sensitivity analysis, presented both as
  tables and diagrams.   It requires a large computer to run.

- MINIKALKYLE is a much simpler and approximate model, with less
  detail. It is in design and complexity adapted to programming
  on a programmable pocket/desk calculator with 25 storage
  registers and 224 program steps. This constitutes a flexible and
  simple tool for the user to experiment with in an interactive way
  in the development phase of the design of a vessel project. It
  can give accurate enough estimates of the economic result, at
  this stage, but its documentation capabilities are of course
  limited. Its strength is that it is available at the user's
  desk and can produce alternative calculations right away.

The basic idea of both models are

- the vessel revenues are expressed as a function of the operation
  of the vessel, depending upon effective fishing time, amount of
  fish gear used per fishing day, catch per unit gear used, and
  fish price.

- The various cost components are associated with the technical or
  operational factors that incur the cost (e.g. fuel costs
  expressed as a function of engine power and operating time, bait
  costs on longlining expressed per gear quantity used etc.)

The main results produced by the models are

- owners net profit before tax
- crew income, annual, and per working hour
- cash flow balance before tax (where loan instalments are taken
  into account rather than depreciation of capital)
- break even revenue and catch rates, with corresponding crew income
  (calculates necessary change in catch rate or fish price in order
  that the vessel breaks even).
- parameter sensitivity analysis.

The reason for bringing working hours and income per working
hour into the models is that new vessels require extremely long
yearly operating times to cover their costs. This of course results
in extreme man-year working hours. The fishermen's man-year is
often very long, 3000-3500 hours in a year are not unusual, and
with new vessels this figure may in some cases approach 4000. Of
course, normally a single fisherman does not work that long, because
vessels operate some sort of crew rotation system where each man
every now and then takes a fishing trip off. The annual income
per man calculated by the models, however, is for a man staying
aboard during the whole operating time of the vessel. To see this
figure in relation to the working hours by computing hourly income
can be useful.

In Norway, there exist three different trade union agreements
between vessel owners and fishermen's associations. They are
all a bit different in structure, but a common outline of these
agreements is adopted as the account structure of the models, making
it possible to use any of the three agreements. A sketch of this
structure is shown in figure 1.

The models are supported by a data base on cost data. These
data are acquired through analysis of 6-800 vessel accounts
collected annually by the Norwegian Fisheries Directorate. This
data base contribute very much to making the models viable and
practical tools.

KALKYLE

The necessary data for this system is structured into a set
of data sheets, which are:

a) Fish species and their processing mode.
   This information is accompanied by a table of conversion factors
   from caught weight (which catch rates are referring to) to
   delivered weight (which fish prices are referring to) specified
   for each specie and processing mode.

b) Fish prices, specified for species and processing modes,
   which can be given for an arbitrary number of price periods

Gross revenue

- sales & product duty (% of revenue)

- common expences (proportional to operating time)
  · fuel costs
  · victuals   (fuel and victuals may be owner's expences)
  · ice, gutting, gear loss

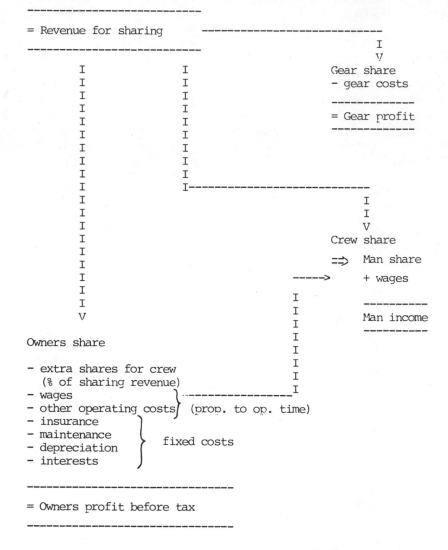

Figure 1. Structure of the Accounts of the Models

per year.

c) Vessel description, containing
   - investment breakdown and corresponding depreciation rates
   - financing plan, including interest rates and payback time of loans
   - other fixed costs
   - gross tonnage and engine power

d) Fishing season descriptions.
   A complete season description has to be made for each seasonal fishery that the vessel operates.
   This description includes
   - plan for vessel operation
     . fishing ground to be used and total operating time
     . no. of fishing days and gear quantity per day
     . catch rate per gear unit used
   These are the basis for calculating revenues. The season may be divided into periods with different catch rates and different fish prices.
   - common expenses for the season, e.g. fuel costs, bait costs, victuals etc.
   - owners operating costs, including gear costs
   - crew shares and wages, plus working hours for the crew.

   The main purpose of this system is to produce an easy to read, clear exposition of assumptions underlying the project and the economic consequences thereof. Much work has therefore been put into the layout of the report, and facilities are provided for the user to choose the volume, of output that he wants. Thus all components of the report are optional, except a very short summary of the results.

   The gross structure of the report, which can be given both for profit calculation and for break even calculation is as follows:

1) Short summary of the results.

   For each season, and summarized for the whole year:

2) Operational data
3) Catch data
4) Seasonal accounts, including crew income, and owners profit
5) Sensitivity analysis tables

   Only on a yearly basis:

6) Complete owner account
7) Sensitivity diagrams.

In addition to this, echo printouts of input data are
optional.

When computing the owner profit for each season, the yearly
fixed costs are distributed according to each season's part of the
yearly operating time.

Sensitivity analysis are carried out on owner profit before
tax and on fishermen income. This is done by varying one parameter
at a time, keeping all others constant. The following parameters
are varied for each season:

i)   Average fish price
ii)   Gear quantity per fishing day
iii) Operating time
iv)   Owner part of sharing revenue.

On a yearly basis:

v)   Cost level of operating costs.
vi)   Vessel investment

You will find in fig.2 and 3 part of a sample report form
KALKYLE, including operational data, seasonal accounts, and
sensitivity diagrams.

MINIKALKYLE

As already mentioned, this model is adapted in size and
complexity to programming on programmable pocket/desk calculators.
We have chosen the Hewlett Packard 67/97 calculators; the job of
course can be accomplished on other calculators of the same capability.
The HP-67/97s are able to read and write programs and data on magnetic
strips, which is a requirement for this application.

The calculations are divided into three steps, each corresponding to one program in the calculator, designed to run in sequence.
One step produces some of the data for the next.   The unit of
calculation is one fishing season.

As an aid in performing the calculations, a data sheet has been
constructed containing a description of all the variables entering
the calculation in each step, their position in the storage
registers of the calculators,and space for filling in values for
actual calculations. The structure of this sheet with a short
summary of data and results is given in fig. 4.

The ACCOUNT step produces the final results from the
calculations. This program is the core of MINIKALKYLE. Data at

OPERATIONAL DATA
================

	GILLNETTING	PRAWN TRAWLING	LINING ON THE BANKS	YEARLY SUMMARY
SEASON				
DURATION FROM	1. JAN	10. MAY	1. OCT	
TO	30. APR	20. SEP	20. DES	
GEAR	GILLNETS	PRAWN TRAWL	LONGLINE	
FISHING GROUND	FINNMARK	BARENTZ SEA	TROMSØFLAKET	
NO. OF MEN ABOARD	10	6	10	
CATCH WEIGHT (KG)	231 000	360 000	198 000	789 000
FUEL QUANTITY (LTRS)	106 329	318 987	80 759	506 076
TOTAL CATCH EFFORT	15400 NET	1500 TRAWLING HOUR	9000 100 HOOKS	
CATCH EFFORT/DAY FISHING	220 NET	20 TRAWLING HOUR	200 100 HOOKS	
CATCH PR. UNIT EFFORT	15 KG/NET	240 KG/TRAWLING HOUR	22 KG/100 HOOKS	
DAYS OPERATING	110	120	80	310
DAYS AT SEA	75	105	55	235
DAYS FISHING	70	75	45	190
CATCH/DAY FISHING (KG)	3 300	4 800	4 400	4.153
FISH PRICE (KR/KG DELIV.)	3.49	5.50	3.22	4.34
MONTHS OPERATING	4	4	3	11

SEASONAL ACCOUNTS
=================

	GILLNETTING	PRAWN TRAWLING	LINING ON THE BANKS	YEARLY SUMMARY
SEASON				
GROSS REVENUE	806 960	1 980 000	637 200	3 424 160
-SALES TAX	24 209	59 400	19 116	102 725
NET REVENUE	782 751	1 920 600	618 084	3 321 435
COMMON EXPENSES				
FUEL COSTS	84 000	252 000	63 800	399 800
BAIT/BAITING COSTS			40 500	40 500
SOCIAL SECURITY TAX	25 016	61 380	19 753	106 149
VICTUALS	27 500	21 600	24 000	73 100
WAGES				
MISC. INSURANCES	2 878	2 691	2 093	7 662
ICE, GUTTING, GEAR LOSS	23 100		18 000	41 100
MISC. OTHER COSTS	5 500	12 000	4 000	21 500
SUM COMMON EXPENSES	167 994	349 671	172 146	689 811
SHARING REVENUE	614 757	1 570 929	445 938	2 631 624
CREW ACCOUNT:				
CREW SHARE (KR AND %)	270 490 44	690 894 44	254 180 57	1 215 564 46
FISHERMAN'S INCOME	27 049	115 149	25 418	167 616
HOURLY INCOME & WORK HHS	20 1 380	65 1 785	25 1 000	40 4 165
AVG. MAN INCOME & WORK HHS	27 725	119 947	26 054	173 726
HOURLY INCOME & WORK HHS	20 1 380	67 1 785	26 1 000	42 4 165
GEAR SHARE (KR AND %)				
OWNERS ACCOUNT:				
OWNERS SHARE (KR AND %)	344 267 56	880 035 56	191 758 43	1 416 060 54
OWNERS OPERATING COSTS	113 544	177 519	51 179	342 242
OPERATING SURPLUS	230 723	702 516	140 579	1 073 818
OWNERS FIXED COSTS	461 681	503 652	335 768	1 301 100
OWNERS PROFIT BEFORE TAX	-230 958	198 864	-195 189	-227 282

Figure 2.   90 Feet Combination Vessel

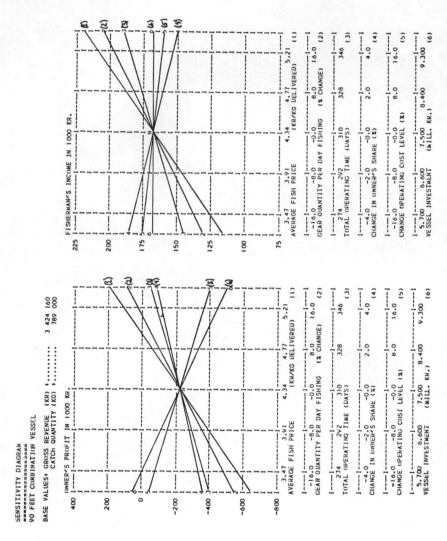

Figure 3.   Program: Kalkyle – V6:0 Run on 29:06:79 for: FTFI-DEMO

this step are aggregated as much as possible with the goal in mind
to still maintain the structure in figure 1.   The function of the
first two steps is to produce the necessary aggregated input data
to ACCOUNT.   In addition they produce some byproducts of interest.
The fourth column of the sheet is identical to the third, and is
used for taking down the break even values, which are calculated
by the ACCOUNT step.

The MINIKALKYLE model is simple, with a limited number of
parameters, and therefore manageable to comprehend.   In addition,
the user has full access to all parameters and he can in an interactive way manipulate them freely to investigate his particular
problem.   For instance he can perform sensitivity analysis with
dependent and independent variables of his own choice.   One example
is a sensitivity diagram of break even catch rate, with fish price,
gear quantity, owner percentage of sharing revenue, and vessel
investment as independent variables, which is shown in figure 5.

The model is easily adapted and/or extended to attack the
problem of operation of fishing vessel economics from different
angles, either by using manual calculations to pre- or post-process
data for the model, or modifying the equations in the formal model.
Examples of this will be given in the next section.

APPLICATIONS AND ADAPTIONS

The models have been used to calculate expected profitability
for several fishing vessel investment projects.   Besides these,
especially MINIKALKYLE has been used on several other applications.

a) A comparison between medium to high speed 50 feet coastal fishing vessels and conventional designs of the same size.

   This was a comparison under the assumption that the time gained
   by higher speed is used to increase fishing time on the grounds.
   Of course the model is not comprehensive to the extent that the
   consequences of a speed and engine change in this way can be
   accounted for directly through the analytical formulae in it.
   The task required manual adjustment of several parameters,
   including
   - speed, engine power, and engine use
   - fishing time on the grounds per trip
   - the crew's working hours
   - gear quantity used per fishing day
   - insurance and maintenance
   - vessel investment and depreciation

   This study required the complete three-step calculation
   procedure of MINIKALKYLE.

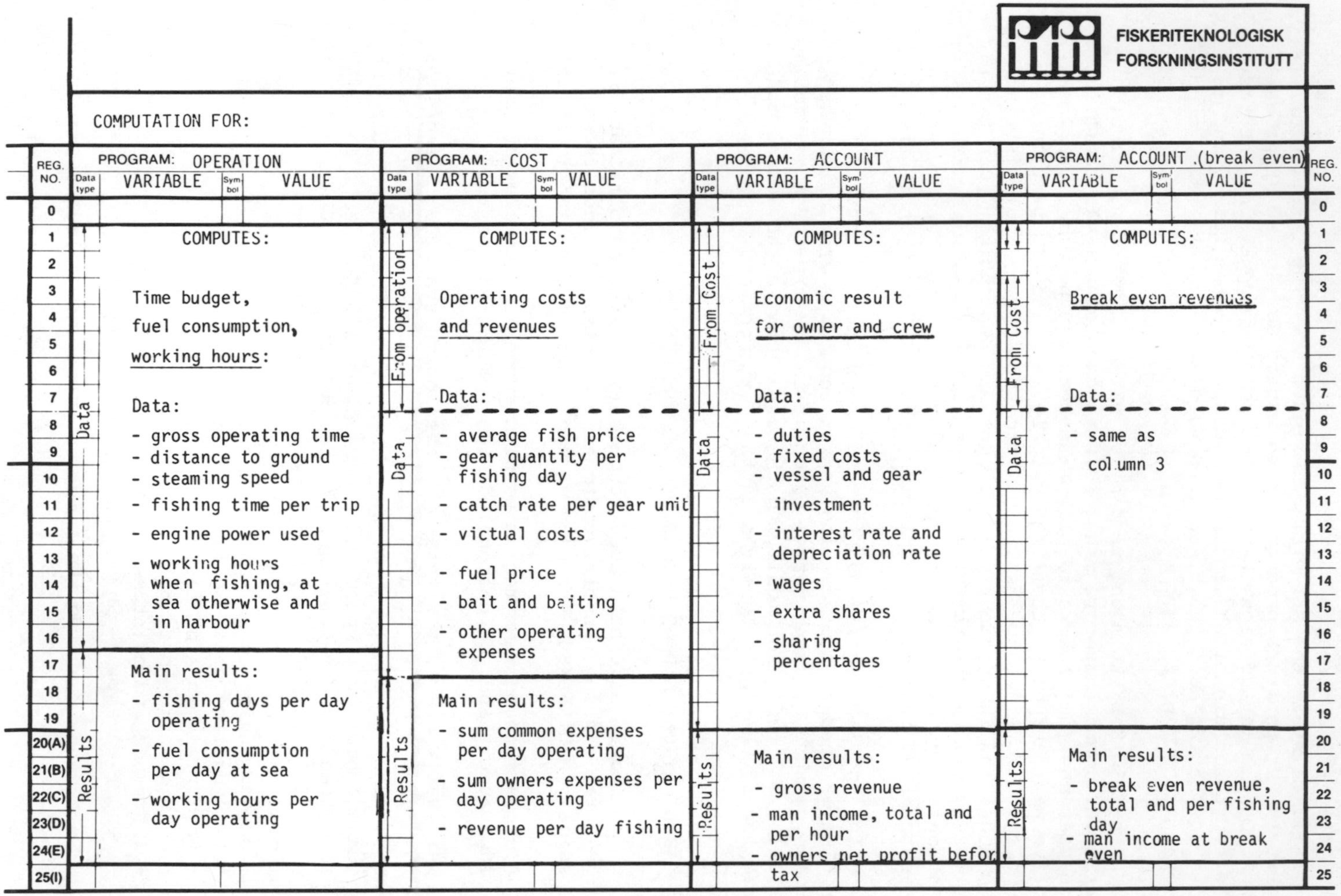

Figure 4. Structure of Data Sheet for Minikalkyle HP-67/97

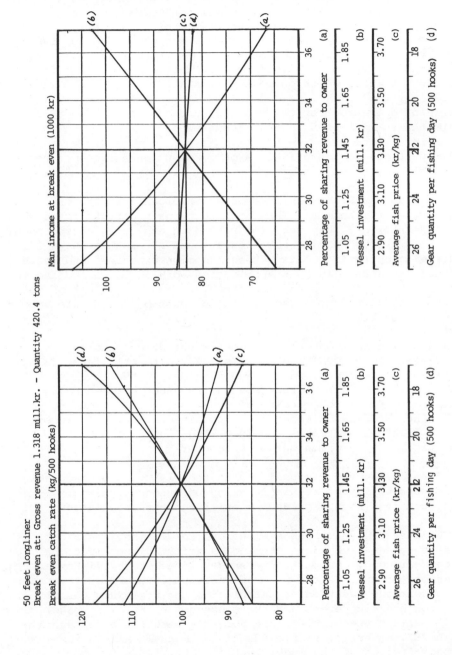

Figure 5. Sensitivity Diagram for Break Even Catch Rate

b) Operating economics for newly built vessels.

This study was done for the Norwegian governmental bank which
finances most of the fishing vessels in Norway.   The background was a sharp increase in vessel building costs and other
costs over the last 2-3 years, and the bank wished to clarify
the vessel operation economics for new vessels of ten
representative vessel types in the fleet. This called for
rough estimates based on collected statistics on operating
revenues and costs. We did this without making any detailed
plan of operation for the vessels, and only step ACCOUNT of
MINIKALKYLE was used in the calculations.

c) Calculations for negotiations on trade union agreements between
vessel owners and crew.

The Norwegian Fishermen's Association organizes both vessel
owners and fishermen. These two sections negotiate trade union
agreements, and they were interested in exploring consequences
of alternative sharing percentages, and alternative ways of
dividing operating costs between "common expenses" and vessel
owner expenses. In this application working hours and income
per working hour was an important part, as the amount of work
associated with the vessel operation must be taken into account
in this context.

In addition to these pure parameter variation studies, we
modified the ACCOUNT step of MINIKALKYLE to calculate "fair"
sharing percentages, based on the principle that each party
should have a part of the "sharing revenue" corresponding to
his part of the total costs of the fishing operation. Costs
for the owner is calculated the traditional way, while cost of
labour is calculated as total working hours for the crew,
multiplied by a price per working hour.

Presently, we are discussing with the Fishermen's Association
the possibility of regularly using both these types of
calculations as background material for the negotiations.

The last 3-4 years the serious decline in many important fish
stocks, together with high catching capacity in the fleet, has
made catch regulations part of everyday life for the fishermen. The
regulations often take the form of individual vessel quota, and in
this context the question of calculating the necessary vessel
quotas for breaking even, and the consequence of fixed vessel quotas
arise.   Both these problems are solvable within the framework of
MINIKALKYLE, either by simple additional manual calculations, or
by a modification of the model itself.

Consequence of fixed quota

1) By using the estimated catch rate, compute the number of
   fishing days required to fish the quota, and the number of
   operating days corresponding to this fishing time. Use this
   figure as new operating time.

2) Adjust this seasons part of the total operating time to account
   for change in fixed cost chargeable to this season (this is
   done only if the new time is longer than the original, or,
   when shorter, if other fisheries are available in the time
   saved).

3) Perform an ordinary profit calculation.

Necessary quota.

   A necessary quota is only relevant if the profit from a free
fishing situation comes out positive. Then it can be computed like
this:

1) Put all fixed costs to zero (depreciation, interest and other
   fixed costs). The result from a profit calculation then will
   be the operating surplus for the owner, which is divided
   by the operating time to give operating surplus per day.

2) Then calculate all fixed costs for the season (can be done by
   zeroing operating costs).

3) Necessary operating time will then be fixed costs divided by
   operating surplus per day.

4. Perform an ordinary profit calculation with the new operating
   time which now will yield necessary gross revenue, which is
   readily transformed to necessary quota.

CONCLUSION

   Two models for computing fishing vessel profitability have been
developed. They are practical, down to earth, purely exploring
consequences of assumptions and data given by the user. They
contain no optimization elements.

   The models are now being introduced to the fishing community in
Norway, through courses for people who have various aspects of
fishing vessel economics as part of their daily work. Thus
representatives of the Norwegian Fisheries Directorate, the
Norwegian Fishermen's Association, the regional advisory service
for fishermen, and educational institutions have participated, and
the responses from these have been positive.

It is this author's opinion that the people who work with the practical problems should be supplied with comprehendable, not too complicated computation tools.   When they have such tools at their disposal, tedious calculation work is eliminated.   Such work can otherwise provide an obstacle to clarifying the consequences of their assumptions.   With models at their hand they can concentrate on being problem solvers rather than being calculators.   The result should be that the decisions taken are founded on more extensive knowledge.

In closing a few words with a bearing beyond the fishing industry.   The microprocessor age has through the development of programmable pocket calculators made computing power available at everyones working desk at a low price, and without the barrier of having to learn how to handle a computer, with its operating system and other "administrative overhead".   No doubt we have only seen the beginning of the development of the computing capabilities that can be put into such calculators.

This situation opens up the possibility to bring quantiative models into wider use outside of the researchers' workshops.   It provides a challenge for the operations research community to develop simple and well structured models that can be mastered by people not trained in OR-disciplines.   The market for this type of models is bound to be huge.

REFERENCES

1. T. Digernes, MINIKALKYLE - Operating Economics of Fishing
        Vessels on Programmable Desk Calculators. Report - Institute
        of Fishery Technology Research June 1979.
2. T. Digernes, KALKYLE - A System for Computing Operation
        Economics of Fishing Vessels. Report - Institute of Fishery
        Technology Research. Preliminary Edition, June 1979.
3. T. Digernes, Th.B. Melhus, H. Erstad, Operating Economics of
        Newly Built Fishing Vessels - The Cash Balance in First Year
        of Operation. Technical note - Institute of Fishery
        Technology Research, January 1979.
4. H. Aasjord, T. Digernes, Operating High-Speed Fishing Vessels.
        Technical note - Institute of Fishery Technology Research,
        November 1978.

All the referenced material is in Norwegian.

A SIMULATION MODEL OF THE CAPELIN

FISHING IN ICELAND

Pall Jensson

University of Iceland

Reykjavik, Iceland

INTRODUCTION

In the last decade, or since the herring disappeared from the Icelandic fishing waters in 1967-68, the capelin has been by far the most important industrial fish for the production of meal and oil in Iceland.

The importance of the fishing industry for Iceland is clearly expressed by the fact that frozen fish, meal and oil account for more than 70% of the export value.

The capelin fleet of approximately 70-80 boats, ranging in capacity from 130 to 1300 tons, catches about 500.000 tons in the winter season, which is the first three months of each year.

This was in 1977 estimated to be less than 10% of the capelin school which every year migrated around the country during the winter seasons.

Fig. 1 shows the behaviour of the capelin school in 1977. This pattern has been practically the same for the last 5 years. Also the location of the 23 factories equipped for processing of capelin (in 1977) is shown.

The factories are different in terms of processing capacities, ranging from 150 to 1.200 tons/day, capelin storage capacities from 1.200 to 16.000 tons, number of storage tanks from 3 to 8 and the capacity of unloading facilities from 100 to 300 tons/hour.

In total the factories can store up to 150.000 tons of capelin

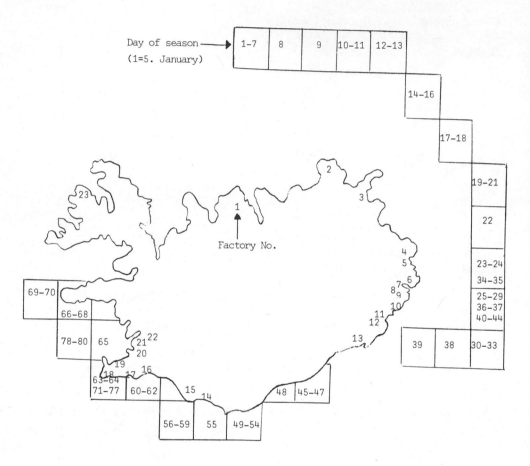

Figure 1. Movement of Capelin School aroung Iceland and Location
          of the Factories

and during the season they are able to process more than 1 million tons. However, the utilization of an average factory is less than 50% over the season, as it is short of raw material before and after the capelin school passes.

Decisions as to where to unload the catch are made by the captains, who usually prefer harbours in the nearest vicinity of the capelin school at any time, even though they can expect long waiting times. The result has been that the capelin fleet has spent up to 30% of its time waiting in harbours.

THE PROBLEM

In 1977 various questions had been raised about the efficiency of the capelin fishery in Iceland, such as where or whether processing capacity should be increased, whether storage capacity should be increased, what the fleet size should be, what the effects would be of using factory ships or specialized ships for transportation etc.

The main question was this: If the decision pattern of the captains could be changed so that they would sail longer distances instead of waiting, what would be the effects on the total catch, on the utilization of the different factories and on the different size categories of boats?

As a centralization of decisions was considered to be out of the question, it was proposed to motivate sailing to factories further away by a new price structure, which should be fixed before the season started: A certain base price plus a certain price proportional to the distance sailed.

This was the main proposal that the fishing authorities had to decide on in the summer of 1977. The various consequences of the proposal had to be evaluated carefully and answers given in great details to the questions mentioned above within a time limit of three months.

To fulfill these requirements a detailed simulation model of the capelin fishery was designed.

THE SIMULATION MODEL

The model was programmed in the GPSS V simulation language. The program is about 830 statements (excluding comments), whereof only 230 statements describe the simulation logic with 600 statements used for initial declarations and output editing.

   The reader should bear in mind that about 75 boats have to be
monitored through the main loop of a fishing trip for 80 days,
resulting in almost 2000 trips.   In each trip a captain chooses
one harbour out of 23, and the storage conditions and the processing
at all factories must be followed in detail.

   By using GPSS, only one trip and only one factory had to be
described (in 130 statements), the others being "duplicated" by
GPSS by use of parameters and indices.

   The simulation logic is outlined in the following sketch of
the model, which also serves to explain the use of GPSS features
(shown by capital letters) in a simplified way.

Program module: Description:

Initialization   GENERATE a "mother"-TRANSACTION,
                                   which initializes a group of SAVE-
                                   VALUES for general assumptions for
                                   the run.

Creation of   SPLIT the "mother" into boat TRANSAC-
the fleet   TIONS and ASSIGN values to their
                                   PARAMETERS describing each category
                                   of boats (6 categories were used).
                                   The "mother" is TERMINATED.

Start of season   TRANSFER part of the fleet to ENTER
                                   the fishery each day and ADVANCE
A   the rest one more day.

Sailing to the   ADVANCE the boats according to a daily
fishing ground   update MATRIX (see later) of sailing
                                   times for empty boats of each category
                                   from each harbour. Here as in other
                                   modules QUEUE and DEPART blocks are
                                   used to gather statistics for each
                                   category of boats.

Weather OK?   The boats may be stopped at a GATE
                                   controlled by a LOGICAL SWITCH (see
                                   later).

Fishing   ADVANCE the boats according to the
                                   distribution of fishing time and
                                   ASSIGN to a PARAMETER a value from a
                                   distribution FUNCTION for the catch.

Choosing a harbour

EXAMINE all factories and LOOP through
the relevant subset to calculate the
revenue and expected time for the trip
for each factory.   SELECT a factory
by using the decision rule assumed.

Updating the
factory

"book-keeping"

A detailed "book-keeping" for the
selected harbour is updated (by use
of VARIABLES and SAVEVALUES).   This
covers f.ex. expected time when
registered boats have all been unloaded,
storage level at that time and a
possible waiting time for an empty tank.

Sailing to
the harbour

ADVANCE the boats according to a
MATRIX of sailing times for loaded
boats.

Waiting in
the harbour

In the QUEUE the boats are LINKED to a
USER CHAIN, possibly even if there are
other boats as a GATE might be closed
indicating factory STORAGE FULL.

Unloading

(Boat)

(Load)

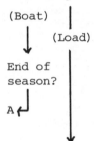

SPLIT the load off in units of f.ex.
100 tons. The unloading facility
keeps the boat until it has unloaded
all the catch. Then the boat UNLINKS
the next boat and DEPARTS.

End of
season?

A

TEST for the time to see if boats shall
be TRANSFERRED to the next trip.
During the last weeks a certain part
of the fleet LEAVES the fishery and
is TERMINATED.

Pump or crane
facility

Each load unit is SEIZED by the unloading FACILITY, ADVANCED and then
RELEASED to the storage.

Capelin tanks

The load ENTERS the capelin STORAGE
and is LINKED to a USER CHAIN.

Processing

Processing occurs from one of the
tanks at any time, and is ADVANCED
according to the capacity of the
factory. Load units are TERMINATED
after the processing.

Daily updating   GENERATE a "day"-TRANSACTION for
                        updating MATRICES of sailing times
                        etc. if the capelin school has moved.
                        Also a LOGICAL SWITCH is regulated
                        according to a FUNCTION for weather
                        conditions.

ASSUMPTIONS

     Although the model description is simplified it should
indicate how close the model comes to reality.   All the earlier
mentioned characteristics of the different factories are included,
as well as those of the different boat categories: Capacity,
speed loaded and empty and fuel consumption.

     As stochastic variables one can include fishing time and the
quantity caught (depending on boat category), sailing times and
weather conditions based on storm statistics. Furthermore the
movements of the capelin school can be regarded as stochastic
although the real behaviour during a given season can be copied
deterministically for comparison reasons.

     The unit of catch is one ton, and the unit of distance one
nautical mile. A matrix was set up for distances taking into
account the coastline between 27 possible fishing grounds and 23
harbours.

     The unit of time was chosen as 1/10 of an hour. This had to
be balanced against the unit of capelin quantity in processing
(which was chosen as 100 tons) to be able to reflect differences
in processing capacities.   The unit of time seems small but it is
not critical for the computing time as it would be in a time-slice
simulation.   Here the number of events is the major factor.

     The most critical assumption is without a doubt the one of
how a captain decides where to unload. Many factors enter this
consideration in real life, such as weather, ocean currents,
possible overloading of the boat, the price for the load and, last
but not least, the expected time to get back to the fishing ground,
estimated from distances and information from a central institution
about length of queues and the status of factory storages.

     On the basis of several interviews it was decided that it would
not be possible at least within the time limit to establish a
detailed model of the preference pattern of the captains. Hence
the following was assumed as a rational decision rule:

     A captain selects the harbour which maximizes his revenue/
time unit = {price x load - fuel cost} divided by the time for the

trip.   This means that a captain is "price-conscious", such that
when no price-motivation is present (same price at all factories)
he minimizes the time for the trip.

RESULTS

    Figures 2-4 show some results from simulation runs based on
data from the 1977 season and with the following price-structures
(Iau = Icelandic "aurar" pr. kg. of capelin):

    I. Base price 1.000 Iau + 0 Iau pr. mile sailed(pr.kg.)
    II.   "   "   700 " + 3 " "   "   "   " "
    III.   "   "   500 " + 5 " "   "   "   " "

    Figure 2 shows the accumulated total catch as predicted by
the simulation model during the season of 80 days.   Also the
actual result of the 1977 season is shown.

    When validating the model a difference like the one in Fig. 2
between the 1977 season and the {1000,0}-run was accepted by all
parts involved as "a natural consequence of the difference between
real and ideal decision making".

    Figure 3 shows how price-motivation moves time from waiting
to sailing with a net decrease in total time for the trip.   It is
clear that further decrease (increase in total catch) cannot be
expected as a result of this price structure and with decentralized
decisions.

    The results in Fig. 4 show that the bigger boats have been
motivated to sail longer distances, leaving space in the nearest
factory storage to the smaller boats.   Still the bigger boats are
not "sacrificed" as they get much higher total price for their load
(not shown here).

    To summarize these results: The simulations revealed clearly
that too much time is spent on waiting. Motivating the captains
away from individually optimal decisions increase the performance of
the system, without leaving anyone worse off.

    The general effect of self- versus social- optimization is well
known among queueing theorists. In the case of the capelin fishery
it is treated analytically in Jensson.[2] (1975)

    However, the simulations also showed that a maximum increase of
10% in the total catch could be expected from the proposed price
structure, and probably less, considering the assumption of "price
conscious" captains. A 10% increase in the total catch would mean
more than $3 millions in export value.

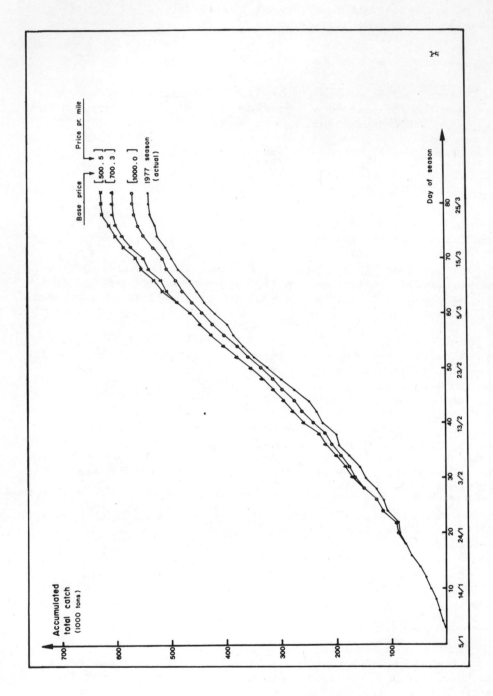

Figure 2. Accumulated Total Catch

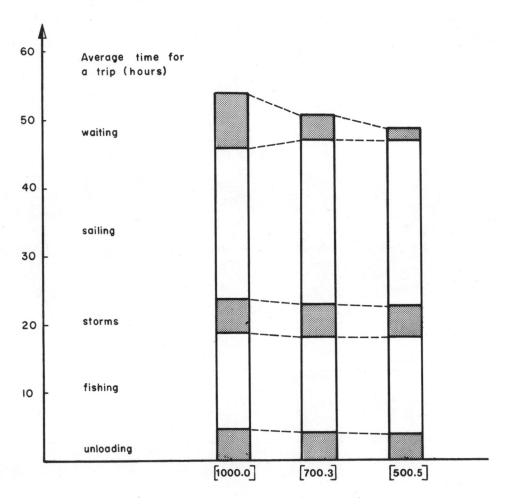

Figure 3. Average Time for a Trip

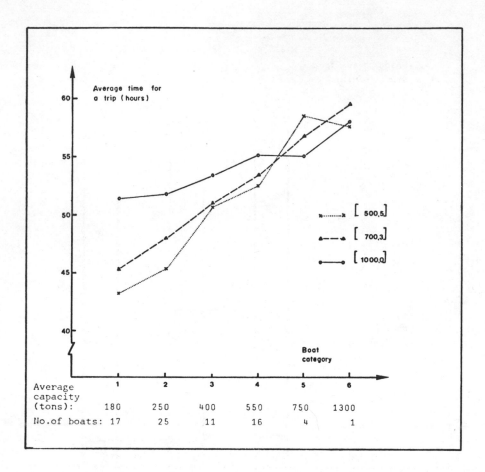

Figure 4. Average Time for a Trip for the Different Boat Categories

Still, the price structure was considered to be a very radical change. The capelin fishing authorities decided on basis of the simulations that the expected return was not enough to justify the proposal.

When this is written, other motivation methods have not yet been evaluated by the fishing authorities by using this model. Optimization models are discussed in the references. Hannibalsson[1] (1978) concludes that up to 30% increase in the total catch could be expected by centrally controlled allocation. However, this looks rather unrealistic in the light of the results in Fig. 3.

Translating results from optimization models into acceptable decision rules and testing by use of the simulation model would be a natural continuation, even though the most serious problems seem to be in the implementation phase.

SOME LEARNING POINTS

In this paper, a case study of applying simulation has been described. The project was run as a consultant task and took about three months. Of the approximately 1 1/2 month spent in GPSS-programming, it took only 1/2 month to make a usable first version of the program. However, this version was very ineffective in terms of computation time.

A month was spent to improve this situation and a seven-fold reduction was achieved by using features in GPSS, which can hardly be utilized except by well-trained GPSS-programmers.

The general experience in the use of GPSS is that it is extremely easy to learn and a detailed model can be programmed with minimal effort. However, in terms of the bill for computer runs the language can be dangerous in the hands of programmers with little GPSS-experience.

An interesting biproduct from the simulation modelling was the design of a "book-keeping system" used in the model to keep track of the various factories and boats. This could be useful for design of an online computerized information system for the capelin fishing authorities.

Since 1977, the model has been used by the capelin fishing authorities for various evaluations mainly concerning capacity expansions. The simulation approach has, like in many other studies, proved its capability of answering "what-if" questions on the requested level of detail. Compared with the alternative of an optimization model, it seems more likely to build up credibility among decision makers and others concerned, a credibility which is necessary for cooperation during the implementation.

REFERENCES

Hannibalsson, I., 1978, "Optimal Allocation of Boats to Factories
    during the Capelin Fishing Season in Iceland". Ph.D. Thesis,
    The Ohio State University.

Jensson, P., 1975, "Stochastic Programming Vol.II: Methodological
    and Applicational Contributions" (in Danish). Lic. Techn. Thesis
    IMSOR, Techn. Univ. of Denmark.

MODELING TECHNIQUES AS A MEANS TO OBTAIN OCEAN WAVE- AND CURRENT-DATA

BY USE OF METEOROLOGICAL ANALYSES OF SEA LEVEL PRESSURE AND WIND

Odd Haug

Meteorologiske Institut
Det Norske Meteorologiske Institut
Blinden, Oslo 3, Norway

As an introduction to this lecture, I think it will be appropriate to look into the variability of the meteorological parameters from year to year. This will quite clearly show the necessity of using data series of considerable length in order to obtain reliable probability statistics for meteorological parameters and oceanographic parameters which are chiefly governed by the wind conditions and/or sea level pressure.

Fig. 1 shows the number of days per year with wind force above 40 kts on a lighthouse station (Utsira) on the western coast of southern Norway. As one can see, there exist continuous periods of several years with storminess either much below or much above the mean conditions. and it is obvious that criteria derived from short series of measurements might be rather misleading if they are taken to represent normal conditions.

In meteorology, data series of close to 30 years of length are regarded as necessary to have acceptable statistical control of the phenomena. The same requirements obviously also apply to series of oceanographic data where meteorological parameters are the generating factors.

We may now look into what kind of data are available which may help us to describe the oceanographic climate in specific waters, and will primarily concentrate on ocean waves.

First of all, there are series of measurements from Waverider buoys and other similar instruments which give a detailed picture of the spectral distribution of wave energy. These series are usually of limited length.

Secondly, fixed ocean weather ships were put into operation shortly after the last world war. The meteorological observational program has included visual observations of significant wave height and period. From these ships positions we have thus rather long series of wave data, but with the inaccuracies one has to accept by visual observations.

Thirdly, coastal observing stations - in particular lighthouse stations - have long series of visual observations of the state of sea, although these data are limited to a rather coarse classification of the sea.

Finally, a potential source of oceanographic information are the long series of sea level pressure maps which are produced several times a day in a regular weather forecasting office. By modelling techniques, as will be described later, ocean wave data and to a certain degree also ocean current data may be derived from the weather maps.

Each of these four categories of wave data have their weakness. The series of measured data are rather short, the observations do not contain information of directions, and they do not separate between winddriven sea energy and swell energy.

The visual observations from ships suffer from systematic errors due to the human factor involved.

The observations from lighthouse stations are rather crude, and are strongly influenced by the coastal geography.

The wave data obtained from sea level pressure maps may contain errors from different sources, like inaccuracies in the analysed isobar maps due to sparse coverage of meteorological observations over the sea, and approximations in the models for computation of wind from the isobaric fields and waves from the computed winds.

It has turned out that inadequacies of the different types of wave data may be greatly reduced by extensive use of the possibilities for intercalibration of the data series at hand. We will come back to some of these possibilities later on.

But first we will briefly describe the modelling techniques for wind and waves, and start by the wind.

The equations of motion for the atmosphere may be written, in vector form,

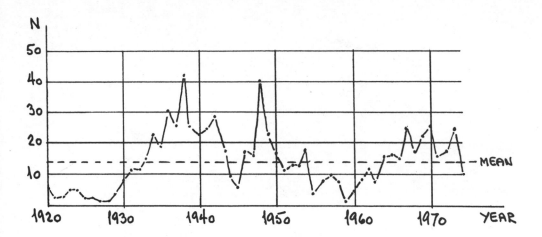

Figure 1. Number of Days with Wind Speed above 40 KTS at Utsira

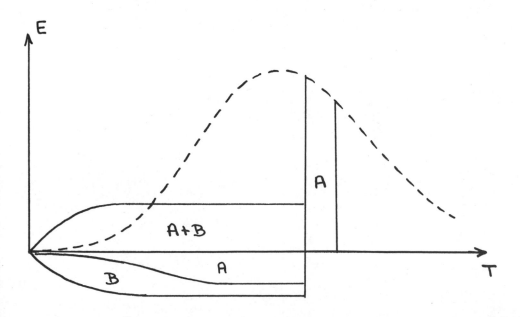

Figure 2

$$\frac{\partial \underline{v}}{\partial t} + \underline{v} \cdot \nabla \underline{v} = - \nabla \phi - f \underline{k} \underline{x} \underline{v} + K \frac{\partial^2 \underline{v}}{\partial z^2}$$

We may, for our particular problem simplify the equation by replacing the individual time derivative of the wind v by the individual derivative of the geostrophic wind, defined by

$$\underline{v}_g = \frac{1}{f} \underline{k} \times \nabla \phi$$

The equation of motion then takes the simplified form

(1) $\quad \dfrac{\partial \underline{v}_g}{\partial t} + \underline{v} \cdot \nabla \underline{v}_g = - \nabla \phi - f \underline{k} \underline{x} \underline{v} + K \dfrac{\partial^2 \underline{v}}{\partial z^2}$

As a first step the wind components u' and v' are computed from the following equation

(2) $\quad \dfrac{\partial v_g}{\partial t} + \underline{v}' \cdot \nabla \underline{v}_g = - \nabla \phi - f \underline{k} \underline{x} \underline{v}'$

The finite difference form of u' and v' derived from this equation is

$$u' = \frac{-(r\Delta_y z + k\Delta_x \Delta_t z)(c+\Delta_y \Delta_y z) - (r\Delta_x z - k\Delta_y \Delta_t z)(\Delta_x \Delta_y z)}{(c+\Delta_x \Delta_x z)(c+\Delta_y \Delta_y z) - (\Delta_x \Delta_y z)^2}$$

$$v' = \frac{(\dot{r}\Delta_x z - k\Delta_y \Delta_t z)(c+\Delta_x \Delta_x z) + (r\Delta_y z + k\Delta_x \Delta_t z)(\Delta_x \Delta_y z)}{(c+\Delta_x \Delta_x z)(c+\Delta_y \Delta_y z) - (\Delta_x \Delta_y z)^2}$$

where

$$k = \frac{d}{\Delta t} \frac{N^2}{N^2+n^2} \qquad\qquad \Delta_x z = \frac{1}{2}(z_{1,0} - z_{-1,0})$$

$$r = 2\Omega d \frac{N^2(N^2-n^2)}{(N^2+n^2)^2} \qquad \Delta_y z = \frac{1}{2}(z_{0,1} - z_{0,-1})$$

$$\Delta_t z = \frac{1}{2}(z_{t+\Delta t} - z_{t-\Delta t})$$

$$c = \frac{r}{g} \qquad\qquad\qquad \Delta_x \Delta_x z = z_{1,0} + z_{-1,0} - 2z_{0,0}$$

$$\Delta_y \Delta_y z = z_{0,1} + z_{0,-1} - 2z_{0,0}$$

$$\Delta_x \Delta_y z = \frac{1}{4}(z_{1,1} + z_{-1,-1} - z_{1,-1} - z_{-1,1})$$

N = distance pole - equator, n = distance pole - gridpoint
d = gridsize, Δt = timestep, Ω = earths angular velocity
Map projection : Polar stereographic

In points with anticyclonic relative vorticity a simplified version of equation (2) is used by replacing v' in the convective term by v_g. This gives the following formulas for u' and v'

$$u' = \frac{-(r\Delta_y z + k\Delta_x \Delta_t z)c - r\Delta_x z\Delta_x \Delta_y z + r\Delta_y z\Delta_x \Delta_x z}{c^2}$$

$$v' = \frac{(r\Delta_x z - k\Delta_y \Delta_t z)c + r\Delta_y z\Delta_x \Delta_y z - r\Delta_x z\Delta_y \Delta_y z}{c^2}$$

In order to take into account the frictional effect, equation (2) is subtracted from (1). Neglecting the terms $(v - v') \cdot \nabla v_g$ and $\frac{\partial^2 v'}{\partial z^2}$ one arrives at the Ekman equation

(3) $K \dfrac{\partial^2 (\underline{v} - \underline{v}')}{\partial^2 z} = f \, k \, x \, (\underline{v} - \underline{v}')$

Assuming a Prandtl layer between sea level and anemometer height and using the 'Prandtl' stress at anemomenter height in the form

$$\zeta_O = \chi \rho |\underline{v}_O|^2 \quad \chi \sim \frac{1}{\sqrt{|\underline{v}_O|}}$$

as boundary condition for the Ekman problem, one obtains the following solution for anemometer height

$$\frac{\sin^2 \alpha}{(\cos\alpha - \sin\alpha)^3} = \frac{A}{f} |\underline{v}_O'|$$

$$|\underline{v}_O| = (\cos\alpha - \sin\alpha) |\underline{v}_O'|$$

where α is the angle between $\underline{v}_O'$ and $\underline{v}_O$.

In order to take into account the static stability of the air in a rough manner, the following empirical formula for $\frac{A}{f}$ is used

$$\frac{A}{f} = (0.8 + 0.4 \frac{v_{ns}}{\sqrt{u^2 + v2}}) . 10^{-2} \quad , \text{ units for wind: m/sec.}$$

where v_{ns} is the north-south component of the wind.

The wave model which was developed at the Norwegian Meteorological Institute in 1967 has with a few modifications been in daily routine operation at the Institute for the last 11 years.

The model distinguishes between the two major states of ocean waves, Sea which is a system of waves under continuous influence of the wind, and Swell which is a system of waves that has escaped the influence of the generating wind field.

In order to work out a numerical scheme for calculation of sea energy, some basic principles have to be established about the nature of the sea.   The following assumptions are forming the basic concepts of the model:

a.   The amount of sea energy and its frequency distribution is at any time uniquely determined as a function of wind velocity and maximum wave period.

b.   The directional distribution of sea energy is at any time uniquely determined as a function of the angle between wind direction and direction of propagation of the waves.

In the following we will assume that the sea energy is given by the Pierson-Neumann-Moskowitz spectrum, which may be written in the following form

$$E = \frac{2}{\pi} KU^5 \int_O^S \int_{-\frac{\pi}{2}}^{\frac{\pi}{2}} s^4 e^{-s^2} \cos^2\alpha \, d\alpha \, ds \ , \quad s = \frac{kT}{U} \ , \ S = \frac{kT_{max}}{U} \qquad (1)$$

where   U = wind velocity
         T = wave period
         α = angle between wave direction and wind direction
     K,k = constants

Integrated over directions, the Pierson-Neumann-Moskowitz spectrum takes the form

$$E = KU^5 \int_O^S s^4 e^{-s^2} ds$$

Under the influence of a wind force, a certain net amount of energy per time increment is added to the wave energy.   In accordance with the basic principles we have stated for our model, this amount has to be added to the spectrum by increasing the upper limit of integration S.

In Fig. 2, an attempt is made to describe in a rough manner the mechanism of energy transfer which is necessary if the assumption of a unique spectral form of the sea energy is going to hold.

Let the area marked by A+B represent the energy input from the wind, distributed over the period range of the spectrum. This area has to be compensated by an equally great area of energy loss in order to maintain the spectral form.

The energy loss will consist of two parts, energy dissipation - indicated by the area marked B -, and transfer of energy by non-linear effects towards the longer-period components of the spectrum, forming the net gain of energy over a time interval.

It should be noted here that the shape of the areas are very much simplified, and do not attempt to give a correct description of the real physical phenomena.

The local change of sea energy per time unit is a result of two effects, namely propagation of energy and transfer of energy from the air to the sea. In the following, these two effects will be treated separately.

We assume that a wave component is travelling with the group velocity for waves on deep water

$$c = \frac{gT}{4\pi}$$

Where T is the period of the wave component.

We may now define a mean speed of propagation for the total sea energy by

$$\bar{c} \int_{O}^{S} s^4 e^{-s^2} ds = \int_{O}^{S} cs^4 e^{-s^2} ds$$

or

$$\bar{c} = \frac{g}{4\pi k} . U . f(S) \quad , \quad f(S) = \int_{O}^{S} s^5 e^{-s^2} ds / \int_{O}^{S} s^4 e^{-s^2} ds$$

We assume, for the purpose of numerical calculations within a short interval of time, that we may approximate the directional distribution of energy by splitting up the energy in two components, one propagating in the x-direction and the other in the y-direction of a Cartesian coordinate system.   The amount of energy and direction of propagation for the two components are given by

$$E_x = \cos\phi |\cos\phi| E$$

$$E_y = \cos(\frac{\pi}{2} - \phi) |\cos(\frac{\pi}{2} - \phi)| E$$

Where ϕ is the angle between x-axis and the direction of wind.

After some calculations, which we will not go into here, we arrive at the following finite difference scheme for the local change of energy due to propagation

$$(E_x)_{t+\Delta t} = (E_x)_t \pm \frac{\Delta t}{d} \left[(\bar{c}|E_x|)_x - (\bar{c}|E_x|)_{x\pm d} \right]_t$$

$$(E_y)_{t+\Delta t} = (E_y)_t \pm \frac{\Delta t}{d} \left[(\bar{c}|E_y|)_y - (\bar{c}|E_y|)_{y\pm d} \right]_t$$

The lower sign is to be used when the direction of propagation is positive along the coordinate axis, and vice versa.

The finite difference scheme is convenient for the solution of our problem, particularly because the sea energy which is reaching a coast line automatically vanishes from the system.

It is easily seen that the scheme is numerically stable provided the following criterion is not exceeded.

$$\frac{d}{\Delta t} \geq \bar{c} \quad \text{or} \quad \frac{d}{\Delta t} \geq \frac{g\bar{T}}{4\pi}$$

In the routine computations made at the Meteorological Institute one is using the values $d = 150$ km and $\Delta t = 3$ hrs, which makes the scheme stable for wave periods up to 18 secs.

In the computations, one has to take into account the effect of change in direction in time or space of the wind which is working upon the propagating energy. A change in wind direction must cause a certain dissipation of energy. This is taken into account in the model by retaining after propagation over one time step the amount of energy corresponding to the overlapping area of the directional distributions at the beginning and end of the time step.

In the course of one time step , some energy is transferred from the air to the sea. The model uses Neumann's results for the growth of sea energy, the growth rate being a function of the energy itself and the wind velocity. The growth function is used

in the model in tabular form, and may therefore easily be adjusted
in accordance with results from later research projects.

The resulting energy after one time step may be written

$$E_{t+\Delta t} = E'_{t+\Delta t} + \Delta E(U_{t+\Delta t}, E'_{t+\Delta t})$$

where E' is propagated energy, and E is growth rate per time step.

The upper limit of sea energy for a certain wind speed is
defined in the model by a cut-off of the energy spectrum at the
wave period for which a wave component has group velocity equal to
the wind velocity.

In applying a computational scheme involving propagation of
some quantity to a grid covering a limited part of the globe, it
is necessary to define some boundary conditions.   One self
evident boundary condition is given by the fact that the energy in
points over land is zero. The second boundary condition used in
the model assumes that wind and sea energy in the normal direction
which is outside boundary points and in open sea is equal to the
values at the boundary.   This condition appears satisfactory
due to the existing correlation in the atmosphere between scales
in time and space.

We will then briefly look into the problem of computing swells
generated in remote areas.

In principle, the sea energy may at any time be regarded as
potential source of swell. In the model, the computed sea energy
over a sufficiently large area, and at every time step for the
last 48 hours, is used as the source of data for swell computations.

In order to handle the swell problem numerically in the model,
the continuous spectrum for the swell-generating sea is replaced
by finite energy elements.   The period range for each element is
chosen equal to 2 sec.

Swell is computed for selected points - chosen according to
needs -, and the swell spectrum is composed element by element for
each point.

For each element, the most significant potential swell energy
fractions from each time step over the previous 48 hours are
computed, and for obvious reasons the largest one is chosen.

Each potential swell energy fraction is reduced by a
dissipation factor during the computation, which in the model is
a function of age and mean period of the energy element.

Having computed the energy spectra for sea and swell, the required quantities like significant wave height and wave periods may be computed.

In Fig. 3 an example of computed waves is shown, as well as the corresponding measured values.

In order to take into account the effect of small scale features of the geography in coastal regions, fjords or even lakes, a special version of the model has been designed, where the mesh width can be chosen arbitrarily. The time step is chosen by the program to secure numerical stability.

For reasons of economy, the program is limited to compute waves for one wind direction and speed at a time.

Fig. 4 shows an example of computed wave conditions over Skagerrack and Kattegat.

We will then briefly look into the possibilities of deriving ocean currents in different depths for the purpose of computing wind-generated drift of different particles like oil, or for example plancton.

The sub-surface current will in general be a resultant of several components, like the Ekman-current, tidal currents, coastal or other semi-permanent currents.

Having at our disposal at the moment 5 years of wind data derived from historical weather maps, we have at the meteorological institute computed trajectories at different depths, starting every 6 hrs during the 5 years. The resulting current was assumed to consist of the purely wind-driven Ekman current, and a semi-permanent current along the coasts of the North sea and Norwegian sea.

On the basis of the approx. 7000 trajectories for every depth, it has been possible to derive some probability statistics for the drift, as shown in Fig. 5. The statistics has been derived for oil dispersed in the water, but could as well be used for plancton.

Finally, I will say a few words about a project which has been going on for 1½ years at the Institute, namely the so-called Hindcast project.

The main goal of the project is to establish, as described earlier in the lecture by hindcast techniques, data series for wind and waves based upon weather maps every 6 hrs for a period of 25-30 years for the Norwegian continental shelf and adjacent waters. This involves digitizing approximately 40.000. maps, and a large

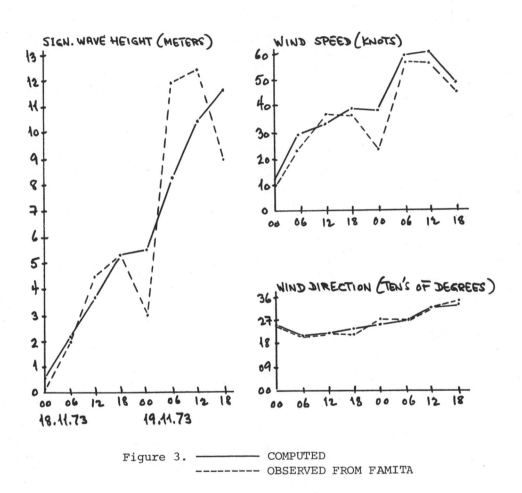

Figure 3. ——————— COMPUTED
           - - - - - - - OBSERVED FROM FAMITA

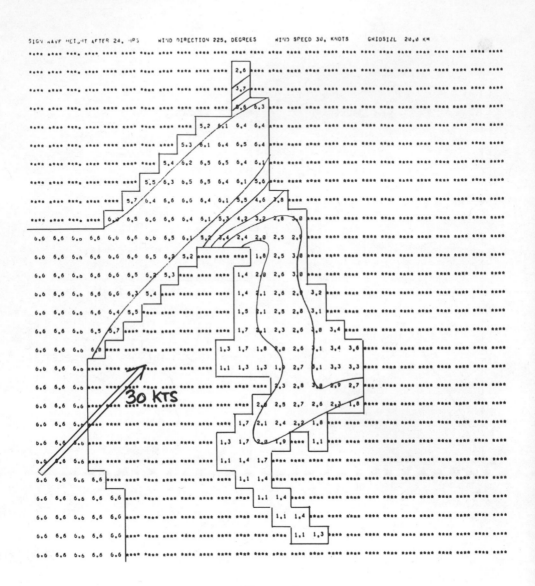

Figure 4

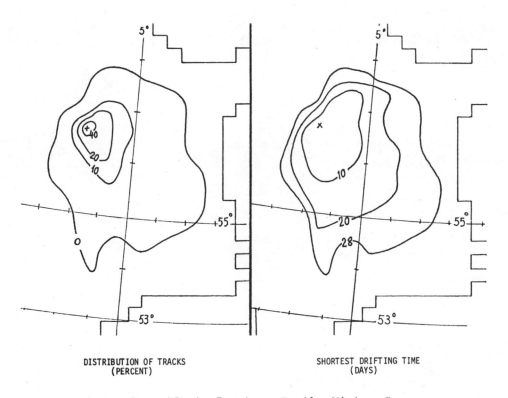

DISTRIBUTION OF TRACKS
(PERCENT)

SHORTEST DRIFTING TIME
(DAYS)

Figure 5. Drift in 5 Meters Depth, Winter Season

amount of computer work to establish the final data.

In order to make best possible use of these data, it is important to study the systematic errors caused by inaccuracies in the maps or approximations in the models. It is possible to do this by comparing computed data to measured data in overlapping periods. Such work is in progress in the Hindcast project parallel to the ongoing digitizing of maps.

The systematic deviations will probably not be useful for corrections in individual cases, but can certainly be used to modify the statistical results which will be derived from the long data series of hindcast wind and waves.

Furthermore, the hindcast series may be used to extend the value of detailed but short series of measurements by placing the short series correctly in the climatological picture. In this way, it may be possible to obtain an adequate climatology also for measured elements which the hindcast series is not capable of describing.

A REVIEW OF MODELS OF FISHING OPERATIONS

Emil Aall Dahle

The Norwegian University of Fisheries

N-7034 Trondheim, Norway

INTRODUCTION

In January 1975, led by L.O. Engvall, FAO held its "Expert Consultation on Quantitative Analysis in Fishery Industries Development", the discussions and conclusions being contained in[1]. At the consultation, the discussions on industry were divided into the following areas,
- Vessels and Fishery operations
- Fishery harbours
- Fish processing plants
- Fish distribution and storage
- Marketing management

A review of application of operational analysis of vessels and fishing operations until 1974, is given in[2].

In summing up of the consultation, J.J. Foster and C. Hamlin[4] recommended that the harvesting system should be divided into,
- Fish biology
- Fishing gear and its catching efficiency, and
- Vessel.

This recommendation is followed below.

FISH BIOLOGY

In any model of fishing operations, income and costs must be quantified if the study undertaken is to have meaning.

213

Fishery Gear	Specified	Fish Distribution	Catch (C) Formula	Catch pr. unit effort (CPUE)	Catch coeff. (k) from logbook
Trawl	Bottom Shrimp	Even	$C = \rho_A (\Sigma \ell_1 \cdot \varepsilon_1) v \cdot t$	$CPUE = \dfrac{C}{(\Sigma \ell_1 \varepsilon_1) \cdot v \cdot t} = \rho_A$	$k = \dfrac{C}{v \cdot t \cdot \Sigma \ell_1}$
	Pelagic	School	$C = \rho_V \cdot \varepsilon \cdot Sgear \cdot v \cdot t$	$CPUE = \dfrac{C}{\varepsilon \cdot Sgear \cdot v \cdot t} = \rho_V$	$k = \dfrac{C}{Sgear \cdot v \cdot t}$
Seine	Purse	School	$C = \rho_V \cdot \varepsilon \cdot Vgear$	$CPUE = \dfrac{C}{\varepsilon \cdot Vgear} = \rho_V$	$k = \dfrac{C}{Vgear}$
	Danish	Even	$C = \rho_A \cdot (\Sigma \ell_1 \varepsilon_1 \cdot d_1)$	$CPUE = \dfrac{C}{(\Sigma \ell_1 \cdot \varepsilon_1 \cdot d_1)} = \rho_A$	$k = \dfrac{C}{\Sigma \ell_1 \cdot d_1}$
Longline	Short	School/Even	$C = \rho_V \cdot \varepsilon \cdot N \cdot f$	$CPUE = \dfrac{C}{\varepsilon \cdot N \cdot f} = \rho_V$	$k = \dfrac{C}{N \cdot f}$
	Long	Even			
Gillnet	Bottom	Even	$C = \rho_V \cdot \varepsilon \cdot Sgear \cdot f$	$CPUE = \dfrac{C}{\varepsilon \cdot Sgear \cdot f} = \rho_V$	$k = \dfrac{C}{Sgear \cdot f}$
	Pelagic	School			
Handline	-	Even	$C = \rho_V \cdot \varepsilon \cdot N \cdot f$	$CPUE = \dfrac{C}{\varepsilon \cdot N \cdot f} = \rho_V$	$k = \dfrac{C}{N \cdot f}$
Traps	-	Even	$C = \rho_A \cdot \varepsilon \cdot Vgear \cdot f$	$CPUE = \dfrac{C}{\varepsilon \cdot Vgear \cdot f} = \rho_A$	$k = \dfrac{C}{Vgear \cdot f}$

Nomenclature:

ρ_A = fish area density
ρ_V = fish volume density
ℓ_1 = gear component length
ε_1 = probability of fish capture of characteristic gear (component 1) parameter
d_1 = displacement of ℓ_1
v = fish gear speed

t = effective fishing time (usually $t < t_{sat}$)
Sgear = effective catch area of gear (f(v))
Vgear = gear volume
N = no. of hooks
t_{sat} = time for saturating the gear ($f(\rho_v)$)
$f = 1 - \exp(-\dfrac{t}{t_{sat}})$

Figure 1. From Reference [2]

The gross income is the product of catch and unit price.   The
fish price depends on several factors, which are dealt with elsewhere. Apart from various Governmental regulations the catch depends
on biological factors, on the gear and on the vessel. There are two
aspects in catch estimation.

Firstly, information on stock size and distribution is needed
for strategic purposes, e.g., decisions on fishing positions during
the year.

Secondly, information on the behaviour of fish (or schools)
during the course of a year in relation to the fishing gear
alternatives is needed for decisions on fishing tactics.

The former problem is dealt with continuously at biological
institutions, and also reported in several papers of the Symposium.
The latter problems are discussed in the ICES's "Gear and Behaviour
Committee".   The "tactic" behaviour depends on the species, and on
several factors which are mutually dependent, and partly difficult
to quantify.   Consequently, the behaviour has to be presented
simply as a probability of capture (ε) in mathematical models.

Nevertheless, fishing vessels are continually being designed
and built all over the world, and some basis for estimation of
catch rates has to be used. In[2], a summary of catch estimation
was given, and it is presented here as figure 1.

In the figure, the fish density $\rho_{A'v}$ can be classified as

strategic information, whereas gear characteristics (ℓ,v, S etc.)
and the probability of capture (ε) provide the tactical information
concerned with capture.   Although the equations in figure 1
obviously are over-simplifying the estimation of catch rates, it is
nevertheless providing a link between the biological sector and the
catching unit, which is necessary for a mathematical model where
the catch is considered as an exogenous variable.

FISHING GEAR AND ITS CATCHING EFFICIENCY

A more detailed discussion on the various fishing gear and its
efficiency is given in[2].   Since 1975, research on fish behaviour
has received continued consideration. One well-known practical
result is the introduction first by DDR and later by other nations
of large meshes in the front part of pelagic trawls. From figure 1
the catch of a pelagic trawl can be expressed as

$$C = \rho_v \cdot \varepsilon \cdot \text{Sgear} \cdot v \cdot t \qquad\qquad (1)$$

where C = catch, ρ_v = volume density of fish, Sgear = catch area

of gear, v = trawling speed, t = effective fishing time pr. haul.
Research has shown that some fish species are herded by such meshes,
(i.e. ε is not affected) while others are not.   Thus, for trawlers,
in equation (1), (v) can be maintained, or lowered and (Sgear)
increased substantially, using the same main engine as before.
Although (t) for each individual haul may have to be shortened, and
hence more time will be used for hauling and emptying, a substantial
increase in (C) can be the result.

However, with conventional trawl gear, no improvements regarding the catching efficiency of the gear have taken place since 1975.

Apart from the catching efficiency, the most important factor
when modelling the catching unit is the hydrodynamic resistance.
The reason is that the power needed to overcome this resistance
decides the size of the fuel-consuming machinery, and thus one of
the most important cost components.   For trawling this is rather
obvious, but it might also be the case for purse seiners and Danish
Seiners.

For trawlers, the estimation of resistance of the gear has
received continued consideration since 1975.   Nevertheless, a
realistic calculation of fishing gear drag from first principles is
still difficult or impossible to perform.   The only available
method may therefore be to extrapolate from often inaccurate
full-scale measurements of gear of related geometry.

The general formula for active gear resistance is:

$$R = \frac{1}{2} \cdot \rho \cdot v^2 \cdot \Sigma \, C_{Di} \cdot A_i \qquad\qquad (2)$$

where R = gear resistance, ρ = density of sea water, v = towing
speed, C_{Di} = drag coefficient of the gear component i, A_i = area

(projected, or net area) for the gear component i.

For the new large-mesh pelagic trawls discussed above, considerations regarding the choice of (v) may have several aspects, such as

- (v) must be above a certain v_{min} to catch fish at all.

- When (v) is increased, the pressure wave and turbulence
  noise may frighten fish away from the trawl opening (e.g.
  influence on (ε) in equation (1)).

- (v) has a strong (third power) influence on the fuel
  consumption.

- Increase in (v) may increase the volume of water swept by
  the gear, even if(Sgear) is decreasing with speed, and

thus increase the catch if other factors are unaltered.

From this list, it is evident that caution is needed when modelling the fishing operation. For trawl gear with deviations from investigated configurations, equations (1) and (2) cannot be expected to be realistic. They can on the other hand be useful for more conventional purposes.

VESSEL

For several reasons the design of a fishing vessel is a task which is different from design of merchant vessels. The most important is that the fishing vessel is a combination of a catching unit and a transport unit (except in certain fleet operations, where the vessel is only a catching unit). Consequently, the fishing gear to be used has a strong bearing on the vessel design. This is particularly the case for vessels fishing with active gear, where also machinery installation may be determined by the gear. Also the minimum crew size depends strongly on the gear (but also on the degree of processing on board).

The transport length of the catch (fishing ground-processing plant) may differ to a large extent, and may have impact on the design, mostly on catch preservation aspects and on fuel capacity considerations. Fishing vessels are normally designed with a conventional under-water hull regardless of fishing technique. Thus, even if the transport length is important for the economy of the vessel and the fleet, it is less important for the designer.

From this it can be concluded that a simple model of the fishing operation can be set up without too many difficulties, provided knowledge from conventional naval architecture is used together with specific knowledge of fishing.

Provided a detailed research on the operations of a vessel type has been undertaken, the time allocation over a year can be given as follows:

$$T = \sum_{1}^{N} T_n = \sum_{2}^{N} T_n + T_1 = \sum_{2}^{N} T_n + \sum t \qquad (3)$$

Where $T = 1$ year, $N =$ time divisions (e.g. steaming, unloading, waiting in harbour etc.), $T_1 = t =$ effective fishing time.

Through the analysis it can be supposed that the costs imposed during the time period $\sum_{2}^{N} T_n$ in (3) have been found. Then, by

using equations (1) and (2), the cost and income of each of the
time periods (t) can be set up for a pelagic trawler as follows:

Profit = Income - Costs = Income - Fuel costs - Other

costs.   (4)

The "Other costs" involve capital costs, crew costs, maintainance
etc., and can also be found through our analysis.   Then

$$\text{Profit} = C \cdot P_c - \frac{1}{\eta_{TOT}} \cdot R \cdot v \cdot f \cdot t \cdot P_f - \text{Other costs} \quad (5)$$

Where C = catch, P_c = catch price, P_f = fuel price, η_{TOT} = pro-

pulsion coefficient, R = gear resistance, v = trawling speed, f =
specific fuel consumption.   Finally, using a total drag coefficient
for the trawl, a total projected area, and neglecting the small
hull resistance, we obtain,

$$\text{Profit} = \text{Sgear} \cdot v \cdot t \cdot (\rho_v \cdot \varepsilon \cdot P_c - v^2 \cdot \frac{0,5 \cdot q}{\eta_{TOT}} \cdot C_D \cdot f \cdot P_f)$$

$$- \text{Other costs} \quad (6)$$

The equation illustrates the influence of some important parameters
as well as the connection of the fishing operation model with the
exogenous variables as follows

ρ_v, ε - biology

P_c, P_f - market

The conflicting influence of increase of (v) is also evident.
There are two trawling sppeds for which the profit is
zero. (v = 0 being one of them), and one trawling speed for which
the profit is maximized.   The latter speed can of course be found

by setting $\frac{\partial \text{Profit}}{\partial v}$ = 0.   These trawling speeds, which are of

interest from a purely economical point of view, must be seen in
conjunction with the biological factors mentioned in connection
with equations (1) and (2).

Although this example deals with the goal of maximizing the
profit, other goals could well be pursued, still using operational
research.

MATHEMATICAL MODELS

In (3) the use of a mathematical simulation model in practical design of a catching unit is described.   The vessel has been built, and has operated successfully for 4 years. The vessel is an automated long-liner of 85 feet length.   The model was written in the computer language SIMULA 67, as shown in principle in figure 2, and in more detail in Table 1 and on figure 3.

Generally in simulation models the physical unit (vessel, gear) has to be represented as a separate entity (element), and the outside world (physical system) must be represented by exogenous variables.   These variables may normally change over the time period studied, and may be represented stochastically or deterministic.

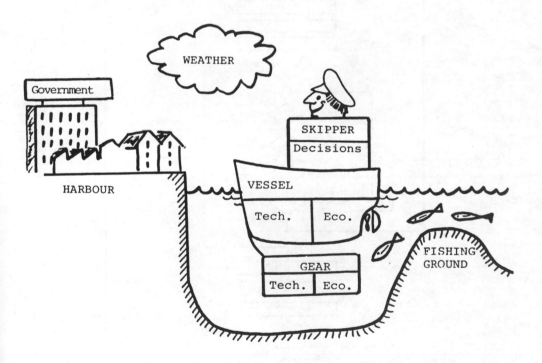

Figure 2. General Concepts of an OA-Model of a Catching Unit

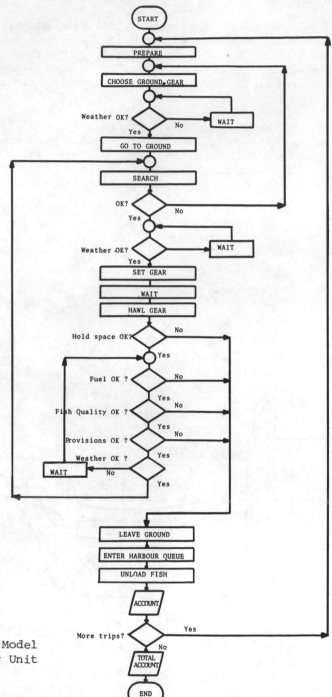

Figure 3.
General OA Model
of Catching Unit

Table 1. Terms used in description of an OA-model of a
catching unit

Model parts:	Consisting of:	Described mathematically by statistics (S) or deterministics (D)
Physical system	Fishing grounds, weather, prices and costs	S and/or D
Physical element	Vessel and gear	D
Operation rules	Governmental regulations. Skipper/owner decisions	S and/or D

In a dynamic simulation model, the use of Monte Carlo drawings from statistic distributions is often employed to describe the physical system, whereas deterministic representation is sufficient in models of a static nature. Also dynamic operations rules belong in dynamic models.   In Table 1, possible descriptions of the various elements in a model of a fishing unit are given.

In[4], an example is given of how to arrive at the best matching of a processing plan with a buffer store and a composite fishing fleet using the heuristic approach associated with simulation. Other applied OA investigations are reported by Haywood, Hamlin, Curr, Digernes and Jennsson. The modelling of the weather is important when simulating the detailed operations of the vessel in the physical system, see table 1. (See Haug).

CONCLUSIONS

As is evident from this Symposium, operations research is undertaken for the purpose of evaluating design and operation alternatives of fishing vessel and fleets to an increasing extent. The crucial problem, is still to arrive at realistic estimates of catch rates, for which no ready solution can be given. The matter has become even more complicated in recent years due to increasing

exploitation of some traditional fishing stocks, sometimes leading
to drastic changes in catch caused by decline in stock and/or
regulatory measures taken.

For evaluating fishing of under- or unexploited stocks, and
especially those within the 200 miles economic zones, operations
research seems to be a very attractive tool.

The normal case is to study an individual catching unit,
assuming that it does neither influence the fish stock, nor the
market price of fish.

Investigations may also encompass fishing fleets. OA investigations should be relevant for both developed, and developing
nations. The former have to change their traditional fishing
pattern, and the latter will have to develop their fishing.
industries to cover indigenous as well as ocean fishing. Several
applied cases have been presented at this Symposium, and the
application of operational research will probably continue in the
future.

REFERENCES

1. Expert Consultation on Quantitative Analysis in Fishery
       Industries Development. Report. FAO Fisheries Report
       No. 167. Vol.1, 1975.
2. E. Aa. Dahle, A Review of Quantitative Analysis of Vessel and
       Fishing Operations. FAO Fisheries Report No. 167. Vol.2,
       1975.
3. E. Aa. Dahle, Ship Design of Mechanized Longliner. Proceeding
       PRAD-Int.Symp. on Practical Design in Shipbuilding, Tokyo
       Oct. 1977.
4. E. Aa. Dahle, T. Digernes and J.P. Irgens, A SIMULA Model for
       Matching a Fishing Fleet and a Production Plant. Proceeding
       of 1976 SCSC-Summer Computer Simulation Conference.
       Washington DC 1970.

A SHORT TERM PRODUCTION PLANNING MODEL IN FISH PROCESSING

Bjørnar Mikalsen[1] and Terje Vassdal[2]

1. Institute of Fisheries
   University of Tromsø
2. The Institute of Fishery Technology Research
   Tromsø

INTRODUCTION *

The Norwegian fish processing industry has clearly less profit-ability than the average Norwegian industries. This is the case for the modern freezing industry as well as the traditional product-ion of stockfish and salted fish. There is hardly one conclusive reason for this, but the explanations are usually sought among the following conditions:

a) A general overcapacity in this industry. The net operating surplus is consequently unable to cover the fixed costs.

b) The prices of raw fish (as input to the industry) are fixed by means of central negotiation based on hypotheses on world market prices. More often than not these price-estimates have overvalued the actual prices, and thus eroded the finances of the manufacturers.

c) In some coastal regions there are great seasonal variations in landings of fish. For long periods the production equipment is hardly utilized.

d) The managerial skills are insufficient, especially in the smallest companies.

The industry has several times been in a poor financial state.

* Financial Support from the Norwegian Fisheries Research Council
  for this Project is gratefully acknowledged.

The Government has often helped the industry, mainly with credits and grants to avoid immediate bankruptcy. These aids, though they are fairly regular, are not meant to be permanent. The Government has as its explicitly stated goal to make this sector financially sound and ultimately withdraw the subsidies and other kinds of financial support.

This sector must, in one way or another, improve its profitability. Our work has been to develop a planning model which may be of some help to the companies in this respect.

OUR PROBLEM

One way to improve the profitability of the fish manufacturing sector, will be to raise the average net operating surplus through efficient production planning. To a certain degree the producers have already, by trial and error, established decision rules that seem to be rather well-founded economically.

However, most of these rules of thumb assume that there is only one bottleneck in the production process, namely manpower. This assumption might be correct in the traditional big seasonal fisheries (for example the Lofoten spring fisheries and the Finnmark summer fisheries) producing stockfish and salted fish. Today the production planning problem has become far more complex, and the old, traditional decision rules are insufficient, if not downright erroneous. Let us give some of the reasons why:

a) It has become increasingly difficult to lay off workers during the low seasons. There are two reasons for this; (1) the unions demand a steady work load throughout the year in this industry, similar to other industries, and (2) the occurrence of a general tight labour market which makes it difficult even to find people willing to work in the traditionally rather unfavourable conditions.

b) The freezing technology makes it possible to freeze gutted unprocessed fish for processing in later periods. Traditionally the processing had to be done with fresh fish. The freezing of unprocessed fish makes it technologically possible to run a processing line quite independent of the actual landings of fish. How much to store, and for how long time, must be solved as economic problems.

c) The markets demand a steady flow of frozen proeucts, and preferably a product mix reflecting consumer demands rather than the availability of fish in the sea. This also represents a new production planning problem for the managers in the fish processing industry. Traditionally, in its planning process, this industry can hardly be said to be market oriented.

d)   The market prices may vary seasonally, and within certain limits the manager can make an extra profit by buying and selling optimally in accordance with the price variations.

We have constructed a linear programming model to help solve this decision problem. The model will be described in the following chapters.

THE MODEL

Static vs. dynamic models

Any linear programming model is basically static, while the problem we are analysing here is dynamic. Partly to overcome this problem, our linear programming model has been designed as a one year multi-period model. The year is parted into five periods. Admittedly this solution is not without shortcomings. The terminal period must allow planning periods beyond the one year time horizon of our model. How much to produce and store of processed products and raw materials in the last period, should ideally have been solved as part of the optimization model. In our model, in its present stage of development, these quantities are set as restrictions exogenous to the optimal solution of the model.

Why use linear programming

Linear programming is undoubtedly among the most widely used methods in applied operations research. There are several reasons for this:

a)   The results of a linear programming model are in accordance with accepted economic theory, and therefore well suited for further economic analysis.

b)   The formulations of complex and interrelated problems often turn out to be surprisingly simple.

c)   Many practical problems, which admittedly are not linear, can be reformulated to fit into a linear programming model with only modest losses in realism.

d)   Well established computer programmes make it possible to solve problems with several thousand equations and unknowns. The computation time for problems of this size may be from a few minutes and upwards.

The actual theoretical foundations of linear programming will not be elaborated in this paper. Those unfamiliar with the method may just as well consult one of the many good textbooks on the subject.

Notation used in this model

Decision variables or unknowns:

R_{ijk} - kilos of raw fish frozen and stored, from period i to period j of species k.

F_{ijp} - kilos of frozen products stored i period i for sale in period j of product p.

Y_{ik} - kilos bought raw fish in period i of species k.

X_{ip} - production in period i of product p.

T_{im} - total hours used in period i in "bottleneck" m.

S_{ip} - sales in period i of product p.

Constants or knowns

C_{ijk} - variable cost of storing one kilo of species k from period i to j.

C_{ijp} - variable cost of storing one kilo of product p from period i to j.

Q_{ik} - net buying price per kilo in period i for species k.

W_{im} - hourly wage in period i in place m.

V_{ip} - variable production cost (except for raw fish and labour cost) for each kilo produced; product p, period i.

P_{ip} - the revenue from product p, period i.

a_{kp} - kilos of species k needed to produce one kilo fillets of product p.

b_k , b_p - hours used to land one kilo of species k,or process one kilo of product p.

Subscripts used

i - denotes time periods ($i=1$, .. ,5)

j - denotes time periods (Usually i denotes time for entering stock and j time for leaving stock.)

k - denotes species bought. In this model we use

             $k = 1$: cod

             $k = 2$: haddock

$$k = 3 : \text{saithe}$$

p - products ($p=1, \ldots, 8$)

m - denotes physical bottlenecks in the processing line ($m=1, \ldots, 4$)

$$m = 1 : \text{landing}$$
$$m = 2 : \text{skinning}$$
$$m = 3 : \text{packing}$$
$$m = 4 : \text{freezing}$$

The objective function

The objective function to be maximized is the net operating surplus during a one year period.

$$(1) \quad \text{Maximize} \quad Z = - \sum_{i=1}^{4} \sum_{j=i+1}^{5} \sum_{k=1}^{2} C_{ijk} \cdot R_{ijk} - \sum_{i=1}^{5} \sum_{j=i+1}^{6} \sum_{p=1}^{8}$$

$$C_{ijp} \cdot F_{ijp} \sum_{k=1}^{3} \sum_{i=1}^{5} Q_{ik} \cdot Y_{ik} - \sum_{m=1}^{4} \sum_{i=1}^{5} W_{im} \cdot T_{im} - \sum_{i=1}^{5} \sum_{p=1}^{8} V_{ip} \cdot X_{ip}$$

$$+ \sum_{i=1}^{6} \sum_{p=1}^{8} P_{ip} \cdot S_{ip}$$

The costs and revenues when split into separate parts show that there is only one kind of revenue; the product price with VAT deducted.   All prices (and costs) are measured with subsidies included.   The costs in the objective function are primarily of four types:

a) Variable costs in storing frozen raw fish and processed products.   This item includes capital costs for the goods in stock but naturally not capital costs for the store-building itself.

b) The costs of buying fish, this price being exclusive VAT, but inclusive subsidies.

c) The labour costs in processing, fixed labour cost such as payment to foremen, administrative personnel and so on not included.

d) Other variable production costs proportional to the production activities.

The restrictions

The restrictions are basically one of two types:

a) Balance equations, stating for example that fish
   bought must either be used for processing or put in
   freezing store; that processed products could be sold
   immediately (i.e. sold within the same period) or put in
   stock for sale in periods to come.

b) Equations giving the upper or lower limits for certain
   activities, like sales in each period, available production
   capacities and the availability of the different types of
   fish.

Let us briefly put down these equations.

Balance equations

The bought fish are produced or stored as frozen raw fish

$$(2) \qquad \sum_{j=1+1}^{5} R_{ijk} + \sum_{p=1}^{8} a_{kp} \cdot X_{ikp} = Y_{ik} \qquad ; \qquad \begin{matrix} i=1,\ldots,4 \\ k=1,3 \end{matrix}$$

$$(3) \qquad \sum_{p=1}^{8} a_{kp} \cdot X_{ikp} = Y_{ik} \qquad ; \qquad \begin{matrix} i=1,\ldots,5 \\ k=2 \end{matrix}$$

Equations (2) and (3) indicate that haddock can not be frozen
before processing, while this is without problems for cod and
saithe.

Production from frozen fish

$$(4) \qquad \sum_{i=1}^{j-1} R_{ijk} = 1.1 \sum_{p=1}^{8} a_{kp} \cdot X_{jkp} \qquad ; \qquad \begin{matrix} k=1,3 \\ j=2,\ldots,5 \end{matrix}$$

The constant 1.1 indicates a decrease in yield of 10% for frozen
raw fish compared to fresh fish when used in the filleting process.

The use of labour in the production process: the landing stage

$$(5) \qquad \sum_{k=1}^{3} b_k \cdot Y_{ikm} = T_{im} \qquad ; \qquad \begin{matrix} i=1,\ldots,5 \\ m=1 \end{matrix}$$

Skinning, packing and freezing

$$(6) \qquad \sum_{p=1}^{8} b_p \cdot X_{ipm} = T_{im} \qquad ; \qquad \begin{matrix} i=1, \\ m=2,\ldots,4 \end{matrix}$$

Sell or store

$$(7) \quad \sum_{i=1}^{j-1} F_{ijp} + X_{jp} = S_{jp} + \sum_{i=j+1}^{6} F_{ijp} \qquad \begin{array}{l} p=1,..,8 \\ ; j=2.,5 \end{array}$$

$$(8) \quad \sum_{i=1}^{j-1} F_{ijp} \leq S_{jp} \quad ; \quad \begin{array}{l} p=1,..,8 \\ j=2,..,6 \end{array}$$

Equation (7) says that for each period and every product, what is produced and taken from stock in this period (left hand side of the equation) must equalize sales plus products stored for later periods. Equation (8) tells us that the sale of any product in one period cannot be less than the total quantity stored in previous periods and intended for sale in this period. This last equation may at first glance look redundant. Its function is to secure an actual sale in periods for which sales from stock are intended.

Maximum and minimum equations

Available fresh fish

$$(9) \quad Y_{ik}^{min} \leq Y_{ik} \leq Y_{ik}^{max} \qquad \begin{array}{l} i=1,..,5 \\ k=1,2,3 \end{array}$$

There may be restrictions on available fish. Usually there are no explicit lower bounds specified, only upper bounds. In some periods, however, the upper limits for some species may actually be zero, meaning no available fresh fish in this period.

Sales

$$(10) \quad S_{ip}^{min} \leq S_{ip} \leq S_{ip}^{max}$$

Sales are usually stated between upper and lower bounds. Since each producer is assumed to sell in a competitive market, such restrictions may seem to be at odds with economic theory. The actual organization of the Norwegian frozen fish export, forces the sales organizations (there are three of them) to give each member processing plant "recommended production quotas" in order to achieve a total production as near total market demand as possible. Season-ally varying prices may also contribute to this goal.

Maximum stock

(11) $\qquad \sum_{k=1,3} \sum_{i=1}^{1} \sum_{j=l+1}^{5} R_{ijk} \leq R_l^{max}$ $\qquad$; $l=2,\ldots,4$

(12) $\qquad \sum_{p=1}^{8} \sum_{i=1}^{1} \sum_{j=l+1}^{6} F_{ijp} \leq F_l^{max}$ $\qquad$; $l=2,\ldots,5$

Equation (11) refers to frozen raw fish, equation (12) to saleable products. Maximum stock is measured at the end of every period. We are thus measuring what is stored from one period to another, and not what is put in stock and removed from stock within a period.

Availability of input data

The estimation of the known constants has not caused any substantial problems. Most of the figures needed will be found in the ordinary company cost accounting system. The yield figures (a_{kp}) are, however, rather roughly measured in most factories. The reduction factor applied when using frozen raw fish as input, has been estimated by us to be 10%. It is quite clear that this factor varies considerably among the companies we have studied. We think the reason for this may be found in an insufficient mastery of this process. Likewise we have observed a wide range regarding the use of labour (b_k and b_p). There seems to be a surprisingly great difference in use of input in the production process between efficient and inefficient companies.

EXPERIENCES - RESULTS

The work on this model has not yet been concluded. We expect that the model has reached its final form, but minor changes may still be implemented. At the time of the writing of this paper, the experiments with the model are about to begin, and the results from these experimental runs cannot be reported here.

We set out to build a short term planning model which could be used in the white fish filleting industry. Among the questions such a model ought to be able to answer, we can mention:

a) What is the optimal product mix to produce?

b) What is the stability of the solution; i.e. would it be optimal to have one product mix in the high seasons and another in the low seasons?

c) Would it contribute to the financial results to use the stock more actively, and what increase in revenues are necessary to compensate for the extra costs if so done?

d) If the workers are paid a fixed monthly salary, which makes the wage costs no longer variable costs, would this dramatically change the production mix, the optimal storing and so on?

e) What would be the consequences of drastic, unforeseen changes in demand and supply of fish, prices on inputs, etc?

f) Would it be desirable to expand the capacities, and in case, how much to pay for such an expansion?

As already noted, many of the questions do not have a numerical answer. Our model is especially well suited to manipulate the cost components, and study the results of any such changes. Since uncertainty has not been formally incorporated into our model, this characteristic may turn out to be very useful.

Any such model must have among its goals that it can be used to solve practical decision problems. The manager must understand the model and be able to use it to solve his planning problems. Today we can hardly report having achieved this. The model has to be run on a fairly large computer, and outputs from the computer runs will most likely look unintelligible for the practitioner. Reporting the results in a comprehensible way, still remains to be solved. We think this can be done in one of several ways:

a) From our experiments with the model we can verify robust decision rules, and communicate just these rules as reliable "rules of thumb".

b) Write a postprocessor which will make the output more readable.

c) "Educate" the manager so that he can more easily understand the implications and limitations of the model, and be able to interpret the results. This may not be an easy task.

CONCLUSIONS

The work with this model has convinced us that linear programming is a very suitable tool to analyse production planning problems.   Our model has been formulated as a multiperiod model to overcome the static concept of linear programming.   This may be a feasible way to solve this problem, if the periods are not too many.   The implementation of our model cannot yet be reported to be successfully done.   It is our belief that problems related to implementation of formal planning models are underestimated among operations researchers.   If we are right, the most difficult part of our work has not yet begun.

A STUDY ON THE DEMAND FUNCTIONS OF THE SACCA DEGLI

SCARDOVARI FISHERY

M. Gatto[1], E. Laniado[1], S. Rinaldi[1] and R. Rossi[2]

1. Centro di Teoria dei Sistemi-C.N.R.,
   Milano, Italy
2. Istituto di Zoologia e Biologia Generale,
   Universita di Ferrara, Ferrara, Italy

DESCRIPTION OF THE PROBLEM

The "Sacca degli Scardovari", located in the southern part of the Po river delta, is a 3200 brackish water inlet (Fig.1). Its mouth is partially blocked by sandbanks and the water depth is, on the average, 1.5 to 2 m. with a maximum of 4 m. in the centre. Salinity is quite variable, ranging typically from 4‰ in the southern part of the "Sacca", where the influence of freshwater coming from the Po is significant, to 25‰ in the northern part.

A co-operative society of fishermen operates a fleet of about 170 small motor-boats, 5 to 8 meters long, with a flat bottom, using different fishing gears (eel-pots, trammel-nets, trawl-nets, etc.).

The main harvest consists of eels (Anguilla anguilla), the rest being grey mullets (Liza saliens, Liza ramada, Liza aurata. Chelon labrosus, Mugil cephalus), sandsmelts (Atherina boyeri), flounders (Platychthys flesus), mussels (Mitilus galloprovincialis) and minor quantities of other species (for further details see the paper by Rossi[1]).

It is worth while to briefly describe the life cycle of eels, which mostly contribute to the catch value. These fish spawn in the open sea and the young migrate inshore to mature. During the spring months the "Sacca" represents a favourable environment for eels, that immigrate to find better feeding conditions. An important event is the transformation from yellow to silver eel, which occurs when the animals are ready to gonad maturation. At this stage eels migrate to the open sea and are therefore easily catchable.

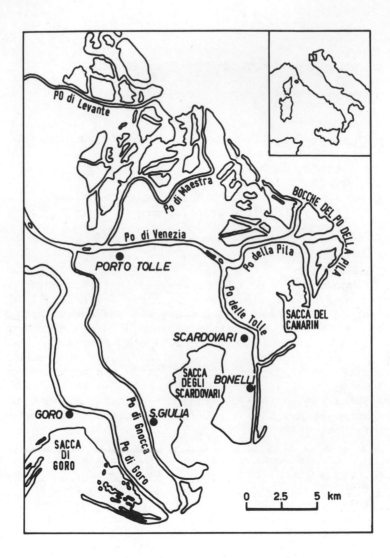

Fig. 1 - The "Sacca degli Scardovari" and the surrounding region.

Fishing takes place all the year round; however the effort concentrates in the period from the end of September to the middle of December, during the emigration of silver eels to the open sea. These eels are bigger than the yellow ones and are very much appreciated by consumers, who are willing to pay exceptionally high prices for eating them during the traditional dinner of Christmas eve.   Fishermen use a particular technique to catch migrating silver eels: by placing fixed nets side by side, they create a sort of enormous barrier in the narrowest part of the "Sacca", so as to trap most animals in the eel-pots which are located beyond such a barrier.

The greatest part of the fish catch is locally auctioned in the Scardovari market and sold to a few wholesalers, who take care of the transport to the fish markets of the largest towns in Northern Italy (these markets will be indicated from now on as terminal markets). The number of wholesalers never exceeds twenty people and varies according to fish availability.

This paper aims at giving a quantitative description of the present exploitation policy, which is based on traditional rules as a first step towards a rationalization of the fishery management.

After a brief exposition of some simple statistics of catches and catches per unit effort, which shed some light on the seasonal fluctuations of the abundance of some species in the "Sacca", the attention will mainly be focused on eels, since they constitute, as mentioned above, by far the most valuable part of the total catch. Price data will be analysed and a model explaining the mechanism of auction built.   Moreover, this will allow us to obtain some information on the demand function characterizing the terminal fish markets, which is not known a priori.

AVAILABLE DATA AND SIMPLE STATISTICS

The available data are: annual numbers of licensed boats through the period 1960-1978; monthly averages of the auction prices from 1968 to 1978; monthly fish landings sold at Scardovari through the period 1960-1978 with the exception of 1967, when the auction was closed because of a disastrous flood.   The data on landings are available separately for eels, mullets, sandsmelts; data on the other species, with the exception of mussels, which are not auctioned in this market, are known only as an aggregate.   Moreover, a rough estimate of monthly fishing effort has been obtained by means of a recent survey, which has yielded the average number of boats operating in each month.

A problem to be taken into account is that about 25% of the fish auctioned in the Scardovari market is not caught in the "Sacca",

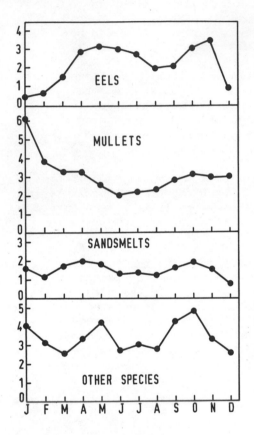

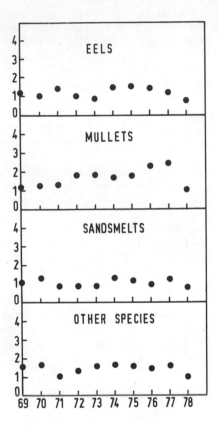

Figure 2. Average over years by
months of catches per unit effort (kg/boat-days)

Figure 3. Annual averages of
catches per unit effort (kg/boat
days)

but conveyed by other fishermen to the market from elsewhere.

Furthermore, in the first 7 months of 1968, due to the drain-
ing of some neighbouring ponds, a big and unknown quantity of fish
was poured into the market. Therefore, the data on the auctioned
landings for the first 7 months of 1968 have not been considered,
while the other available landings have been curtailed to 75% so
as to obtain estimates of the "Sacca" catches.

A well-known and accepted indicator of a population abundance
(as shown, e.g., by Beverton and Holt[2]) is the catch per unit
effort (CPUE).   The averages over years by months and the annual
averages have been computed and are displayed respectively in
Fig. 2 and 3.   Annual averages are shown only starting from 1969
because of the above mentioned lack of data in 1967 and 1968;
furthermore, annual CPUE's from 1960 to 1966 are not comparable
to the ones of the following years since the 1967 flood changed
the shape and the hydrological regime of the "Sacca".

Fig. 2 suggests the following comments:

a) the autumn peak, which is common to all the four groups of
   species, might be due to the emigration towards the open sea.
   This phenomenon is particularly significant for eels;
b) the spring peak, which is not remarkable for mullets, may be
   caused by the immigration towards the "Sacca", whose waters get
   warm earlier than the open sea and offer better feeding
   conditions;
c) January is characterized by an absolute minimum for eels, an
   absolute maximum for mullets and relative maxima for the other
   two groups. A possible explanation is the very low catchability
   of eels during the cold season, which is due to their scarce
   mobility on the bottom of the "Sacca".   This fact leads fishermen to fish mainly mullets.

Notice that CPUE's for the fourth group (other species) is not
very significant, as it collects different species with quite
different behaviours.

Annual CPUE's, as shown in Fig. 3, seem to indicate that
abundances have remained at the same levels in the last 10 years
and no significant deterioration has occurred after the 1968 flood.

ECONOMIC DATA FILTERING

The economic analysis will be restricted to eels, because the
annual revenue from this fish usually amounts to more than 50% of
the total, reaching 65% in some years. A trend towards a slightly
lower share of the revenue can be remarked in the last 3 years, due

to fast increasing prices of the other species.

In order to analyse the data on auction prices, the first
task to be accomplished is to extract the trend caused by inflation
and by long-term cycles so as to obtain a time series of prices,
which reflect only brief-term cycles and are, therefore, comparable
to one another. Brief-term cycles are to be meant as 1 year
cycles, i.e. with periods of 12, 6, 4, 3, 2 months. A standard
approach, illustrated by Faliva[3], is to design a filter which is
capable to cancel the corresponding frequencies $\frac{\pi}{6}, \frac{\pi}{3}, \frac{\pi}{2}, \frac{2}{3}\pi, \pi$.
Let p^i be the original time series of prices and consider the
following weighted moving average

$$\mu^i = \frac{1}{144} \sum_{h=-11}^{+11} (12-|h|)\; p^{i+h}$$

It is possible to prove that the corresponding transfer function
of such a filter is given by

$$G(\omega) = \frac{\cos^2\omega(1+\cos\omega)}{18}\,(3-16\;\sin^2\omega\cos^2\omega)^2$$

where ω is the angular frequency. It can easily be checked that
$G(\omega)$ is a non negative function such that

$$G(0) = 1$$

$$G\left(\frac{\pi}{6}\right) = G\left(\frac{\pi}{3}\right) = G\left(\frac{\pi}{2}\right) = \left(\frac{2}{3}\pi\right) = G(\pi)=0$$

A plot of the trend μ^i is reported in Fig. 4a. A slowing down with
respect to inflation can be remarked in 3 distinct periods (1970,
1972-73, 1977). As far as we know, the causes of this phenomenon
can be pointed out only for two of these periods: a cholera epidemic
can explain the price drop in 1972-73, a massive eel import from
abroad the one in 1977.

Dividing p^i by μ^i, and suitably scaling the resulting time
series, we obtain the seasonal fluctuations of prices, which will
be referred to from now on as unitary auction prices. The averages
over years of such auction prices are reported in Fig. 4b for every
month together with the average eel catches.

ANALYSIS OF THE DEMAND FUNCTION

This section is devoted to illustrating a model which explains

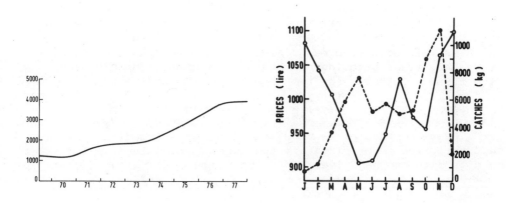

Figure 4. a) Trend of ell unitary price in the Scardovari auction
             (lire/kg)
          b) Averages over years by months of eel auction prices
             (solid line) and ell catches (dotted Line)

the mechanism of eel price formation and to obtaining some knowledge on the terminal markets demand function.   The model is subsequently tested against data.

Given a particular month in a particular year, define
Q   = month's catch
E   = month's effort
P_1 = unitary auction price
P_2 = unitary price in the terminal markets
T(E) = transportation costs
C(E) = fishing costs

Notice that transportation and fishing costs are supposed to depend only on the effort; while this is a classical assumption for fishing costs, in this case it holds approximately true   also for transportation ones.   In fact these costs depend mainly on the number of wholesalers operating in a particular month which in turn depends on the expected month's catch.   Since the expected catch is related to the month's effort, the conclusion   can be drawn that transportation costs are a function of effort only. With these hypotheses the two groups' profits are given by
Π_F = Fishermen's profit=p_1Q - C(E)
Π_W = Wholesalers' profit= p_2Q - p_1Q - T(E)

Assuming that p_2 is known in advance, an immediate remark is that the price p_1 is fixed, in the auction, at a level which makes both profits non negative.   This implies the following inequality

$$\frac{C(E)}{Q} \leq p_1 \leq p_2 - \frac{T(E)}{Q}$$

Another assumption is that p_1 is a convex combination of prices $\frac{C(E)}{Q}$ and $p_2 - \frac{T(E)}{Q}$ at which, respectively, the fishermen's profit and the wholesalers' one vanish, namely

$$p_1 = \frac{C(E)}{Q} + (1-\lambda)\ (p_2 - \frac{T(E)}{Q})\qquad\qquad (1)$$

where λ is independent of Q.   Notice that this assumption entails that the fishermen's and the wholesalers' profits are given by

$$\Pi_F = \{\lambda\ \frac{C(E)}{Q} + (1-\lambda)\ (p_2 - \frac{T(E)}{Q})\} Q - C(E) =$$

$$= (1-\lambda)\ \{ p_2Q - C(E) - T(E) \}$$

$$\Pi_W = \lambda\{p_2Q - C(E) - T(E)\}$$

Therefore

$$\Pi_W = \frac{\lambda}{1-\lambda}\ \Pi_F$$

i.e. the wholesalers' profit is a fixed fraction of the fishermen's.
Thus the coefficient λ reflects the relative importance of the two
groups, $\lambda=0$ meaning that the number and/or the interests of the
fishermen are prevailing.   Another interesting property entailed by
the above assumption is relevant to the maximization of fishermen's
and wholesalers' profits.   Notice that these profits are to be
considered in a statistical sense, because the monthly catch Q
cannot be decided a priori.   Therefore the real objective functions
of fishermen and wholesalers are the average profits.

$$\bar{\Pi}_F = (1-\lambda) \ (P_2\bar{Q}-C(E)-T(E))$$

$$\bar{\Pi}_W = \lambda(p_2\bar{Q} - C(E)-T(E))$$

where averages have been indicated by the symbol " $\bar{}$ ".   Moreover the
real decision variable is the effort E.   In fact it can be assumed
that

$$Q = N \ f(E)$$

where N is the fish abundance and f(E) a suitable function.   It
follows that

$$\bar{Q} = \bar{N} \ f(E)$$

Therefore the two groups face the following optimization problems
for each given p_2

$$\max_{E} \ \bar{\Pi}_F \ = (1-\lambda)\max_{E} \ (p_2\bar{N} \ f(E)-C(E)-T(E))$$

$$\max_{E} \ \bar{\Pi}_W \ = \lambda \max_{E} (p_2\bar{N} \ f(E)-C(E)-T(E)) .$$

Notice that both profits are maximized by the same effort and
therefore our assumptions lead to the conclusion that the two
groups behave like a unique co-operative society with respect to the
external market.

A subsequent step is to try to get information on the external
demand function.   It is reasonable to assume that this function is
given by

$$p_2 = a - b \ Q^{TOT} \qquad a>0 , \qquad b \geq 0$$

where Q^{TOT} is the total eel harvest that flows into the terminal
markets.   This is not the demand function which is faced by the

Scardovari people, because they do not produce Q^{TOT}, but only a fraction Q of such a harvest.   However, since the total eel production Q^{TOT} is mainly harvested in the Emilia-Romagna and Veneto fisheries located not far away from Scardovari, so that they are influenced by the same climatic changes, it is conceivable that there exists a significant correlation between Q and Q^{TOT}.   A realistic assumption is therefore that

$$Q_t = \alpha \, Q_t^{TOT} + \nu_t \qquad 0<\alpha<1$$

where the subscript t indicates the t-th year and ν_t is a random noise with null mean value. Thus, it turns out that we must allow for a certain variability of the external demand function faced by Scardovari, which is given by

$$P_{2t} = a- \frac{b}{\alpha} Q_t - \frac{b}{\alpha} \nu_t \stackrel{\Delta}{=} a - d \, Q_t + \varepsilon_t \qquad\qquad (2)$$

In the above equation, ε_t is again a random noise with null mean value. Eq.(2) implies, in view of eq.(1), that

$$P_{1t} = \lambda \frac{C(E_t)}{Q_t} + (1-\lambda)\left(a-dQ_t+\varepsilon_t - \frac{T(E_t)}{Q_t} \right)$$

By averaging over N years, it results

$$\bar{P}_1 \stackrel{\Delta}{=} \frac{\frac{1}{N}\sum\limits_{1t}^{N} P_{1t}Q_t}{\frac{1}{N}\sum\limits_{1t}^{N} Q_t} = \frac{1}{N}\frac{1}{\sum\limits_{1t}^{N}Q_t} \{\sum\limits_{1t}^{N}(\lambda C(E_t)-(1-\lambda)T(E_t))+$$

$$+(1-\lambda)\sum\limits_{1t}^{N}\left[(a-dQ_t)Q_t\right] +(1-\lambda)\sum\limits_{1t}^{N}\varepsilon_t Q_t\}$$

Since it is reasonable to assume that ε_t is independent of Q_t, we get

$$\bar{P}_1 = \frac{\bar{C}}{\bar{Q}} - (1-\lambda)\frac{\bar{T}}{\bar{Q}} + \frac{(1-\lambda)}{N} \frac{\sum\limits_{1t}^{N}\{(a-dQ_t)Q_t\}}{\bar{Q}}$$

$$\text{where } \bar{C} = \frac{\sum\limits_{1t}^{N}C(E_t)}{N}$$

$$\bar{Q} = \frac{\sum\limits_{1t}^{N} Q_t}{N}$$

$$\bar{T} = \frac{\sum\limits_{1t}^{N} T(E_t)}{N}$$

Introducing

$$Var \ Q = \frac{\sum\limits_{1t}^{N} Q_t^2}{N} - \bar{Q}^2$$

The final result is

$$\bar{p}_1 = \lambda \frac{\bar{C}}{\bar{Q}} - (1-\lambda)\frac{\bar{T}}{\bar{Q}} + (1-\lambda) \ a-(1-\lambda)\,d\bar{Q}-(1-\lambda)\,d \ \frac{Var \ Q}{\bar{Q}} \qquad (3)$$

Notice that eg. (3) refers to a particular month. Eq.(3) can be approximated in the case we are dealing with by making the following considerations:

a) λ, as we said above, reflects the relative importance of fishermen versus wholesalers. Since in the "Sacca degli Scardovari"
case the average number of wholesalers is about 1/15 of the
average number of fishermen operating, the coefficient λ is
quite small.
b) Since the average unitary transportation costs are negligible
with respect to the unitary auction prices, the term $(1-\lambda)\bar{T}/\bar{Q}$
is very small.

As a consequence

$$\bar{p}_1 \simeq a-d \ (\bar{Q} + \frac{Var \ Q}{\bar{Q}} \ ) \qquad (4)$$

This equation again refers to a particular month and the averaging
operation makes it impossible, of course, to estimate a and d. However, it must be reminded that from January to September mainly
yellow eels are caught by the fishermen and auctioned at the Scardovari market, while in the remainder of the year the catch consists
almost toally of the more appreciated silver type. A graph of $\bar{p}_1$
versus $\bar{Q} + \frac{Var \ Q}{\bar{Q}}$ in the twelve months (Fig. 5) suggests that a
same average demand function can be assumed for the period of yellow
eel catch (9 months). A regression line has been fitted through the
9 points and is shown in Fig. 5. Shifts from the regression line can
be observed for August, in accordance with the high tourists' demand
and June, in accordance with the risk of eel perishableness due to
the concurrence of the beginning of summer season and the underfeeding
of eels.

Notice, moreover, that the 3 points relative to October, November
and December (silver eels) are well above the regression line as it

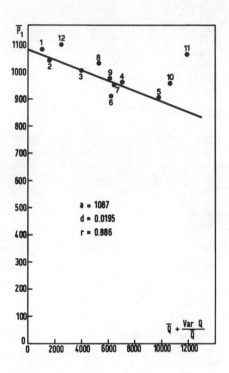

Figure 5. $\bar{P}_1$ (lire) vs. $\bar{Q} + \dfrac{\text{Var } Q}{\bar{Q}}$ (kg) in different months (numbers near crosses refer to month) and regression line through the first 9 months of the year. Parameters a and d of the regression line (see eg. 4) and the correlation coefficient r are also given.

is to be expected since the demand is higher for such eels. However, the 3 points do not specify a demand curve for silver eels.
This is due to the fact that demand concentrates especially around
Christmas time and therefore wholesalers are obliged to stock the
October and November production. The location of the 3 points is
in agreement with this fact, because the highest price is that of
December, the lowest one that of October. Anyway, it seems
reasonable to deduce that the external demand is highly unelastic
during these months, so that the Scardovari fishermen could double
the silver eel production without substantially affecting the
auction price. Therefore an obvious management suggestion might
be to decrease the less valuable yellow eel catch in order to
increase the harvest of silver eels.

CONCLUDING REMARKS

The paper has examined some aspects of the mechanism of eel
supply and demand in the "Sacca degli Scardovari". Though the
Scardovari eel catch is only 2-3% of the national eel production,
this study is certainly of interest, since the above described
methodology can be applied to other similar fisheries.

A final remark is that the analysis has been limited by the
lack of data. On the one hand, unavailability of reliable national
eel market data has been a severe restriction to a deeper economic
study, on the other the lack, up to now, of a regular collection
of effort data, month by month, has not allowed the building of an
accurate biological model, which is, of course, the basis for
sounder management suggestions. It is to be hoped that such data
will be made available in the future, so that this study can be
further developed.

ACKNOWLEDGEMENT

This work has been supported by Centro di Teoria dei Sistemi,
C.N.R., and by Progetto Finalizzato C.N.R. Oceanografia (Risorse
Biologiche).

REFERENCES

1. R. Rossi, Analisi di una serie pluriennale di dati di pesca
      nella Sacca degli Scardovari, 10th Congress of the Italian
      Society of Marine Biology (1978) (in Italian).
2. R.J.H. Beverton and S.J. Holt, On the Dynamics of Exploited
      Fish Populations (Ministry of Agriculture, Fisheries and
      Food, London, 1957).
3. M. Faliva, Optimal Filtering for Seasonal Adjustment of Quaterly
      Data, Rivista Internazionale di Scienze Sociali 1 (1978) 55-86.

ANALYSING THE DEMAND FOR FISHMEAL BY A LINEAR PROGRAMMING MODEL

Terje Hansen

Norwegian School of Economics and
Business Administration
Bergen

INTRODUCTION

Fish meal is together with fish oil the end product of cooking pressing and drying of various types of fish. Because of the relatively low value of the end products, the fish meal industry is typically based on types of fish with limited markets for human consumption. In Norway fish production is thus mainly based on capelin, in Peru on anchoveta and in the United States on menhaden.

The main use of fish meal is in the production of mixed feed. Whereas fish meal in the U.S. for all practical purposes only is used in compounds for the broiler chicken and turkey industry, a considerably wider market is found in West Germany and the United Kingdom. The fact that makes fish meal interesting in the production of mixed feed is its high content of important amino acids, such as lysine and methionine, which are required at relatively high levels in certain compounds. Fish meal is also valued for its high content of protein.

Until recently the closest substitute of fish meal has been soyabean meal, which also has a high content of protein. Presently a synthetic meal, the so called single cell protein, is being produced in small quantities. Single cell protein has for all practical purposes the same nutritional composition as fish meal and represents thus a perfect substitute for fish meal. The supply of single cell protein may therefore, from a modelling point of view, be treated as an additional supply of fish meal. Because of high production costs there has been some doubt that single cell protein will be produced in large quantities in the

247

foreseeable future.

An important aspect of the fish meal market is the general
belief that fish meal contains an unidentified growth factor.
As a consequence the inclusion of a minimum amount of fish
meal in a compound promote growth in young fowles. At high
prices of fish meal the degree of substitution of soyabean meal
for fish meal therefore becomes quite inelastic.

The fish meal market has been previously studied by Segura [6]
As in the present study Seguras analysis if focussed on the two
major fish meal markets, the United States and Western Europe
Segura develops an econometric model of the fish meal market.
The parameters of the model is then estimated from time series
of the relevant variables.

The approach used in this paper is entirely different.
Essentially, we simulate the behaviour of the mixed feed producers,
who are assumed to produce mixed feed at minimum cost. For each
type of mixed feed a linear programming model is specified. The
objective is to minimize costs subject to a number of constraints,
reflecting the nutritional specifications of the specific compound.
By varying the price of fish meal, a demand schedule for fish meal
is generated. Aggregating over different types of mixed feed and
different markets an aggregated demand schedule is derived. The
market price is then determined by equality of demand and supply.
We shall be concerned with the character of the aggregate demand
curve, for example whether aggregate demand is elastic or
inelastic, and factors that may cause a significant shift in the
demand curve and as result the market price.

The paper is organized in the following way. In section 2,
a characterization of the world fish meal market is given. In
section 3 linear programming and least cost production of mixed
feed is discussed. A world model of the fish meal market is
presented in section 4, and a numerical illustration of the
application of the model is given in section 5.

THE FISH MEAL MARKET

From 1955 to 1970 world production of fish meal increased from
1.3 million tons to 5.5. million tons. Since 1970 world production has
been decreasing and in the period 1975 to 1977 it averaged 4.6 million
tons per year. The principal producers of fish meal are given
in the table overleaf:

Table 1 Production of fish meal (1000 tons)

	1968	1971	1974	1977
Japan	484	676	767	461
USSR	326	406	542	635
Peru	1922	1935	878	498
Norway	402	384	320	498
Denmark	244	248	307	320
United States	258	306	327	304
Chile	235	256	192	255
S & SW Africa	469	273	259	176
Iceland	51	64	101	162
Others	557	627	778	889
Total	4948	5175	4471	4463

Source : Statistics published by the International
         Association of Fish Meal Manufacturers

As the table indicates, the reduction in world production of
fish meal from the beginning to the late 70'ies is mainly due
to a reduction of production in Peru.

Whereas all the production in important producer countries
like Japan, USSR, the United States and S & SW Africa are consumed
domestically, the bulk of the production of the remaining
important producer countries is exported. Together these five
countries account for some 75% of total world exports of fish
meal as illustrated by the table below. (Re-export of fish meal
is included in export statistics. The table therefore underestimates
these countries' share of world exports).

Table 2   Exports of fish meal (1000 tons)

	1968	1971	1974	1977
Peru	2078	1760	616	412
Chile	176	199	109	199
Norway	435	319	282	492
Denmark	184	184	258	267
Iceland	62	60	82	146
Others	571	484	613	517
Total	3506	3006	1906	2003

Source : Statistics published by the International
Association of Fish Meal Manufacturers

The export market is thus dominated by 2 distinct geographic
blocks; the South American producers, Chile and Peru on one hand,
and the Nordic countries, Denmark, Iceland and Norway on the
other hand. The table illustrates the Peruvian dominance of fish
meal exports in the 60's and beginning of the 70's, and the increased
relative importance of the Nordic countries as fish meal exporters.
As with exports, imports of fish meal is equally concentrated, namely
to USA and Europe.

Table 3   Imports of fish meal (1000 tons)

	1968	1971	1974	1977
USA	776	257	62	74
EEC	1547	1245	627	742
Of which:				
UK	(499)	(301)	(198)	(215)
W Germany	(520)	(516)	(337)	(248)
Rest of W Europe	349	390	237	239
E Europe (excluding USSR)	388	618	484	391
Rest of World	419	460	501	490
Total	3439	2970	1911	1936

Sources : Statistics published by the International
          Association of Fish Meal Manufacturers

Note : Total imports and exports do not coincide because
       of statistical discrepancies

As the table illustrates, United States and Europe together
represent 75% of world imports and exports into these two
geographic areas, roughly equals total exports of the five
major exporters. As a simplification one may thus say that the
world fish market for all practical purposes is characterized
by two major blocks of exporters, Chile/Peru on one hand,
and the Nordic countries on the other hand, and two major
markets, the United States and Europe.

LINEAR PROGRAMMING AND MIXED FEED PRODUCTION

The function of the mixed feed producer is to combine
various ingredients such as maize, wheat, soya meal, fish meal,
etc. into a compound which meets certain specified nutritional
requirements such as the content of protein, amino acids and
energy. The mixed feed producer, being a profit maximizer,
obviously wants to produce the compound at the least possible
cost.

Consider a specific compound that is produced from n
ingredients. The compound is characterized by m nutritional
specifications. Consider the production of 1 lb of the compound
and let us introduce the following notation:

x_j will denote the quantity of ingredient j

p_j will denote the price per lb of ingredient j

a_{ij} will denote the content of nutrient i per lb of
                 ingredient j.

$b_{i,max}$ will denote in the maximum permissible content
                 of nutrient i in the compound.

$b_{i,min}$ will denote the minimum permissible content of
                 nutrient i in the compound.

$x_{j,max}$ will denote the maximum permissible quantity
                 of ingredient j in 1 lb of mixed feed.

$x_{j,min}$ will denote the minimum permissible quantity
                 of ingredient j in 1 lb of mixed feed.

The compounders optimization problem is then given by

(1.1)   minimize $\sum_j p_j x_j$

subject to

(1.2) $b_{i'min} \leq \sum_j a_{ij} x_j \leq b_{i'max}$, $i = 1, \ldots, m$

(1.3) $\sum_j x_j = 1$,

(1.4) $x_{j'min} \leq x_j \leq x_{j'max}$, $j = 1, \ldots, n$

MODELLING THE DEMAND FOR FISH MEAL

Suppose the prices of all ingredients, but fishmeal, are given and we want to derive the demand/price schedule for fish meal. We shall show that this may be done essentially by solving (1) for each type of feed for alternative values of the price of fish meal. Consider therefore a specific compound and let us solve (1) for alternative values of the price of fish meal. Figures 1 illustrates the results of such an exercise for a specific mixed feed say Broiler grower.

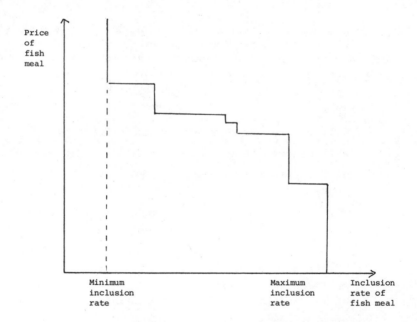

Figure 1. The relationship between the price of fish meal and the rate of inclusion of fish meal in Broiler Grower feed.

Suppose annual production of Broiler Grower feed in the
market in question is 1 million tons. Annual demand for fishmeal
in Broiler Grower feed is then found by multiplying the rate of
inclusion by annual production. Obviously this exercise may
be done for each type of feed and we get a demand schedule for
each type of feed as illustrated by Figure 2.

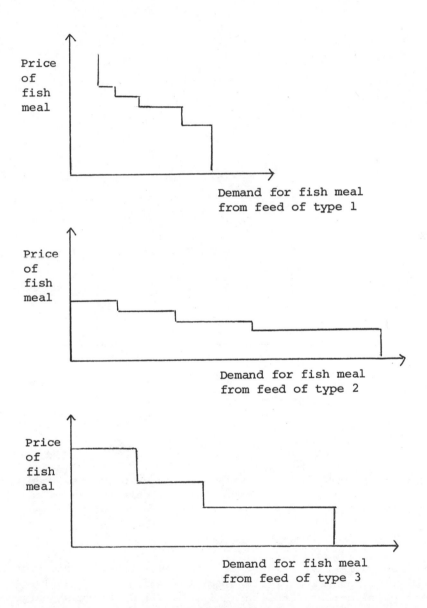

Figure 2. Demand schedules for fish meal for 3 types of feed

The demand schedules for the various types of feed may then
be aggregated to a total demand curve for fish meal for a
specific market. If we for the present disregard transportation
costs the market demand curves may again be aggregated to a "world"
curve as illustrated by the figure below.

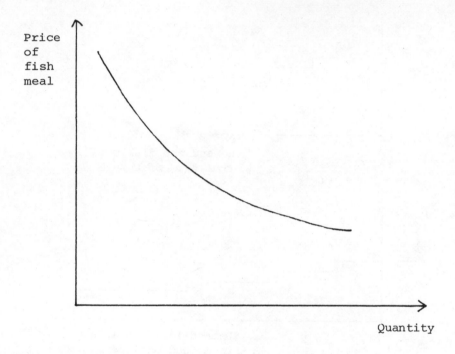

Figure 3 : "World" demand of fish meal

Given the supply of fish meal the market price may then be
determined from the "world" demand curve. This procedure could
obviously be used in predicting the future price of fish meal.

A NUMERICAL ILLUSTRATION OF THE MODEL

For our numerical illustration we shall consider the problem
of predicting the price of fish meal in 1985. In our analysis
of the "world" market for fish meal we shall distinguish between
the following markets:

1.   USA

2.   EEC
          West Germany, United Kingdom and Ireland, Belgium,
          Denmark, France, Holland, Italy and Luzembourg.

3.   Rest of Western Europe

4.   Eastern Europe

     Demand for fish meal in Eastern Europe and Rest of Europe
appears for all practical purposes to be independent of the
price of fish meal. Consequently we may limit ourselves to
modelling demand for fish meal in the USA and EEC.

     Fish meal will be assumed to be used in the following types
of feed in USA and EEC.

     USA : Broiler, layer and turkey feeds

     West Germany : Broiler, layer and swine feeds

     United Kingdom and Ireland : Broiler, turkey, layer and swine
                                  feeds

     Rest of EEC : Broiler and layer feeds

     So far, we have implicitly assumed that all feed is produced
by a compounder. This is to a large extent true as far as broiler,
turkey and layer feeds are concerned. In countries where fish meal
also is used in swine feeds one also finds farmers that buy the
non-cereal ingredients such as fish meal and soyabean meal, and
then mixes their own feed on the farm. We shall therefore
distinguish between the demand for fish meal by compounders and by
farmers that mix their own swine feed on the farm.

     In the mixed feed industry, there is obviously some variation
in the nutrient specifications for a specific compound. Nevertheless
for each type of feed, we use only one nutrient specification, which
is meant to represent an "average" of the industry. We distinguish
between the following mixed feeds:

     1.   Broiler starter
     2.   Broiler grower
     3.   Broiler finisher
     4.   Turkey starter
     5.   Turkey grower
     6.   Turkey finisher
     7.   Layer
     8.   Pig grower
     9.   Pig fattener
    10. Pregnant and lucatating sow.

For each of the four markets, we use the same specifications for the same type of mixed feed with the following exception:

1. For the United Kingdom and Ireland, the inclusion of maize in broiler diets is constrained so as to produce white flesh broilers.

2. The minimum inclusion rates of fish meal in mixed feed differ among the 4 markets as given by Table 4.

Table 4 <u>Minimum inclusion rate of fish meal in mixed feeds</u>

Type of mixed feed	USA	UK and Ireland	West Germany	Rest of EEC
Broiler starter	2	5	4	3
Broiler grower	2	4	3	2
Broiler finisher	1	3	3	1
Turkey starter	2	6	x	x
Turkey grower	2	4	x	x
Turkey finisher	O	1	x	x
Layer	O	1	1	1
Pig grower	x	1	2	x
Pig fattener	x	O	2	x
Pregnant and lacatating sow	x	2	5	x

x Not included

For the swine feeds mixed on farm an inclusion rate of 3% was assumed.

Table 5 gives the quantitites of mixed feed to be produced in the various markets.

Table 5 Mixed feed production (1000 tons)

Type of mixed feed	USA	UK and Ireland	West Germany	Rest of EEC
Broiler starter	1500	200	216	600
Broiler grower	6300	950	1080	2700
Broiler finisher	4700	650	720	1800
Turkey starter	150	15	–	–
Turkey grower	800	100	–	–
Turkey finisher	1600	200	–	–
Layer	10000	2625	2725	7500
Pig grower		900	3000	
Pig fattener		950	3600	–
Pregnant and lacatating sow		850	1060	–
Mixed on farm				
Pig grower		750	600	–
Pig fattener		1150	2400	
Sow		300	450	

Note: The numbers above have been obtained by combining data for mixed feed production from various sources for 1977 and adjusting for an expected change up to 1985.

Table 6 lists the ingredients used in the production of mixed feeds. Fish meal of different origins differ with respect to nutritional specifications. To deal with this problem, we shall convert the different types of fish meal to herring meal equivalents. We shall then assume that 1 ton of South American meal and U.S. produced meal is equivalent to .9 ton of herring meal.

Table 6   Prices in 1979-US$ per ton of various ingredients

Meat and bone meal	270	300
Blood meal	400	500
Lysine	4000	4000
Methionine	3000	3000
Soyabean meal (44% prot)	320	360
Maize	176	250
Barley	not included	260
Wheat	206	290
Wheat middlings	210	260
Maize Gluten 27%	210	240
DG + Solubles	200	200
Tallow	500	600
Molasses	120	140
Dicalsimuphosphate	185	200
Salt	60	60
Limestone	25	25

A demand schedule for each market was then derived.

     In order to derive the joint demand curve for the EEC markets
and the US, transportation costs for South American fish meal
has to be taken into consideration. We have estimated
transportation costs from Peru/Chile to USA and Europe to
$45 and $70, respectively. Since the US and Europe represent
alternative markets for South American meal, one would
consequently expect a differential in price reflecting the
differential in transportation costs. In other words, in
market equilibrium we would expect the price of fish meal in
the US to be $25 below the price of fish meal in Europe.

We may then generate the joint EEC-USA demand schedule.

Table 7   Demand for fish meal in EEC and USA for alternative
          prices for fish meal (1000 tons)

Price $ per ton of fish meal delivered to mill in EEC	EEC	USA	Demand
650	824	275	1099
625	849	275	1124
600	869	325	1194
575	946	357	1303
550	1001	464	1465
525	1009	503	1512
500	1358	540	1898
475	2000	1671	3671

Demand in 1977 (in herring meal equivalent)			1190

At prices above $600 the aggregate demand for fish meal by
the EEC and USA are thus extremely inelastic whereas the opposite
is the case at prices around $500.

As we shall see the demand schedule derived above may be used
to make a prediction forthe price of fish meal for 1985.   We
then first have to estimate production.

Estimate of production in herring
meal equivalents (1000 tons):

US	265
Chile	270
Peru	900
Ireland	120
Norway	320
Denmark	300
EEC excl Denmark	175
Eastern Europe	100
Rest of Europe	50
Sum	2500

Suppose we estimate production of single cell protein to
100,000 tons and total demand in Eastern Europe and Western Europe
to 1,100,000 tons. 1,500,000 tons will then have to be marketed
on the EEC and US markets.

From the demand schedule we then conclude that the expected
price for 1985 will be $525 per ton of fish meal delivered
to mill. We may also feel reasonably confident that the price
will be at least $500, since 400,000 additional tons could be
marketed at that price. On the other hand, a corresponding
reduction in supply on the US and EEC market would mean an
increase in the price of fish meal to 650 per ton delivered to
mill.

The present examples should illustrate how the market model
fish meal may be used in analysing the market for fish meal.
In the studies that are presently being pursued the sensitivity
of the market equilibrium price for fish meal with respect
to changes in the price of soyabean meal and other ingredients
as well as changes in the nutrient specifications for the various
mixed feed, will be studied.

REFERENCES

(1)   Bjørnerem, Magne: Etterspørsel etter fiskemel. En analyse
       av det engelske marked ved hjelp av fórblandingsmodeller
       og lineær programmering. Arbeidsnotat nr 2, Senter for
       anvendt forskning, Norges Handelshøyskole, 1978.

(2)   Hansen, Terje and Magne Bjørnerem: Sildemel til fór.
       Rapport fra et forprosjekt. Senter for anvendt forskning,
       Norges Handelshøyskole, 1975.

(3)   Houck, James P. et. al.: Soyabeans and their products,
       University of Minnesota, Minneapolis, 1972.

(4)   International Association of Fish Meal Manufacturers:
       Various statistical publications and studies of the fish
       meal market.

(5)   Paarberg, Phillip: The demand for soyabeans and
       soyabean products in West Germany, Purdue University,
       1977.

(6)   Segura, Edilberto: An Econometric Study of the fish meal
       industry, Columbia University, 1972.

(7)   United Kingdom Agricultural Supply Trade Association:
       Do you know? Facts and figures about the UK compound
       animal feeding stuffs industry.

A BUSINESS MANAGEMENT GAME FOR TRAINING

OF FISH MARKETING PERSONNEL

K.H. Haywood[1] and E. Ruckes[2]

1. W.F.A.
    United Kingdom
2. F.A.O.
    Rome, Italy

INTRODUCTION

One of the most important problems confronting the development of fisheries throughout the world is the education and training of personnel.   It is clear that development programmes in fisheries technology must be accompanied by parallel programmes of education and training in many areas and the serious lack of management skills required to spearhead this development also emphasises the need for improved training of management.

The acceptance of this premise leads to a consideration of the nature and type of management training that is required and also of the problems in devising a methodology that is satisfactory for a wide range of personnel associated with the industry in such diverse occupations as fishery officers, personnel from private industry and cooperatives, government officers and development bank employees.

The management role common to all these disciplines can be considered as a responsibility combining four essential elements – planning, motivation,coordination and control and its primary function is to make decisions that determine the future course of action for the appropriate organisation over the short and long term.   An essential basis of all real management problems is the existence of alternative courses of action, with a choice to be made among them, and decisions may be directed in every conceivable physical and organisational area;   they may deal with financial planning, personnel and marketing.

Marketing is a dynamic force in fishery development. However,

Table I Subject matters for Training Fish Marketing Personnel

Category of personnel / Subjects	Merchants		Market Staff			Staff of Public Development Agencies					
	Wholesalers	Retailers	Technicians	Administration	Inspection Service	Private Marketing Firms	Marketing Boards	Banks and Development	Marketing Support Services (extension, regulation)	Marketing Improvement Services	Marketing Policy Formulation
Fish as food	v	v		v	v	v	v	v	v	v	v
Fish as raw material	v	v		v	v	v	v	v	v	v	v
Fishing methods	v			v		v	v	v	v	v	v
Principles of fishery economics	v	v	v	v	v	v	v	v	v	v	v
Principles of marketing	v	v		v		v	v	v	v	v	v
Marketing research	v	v		v		v	v	v	v	v	v
Business administration	p	p				v	v	v	v	v	
Costing and pricing	p			v		v	v	p	v	p	v
Planning				v		v		p	v	p	v
Feasibility studies						v	v	p	v	p	v
Sales promotion		v				v			v		
Exporting	v	v	v	v	v	v	v	v	v	v	v
Food legislation	v	v	v	v	v	v	v	v	v	v	v
Storage	v	v	v	v	v	v	v	v	v	v	v
Packaging	p	p	p	v	p	v	v	v	v	v	
Transport	p	v	p	v	p	v	v	v	v	v	
Handling and preservation	p	p	p	p	p	p			v	v	v
Traditional processing	p	v		v	p	v			v	v	v
Industrial processing	v	p			p	v			v	v	
Product development		p				v			v	v	v
Inspection and quality control	v	v	p	v	p	v			v	v	v
Hygiene and sanitation	p	p	p	v	p	v			v	v	v

Symbols: v - vocational; p - practical exercises

it can play this role effectively only if adequate numbers of
trained personnel at all levels are available.   The shortage of
such personnel is one of the major bottlenecks hampering rapid
improvements in fish marketing in developing fisheries.   Good
general education would already improve the situation enormously,
but technical instruction of larger numbers of staff in clerical
duties, book-keeping and fish handling are a pressing need.
Training in economics, business management and marketing subjects,
as well as sufficient knowledge in fish technology are required at
the intermediate and senior levels, both in the private and the
public sector.

Subjects of training and categories of personnel for which
training is usually required are listed in Table 1.

FAO TRAINING ARRANGEMENTS

Training arrangements commonly provided by FAO are fellowships
and in-service training for individual training, study tours for
individuals and groups and group training activities such as
workshops, seminars and training courses or training centres.[1]

In most cases fellowships are awarded to junior personnel and
government officials who have already concluded their basic studies
and have some practical experience, in order to enable them to
receive training abroad which is not available in the country.
Even though such training is very much needed, it must be born in
mind that only in a very few cases the knowledge gained abroad is
of the same value in the home country as it is in the country
where the training is received.   The reason for this is that the
economic and social conditions in the two countries differ to a
very large degree.   Since marketing is very closely linked with
the social environment, the fellow must have a much greater gift
for adapting his newly acquired knowledge to conditions in his home
country than trainees in other disciplines.   Another drawback is
that fellows with an academic background frequently are not inclined
toward practical marketing improvement work that is very much
needed in developing countries.

The value of in-service training in institutions such as
commercial firms, market news service, wholesale markets, marketing
cooperatives, research institutions, etc. is often limited because
these institutions are not really equipped to give training.   Firms
are often not really interested to show "intruders" details of
their operation and fellows attached to a business enterprise will

_____
1.   There is no clear definition established for these terms in
     FAO, but seminars usually refer to somewhat higher level
     group training.

not be able to learn much if not directly <u>working</u> either in the
office or on the factory floor.   This required that the fellow
adopts the same attitude to work as the labour force in the host
firm.   However, questions of punctuality, adherence to working
regulations, etc. arise and according to experience this is a
major stumbling block in practical training of fellows.

Despite all these difficulties, fellowships will continue to
play an important role in the training of personnel particularly
at a higher level and therefore every effort should be made to
improve the effectiveness of this method.

Participation in seminars and training centres involves
absence from the normal working place.   Hence, there is a disadvantage in this method for certain groups such as established
retailers and other traders, due to their restricted mobility.
However, for government personnel and employees of larger marketing
organizations seminars appear to be an effective method of training.
It will be preferable to have a group the interest of which is
largely the same, since the success will greatly depend on the
precise knowledge of the problems the group members are facing in
their work and which they are expected to solve.

Seminars and training centres offer the particular advantage
to participants of an intensive study of problems with guidance
and under supervision.   Case studies and business games are
teaching methods which appear to be very well suited for this
type of training.   These offer a variety of training opportunities
which can cover almost every level and any field of expertise.   In
view of the extensive needs for training in developing countries
they are probably the most important training arrangement.

Furthermore, with increasing attention being directed to
marketing in fishery development projects, with the establishment
of fish marketing organizations and advisory services at national
level such courses will gain importance for training fish marketing
personnel at national level and to a somewhat lesser degree for
regional training activities.   Since fish marketing personnel,
particularly from private firms can devote only limited time to
training activities, such courses must be of short duration.   This
requires a condensation of the subject matters and a method of
learning by doing.   Therefore, a business management game can be
a most suitable training tool.   In practice it would be comple-
mented by lectures or discussions on items of specific interest and
by case studies which refer to the situation given in the country.

THE DEVELOPMENT OF TRAINING AT THE W.F.A.

The W.F.A. has been concerned increasingly with the provision

of training courses covering a wide range of subjects concerned
with fisheries development.   These now include courses on gear
technology, vessel engineering, acoustic fish detection and electrics
and hydraulics.

Not all the problems are on the catching side however;   there
is much scope for improved training of management in the processing
and marketing sectors of the industry.

Prompted by the obvious need for improved training in this
area, the W.F.A. designed a course for management personnel from
fish processing and marketing companies.

Based on a business management game and supported by lectures,
and discussion periods, the intensive three-day course covers such
topics as production management, corporate finance and marketing
management.   The major areas of importance are covered by expert
contributions from the fish industry.

The main objectives of the course are:-

(a)   To provide a meeting point for executives from fish processing
      and marketing companies of varying sizes, differing policies
      and from different sectors of the industry, so that mutual
      problems can be discussed.

(b)   To introduce the participants to the corporate nature of
      planning and control and instruct in the use of management
      techniques for decision making by means of the business
      management game.

(c)   To bring the course members into contact with prominent members
      of industry by means of lectures and informal discussion.

(d)   To describe the W.F.A.'s current R & D programme and the
      services offered to the fishing industry.

This course has now been running successfully for a number of
years and in addition to the support from British companies, it has
attracted considerable international interest.   This success has
led to the creation of other courses which now include a management
course for participants from developing countries.   A further
development has resulted from the cooperation between FAO and the
WFA.   After an evaluation exercise conducted with the assistance
of FAO projects teams in Peru and Chile, contributions have been made
to courses in Malaysia and Sri Lanka covering the management of
small scale fishery enterprises, in the Philippines covering
licensing agreements and joint-ventures and in Sri Lanka with the
latest course in fish marketing.

THE GENERAL CHARACTERISTICS OF MANAGEMENT GAMES

Management games have developed from the war game technique, the "Kriegspiel", first established in Prussia in 1792. War games spread through the military training establishments of Europe reaching the United Kingdom in 1872 and then later the United States. As a result of a visit by a member of the American Management Association to the US Navy Academy many years later, this method was recognised as a management training tool for industry. The very first business game was designed by the American Management Association in 1957.

Although there are many different types of games, there is sufficient common ground between all of them to distinguish them from the other teaching techniques used in management education.

One of the major distinguishing features is the simulation of the environment within which a manager operates and the interaction between the participants in the game and their simulated environment. The characteristics of the environment, which is simulated, are expressed in a quantitiative form by a set of mathematical relationships. Some of these relationships are known to the participants and are termed the "rules of the game", while other relationships, usually those describing the "structure" of the environment, are not made available to the participants except in a vague, qualitative way. The fact that the environment is characterised by a set of relationships necessarily means that some method of computation is necessary.

One feature of management games which is often misunderstood but which is essential to a correct understanding and use of games, is that all games are simple. However complicated a game may appear to be, the simulation involved is always a simplification from the reality which it purports to represent.

The most important characteristic of management games is probably the concept of "feedback". Many of the previous remarks made in this section could apply to case studies and similar training techniques, but the "feedback" elememt is strongest in management games. They have the virtue that, when a decision has been made by the group of participants which represents an organisation in the game, the consequences of this decision are calculated by the appropriate computing device and the results returned to the group. These then form the basis upon which the next decisions are to be made. In this way, the participants in the game must live with their decisions and the consequences. The distinguishing characteristics, therefore, of a management game as a teaching device, especially when compared to case studies, lies in its distinctive use of time. Indeed, it may be said that:-

A management game is a dynamic teaching device which
uses the sequential nature of decisions as an inherent
feature of its construction and operation.

FIELD EXPERIENCE IN SRI LANKA

The latest event in this field was the recent FAO/SIDA
training course for fish marketing personnel in Sri Lanka. Briefly
the main objectives of the course were:-

a) to improve understanding of the concept and role of
    marketing organisations, management functions and
    responsibilities.

b) to improve knowledge of marketing management tools
    and techniques.

c) to increase the ability to analyse and solve management
    problems.

d) to improve knowledge and skill in the planning,
    supervision and implementation of fish marketing
    development programmes.

The course was attended by 30 participants from Sri Lanka
and included managers from wholesale markets, retailers, marketing/
distribution managers from large organisations, extension officers
and market development officers.

A part of the course was devoted to lectures and case studies
and the remainder to the Marketing Business Management Game.

The game run in Sri Lanka simulated fishery cooperatives in
competition with each other. The teams representing the
managements of the competing fishery cooperatives are required to
make executive decisions on matters relating to marketing, storage,
placement of sites, purchasing, processing and finance.

The teams are expected to improve the performance of their
particular cooperative which is initially in a poor financial
situation due to previous bad management of the marketing and
distribution aspects of the business. Decisions are made once each
period and the exercise will last for several periods, usually 4
or 5, in addition to which, there is an initial period for the
setting up of starting conditions.

The marketing system has been conceived as a depot system with
inland central wholesale markets and a distinction is made between
marketing in rural areas and cities. Decisions on retail patterns

have to be considered and the choice of products include iced, frozen salted and dried products produced from groundfish, tuna and shellfish.   Each cooperative at the commencement of the game owns a freezing and loining plant specifically for export use, an ice-making plant, a processing plant for salting and drying, a coastal wholesale market and coastal cold storage facilities.

Initially, the majority of the marketing activity is to take place on the coast. A small proportion of fish caught is, however, to be transported to the main inland city to be sold direct to retailers as no wholesale and cold storage facilities exist.

Once the game is underway it is possible for each co-operative to set up inland wholesale markets both in the city and in rural areas. Descriptions of three city and three rural sites, with differing potential, are to be given to the participants. The decision as to whether to develop any one of these sites, or none, is to be left to the participants. In this way expansion of each co-operative should be guaranteed.

The administrators have control over the quantities of fish entering the market subject to the proviso that, overall, quantities will increase due to an increase in the size of the fishing fleet. Fluctuations due to seasonal trends, monsoons and other climatic conditions are incorporated.

Each co-operative has a guaranteed supply from local fisher- men in return for loans for fishing gear. Further quantities are, however, available from auctions which are to be held each period.

A marketing model is used to aid the administrators in allocat- ing fish sales to the cooperatives. The model is interactive in order that the competitive nature of the system may be developed. The model is based upon prices offered and promotional expenditure of each co-operative in relation to those of competitors.

Each cooperative controls its own accounts system.   This tends to give participants greater insight into the interactive nature of fish marketing operations.

The assessment of the value of such courses and also business management games requires further consideration and research. Reaction of course participants was monitored by the use of a questionnaire which was completed anonymously at the end of the course.   Only limited conclusions may be drawn from this with regard to long term assessment but it is useful as an aid to improv- ing the course in the short-term.

Participants were asked to judge each section of the course using ranking methods and the outcome was extremely encouraging.

The most interesting conclusion was that the balance of teaching
methods, i.e. lectures, case-studies, films and business management game was considered correct.

THE EDUCATIONAL ADVANTAGES AND LIMITATIONS

For any management training course, the particular educational
approach will depend on the situation and the course participants,
but will involve a carefully-constructed programme to meet the
requirements. It might contain lectures, case-studies, management games, syndicate projects and role-playing. Although this
paper is concerned with management games, it is recognised that
the main advantages of this technique accrue when games are used
as the core of a comprehensive management course. A game used in
a course allows the organisers to illustrate a particular managerial
function, for example, scheduling of fish processing plant, and to
illustrate the way in which managerial functions and departments
interact such as in fish buying and fish marketing. It is very
important to link the playing of the game to the underlying
theoretical management concepts covered in the lecture periods.

The lectures give the opportunity to impart additional skills,
such as management accounting techniques and development of
information systems, with the aim of a transfer of skills into the
real life situation, but the overall course should also enable
participants to think more of the total system in which they are
involved, what courses of action are available, and what facts must
be considered in order to improve their particular situation.

The response to management games has been one of great
enthusiasm arising from a sense of involvement. It is plainly
obvious that in addition to benefiting from the games in a
professional sense, participants have also enjoyed playing them.
This gives the technique a considerable advantage.

Participants are forced to realise the nature and importance
of effective organisation and intelligent cooperation and to
appreciate this in relation to fisheries "business-like" situations.
They must discover for themselves where the problems lie and how
the choices they make, or can make, relate to the solutions of
these problems. This is clearly superior to merely being told
what the problems are and discussing possible solutions.

If a game is well designed it provides a focal point for
thought and discussion. It establishes a common basis for
communication between participant and participant and also between
participant and lecturers. It is a feature of regional fisheries
training courses that participantd are from different backgrounds
and all are different from the experience of the lecturers. A

game provides a new form of communication across international barriers promoting discussion on different aspects of the managerial process.

The educational value of management games has not been proven, but then neither has the educational value of lectures, case studies or other techniques. Their use currently rests on the intuitive feelings of those who, experienced in using them, are convinced of their usefulness. This means that a great deal of emphasis must be placed on the construction of games and decisions on the characteristics which a particular game should or should not possess. It also means that the personality and ability of the lecturers are just as important for this technique as for lectures, case studies and other methods.

In view of the simplistic nature of all management games, they are open to the criticism that certain of their aspects are unrealistic. This is true, but this is source of strength rather than a weakness. The critical analysis of the specific assumptions upon which the game is based is an exercise in analytical skills and also a starting point for a consideration of the way in which real-world managerial problems evolve and impinge upon one another.

A final observation on advantages and limitations is that this method of management training above all others depends on the desire of course participants to acquire knowledge.

It places a particular emphasis on participative learning where course members tend to gain learning in proportion to their contribution to the activities and their use of the resources available.

THE FUTURE

It is important to examine this technique to see if it could make further contributions to management training programmes.

The type of fisheries management game currently in use and developed through the experience in the United Kingdom, South America and Malaysia and Sri Lanka could be extended to other countries and regions.

These management games would have to reflect the nature of the industry of the country involved, the ownership, the structure, the particular problems experienced in cooperatives or vertically-integrated company systems, the species of fish marketed, and so on. It would also include the products prices, costs yields, etc., pertaining to the local situation. These are only problems of detail.

However, there are other applications that could also benefit from this approach, such as long-range planning and policy making, administration, organisation theory and the identification and preparation of fishery investment projects.

The most rewarding area of investigation might result from the development of a management game linking the economic concepts of the games already constructed for fisheries management training with the physical concept of fisheries resource management. The extension of fishery limits and consequently, the reduced international involvement, is leading to a situation where management decision-making of the "business-like" variety is a more realistic proposition than in the past. Games could be useful for training management personnel in this situation and also for the investigation of the economic effect of alternative policy proposals for the exploitation of fish stocks.

Managment games have been constructed and used as integral parts of fisheries business management training courses. They have made an effective contribution and experience indicates potential contribution in other management training areas in fisheries. It now remains for that potential to be fully exploited.

BIBLIOGRAPHY

Brech, E.F.L., 1953, The Principles and Practice of Management, Longmans, London.
FAO, 1976, UNDP/FAO Training Course on the Management of Small-Scale Fishery Enterprises. FI:DP RAS/74/d3/1.
Haywood, K.H. and Wray, T., 1973, Business Management Courses in processing and marketing management for fish industry executives. Fish. Ind. Rev. 3(4):5-6.
Haywood, K.H., 1974, Management training "games" for fishery executives. Fish. News Int. 13(7):67, 68, 71.
Haywood, K.H. and Nicholson, R.J.A., 1974, Business Management Game: FAO South American Exercise. White Fish Authority Tech.Serv.Rep. (253)
Haywood, K.H., 1975, A Review of Quantitative Methods for Marketing Management. FAO Fish. Rep. (167) Vol.6:7p.
Haywood, K.H., 1976, Fisheries Trainees Play a Malaysian Management Game. Fish. News. Int. 15(1):31,32.
Haywood, K.H., 1976, Fish Industry Operational Research Course held in Trondheim. Fish. New Int. 15(7):67,68.
Lovelock, C., 1969, Notes on the Construction, Operation and Evaluation of Management Games. Management Games Ltd. London.
Sculli, D. A Management Game for the Purchasing, Processing and Marketing of White Fish. Unpublished M.Sc. Report 1970. University of Birmingham.

Kingscott, R.P. and Taylor, J.M. A Management Game for the Market-
    ing of Fish in the Far East. Unpublished M.Sc. Report 1978.
    University of Birmingham.

A REVIEW OF MODELS OF HARBOURS, STORAGE, PROCESSING,

TRANSPORTATION AND DISTRIBUTION

E. Page

Department of Operational Research
The University of Hull
U.K.

In considering the parts of the fishing industry included in
the title, work published since 1975 and not covered by other
contributors in this session will be given preference. For work
prior to 1975 reviews are given by Madziar[1] and Haywood.[2] Fishing
News International gives a ranking of the top 30 fishing nations
by weight caught in 1975.   Japan, U.S.S.R., and China being the
top 3, U.S.A. fifth, Norway sixth, Canada 16th, and U.K. 18th. The
amount of published work does not seem to be proportional to the
fishing effort. Some papers describe work which covers more than
one of the aspects given in the title, when this happens the work
will be described under the first appropriate heading.

HARBOURS

An F.A.O. paper by Madziar[1] reviewed work done upto 1975. A
simulation study of the facilities required for the fishing harbour
at North Shields by myself for the White Fish Authority was carried
out in 1975.   The purpose was to study the feasibility of operation
of the harbour over a range of conditions of catch rates, of species
of fish, types of vessel, and fishing methods. Seasonal variation
in catch rates was catered for, and weather conditions on the
fishing grounds. The facilities available at the harbour considered in the simulation included the length of quay available, the
number of unloading points, susceptibility of the harbour to tidal
conditions, the size of fish storage area, and the capacity of the
fish plant. Output gave the maximum amount of storage needed, the
maximum number of ships waiting and the total number of ship hours
waiting per year. By running the simulation for a range of
conditions it was possible to recommend a range of facilities at

273

the port.   Because of the long time scale in implementing this
type of project the development of the port is still in progress
so a comparison of the final design with the simulation results is
not yet possible.

    Haywood and Farstad 1976[3] describe a study of the sardine
fishery in Morocco.   This was a mixture of a linear programming
exercise to determine the catches at each ground, the port of landing, the processing site and the market to which the fish should be
sent, and a simulation of day to day operations at the ports and
plants to determine the facilities required for normal operation.
The simulation covered upto ten ports of landing, forty fish meal
and oil plants, one hundred and twenty canneries and ten other
processing plants.   Transportation to processing plants and
markets was included in the simulation, and allowance was made for
the effect of climate on the storage and transportation of fish.
The L.P. model was used to limit the range of possibilities needed
to be covered by the simulation, and a final costing exercise
enabled firm recommendations to be made.

    Many of the one-year postgraduate M.Sc. courses in Britain
include a project on an Operational Research problem from Industry
or Government, the White Fish Authority has for many years been
helpful in providing projects for students on such courses.

    C. Elwell[4] studied the operations of the British fishing fleet
at sea and in port.   A simulation was developed for the distant
water fleet but allowed extension to the near water fleet merely
by data changes.   The cycle of time at sea was divided into steaming, fishing, and return, while the time in port was divided into
queueing, unloading, and servicing.

    Various numbers of operating ports were considered.   For each
number of ports the annual operating costs were calculated for the
optimum number of facilities at each port.   The costs included in
the operating costs were the unloading facilities, the labour, and
the demurrage on keeping the vessels waiting for unloading.   The
conclusion of the study was that super single port was the most
efficient operating system.

STORAGE

    Coverdale[5] considered the use of chilled water containers for
the storage of fish from catching to processing plant, for the
Scottish inshore fishing industry.   The use of chilled containers
on ship but not in port was also considered as was the sorting of
the fish by species into the containers rather than having a mixture
of species in each container.   The study considered how many
containers would be needed on each vessel in the white fish and

herring seasons, the cost of converting the vessels, the cost of
container unloading equipment at the port, and the costs of the
traditional system of boxed fish.

The conclusion was that the container system was marginally
cheaper if the destination was a nearby processing plant, but the
savings were so marginal as to cast doubt on the feasibility of the
container system.

Eastley[6] developed a stock control policy for a range of
frozen fish products. The project looked into the forecasting
methods available for supply and demand and stock control for the
monitoring service of the White Fish Authority.   As far as availability was concerned the species divided into three groups, those
available all year, those with a high and low season, and those
with a season and a period of non-availability. The major task of
any forecasting system was then to predict when the change points
occurred in the availability patterns.   Reductions in cost of 30%
had been obtained in introducing the monitoring system, and the
study showed a further 25% was possible by use of the forecasting
and stock control system.

Lee and Neilson[7] studied the processing and storage of blue
whiting from landing to despatch. A Linear Programming model was
developed to find the plant facilities required and a simulation
model to determine the storage requirements in process and of
freezer facilities.   While the process structure was not too
complex, the simulation did involve the use of a lot of computer
time which restricted the number of alternatives studied.

Stael von Holstein[8] looked at the economics of marine fish
farming.   His work divided into three sections, an overall costing
model, a feeding model, and a model of holding or storage facilities.
The feeding programme included the maintaining of the breeding stock
as well as growth rates.   The holding model was an L.P. model which
included the shelf life of the finished product. The reviewer was
struck by the similarity between this problem and that of the modern
turkey or chicken producer.

Some technical aspects of fish storage were considered in Botta
and Shaw[9].

The use of management games in training fisheries personnel is
increasing. A fairly common structure is for the game to begin with
an auction followed by a marketing, sales, and distribution section.
The whole activity is costed and suitable accounts produced.   The
game is played for a number of periods and a final summary session
drawing together the lessons to be learned concludes the game.
Sculli[10] developed an early game of this type. Many more such games
have been developed since Sculli's, the fishing industry is

fortunate in being able to be modelled with a high degree of realism
in such games.   Personnel are thus able to appreciate the interactions of the different aspects of the industry from such games.
The early games were manual and involved a fair amount of calculation either by the participants or the administrators, or both. More
recent games have tended to be computer-based in view of the
continued availability and reduction in cost of computers.   In the
marketing and sales section of such games there is usually a
storage facility for the fish of limited capacity or limited
duration.

PROCESSING

     Linear Programming models of processing plants are very common.
Lee and Neilson[7] consider the processing of blue whiting including
operations, plant, and labour.   Up to ten products were considered
each with its own process specification.   Most Fisheries Management Games have a processing section of which an L.P. model could
be made if the teams so wished.   The study of the sardine fishery
of Morocco[3] developed a simulation which allowed a wide variety
of processes to the sardines and calculated the amount of waste
product if the time between landing and processing was in excess
of a specified limit.

     Johnson[11] considered the relationship between the fleet cap-
acity and the capacities of fish meal plants over time, concluding
the growth of plant to be in proportion to the fleet capacity.
This was an early project and the conditions have changed since
the report, whether the same relationship held over the period of
decline is not known.

TRANSPORT AND DISTRIBUTION

     A rich field for published work in L.P. and simulation models.
Coverdale,[5] and Sculli[10] already mentioned, consider distribution
of fish from the port to processors and markets. Hope and Niven[12]
developed an L.P. model of the catching and distribution of fish
for the British fishing industry. This studied the alternatives
available after the ban from Icelandic waters and make recommendations on tonnages from grounds to each port, and from each port to
demand areas. Restrictions on the system were imposed, e.g. all
ports currently being used must still be used.

     Kitchen[14] did a study of the catching, landing, and distribut-
ion of frozen fish in the U.K. as a whole, firstly keeping the
landings at each port as at present, and secondly giving the vessels
a free choice of port of landing.

SUMMARY

All parts of the fishing industry in the title of the paper have been covered by studies in recent years. The nature of the studies has reflected the changing conditions of the industry. Early work on tactical problems and installation capacities is being replaced by strategic studies on how best to manage a limited resource, how should any future catch quotas be divided between the vessels in a fleet?, what sort of fleet should a country have or aim to have in the future, and so on.

Over a period 1975 to the present the reports of studies has moved from the developed countries to the developing countries, Morocco, the Philippines, Saudi Arabia, Sharjah, etc.

Some projects reported would seem ideal opportunities for the application of O.R. techniques and yet no such studies are mentioned. For example, the development of fourteen integrated harbours in India,[15] and a deep sea fishing study in Malaysia involving 220 trawlers and five ports.[16] And more recently "two ports in India developed" - Fishing News International 1976.

REFERENCES

1. J.B. Madziar, "A Review of Quantitative Methods as Applied to Fishery Harbour Planning, Design and Operation". F.A.O. Fisheries Reports No. 167 V.3, Rome 1975.
2. K.H. Haywood, "Export Consultation on Quantitative Analysis in Fishery Industries Development" F.A.O. Fisheries Report No. 167 V.4, V.5, V.6, 1975.
3. K.H. Haywood and N. Farstad, "Sardine Fisher of Morocco" Fishing News International 1976, v.15, p.22.
4. C. Elwell, "Simulation of the Fishing Fleet", 1970. M.Sc. project report, University of Hull.
5. I.L. Coverdale, "Application of a chilled container system to the Scottish Inshore Fishing Industry", 1972. M.Sc. project report, University of Hull.
6. D.R. Eastley, "Stock Policy for the Frozen Fish Industry", 1973. M.Sc. project report, University of Hull.
7. G. Lee and J. Neilson, "An Investment Appraisal Model for Fish Processing" 1977, M.Sc. project report, University of Hull.
8. G. Stael von Holstein, "Economics of Marine Fish Farming" 1969. M.Sc. project report, University of Birmingham.
9. J.R. Botta and D.H. Shaw, "Effect of double freezing and sub-sequent long-term refrozen storage", 1978. J.Fish.Res.Board Can. v.35, pp.452-456.
10. D. Sculli, "Management Game for Purchasing, Processing and Marketing of White Fish". 1970. M.Sc. project report University of Birmingham.

11. B.G. Johnson, "Industrial Fishing" 1969. M.Sc. project report, University of Birmingham.

12. C.E. Hope, and F.N. Niven, "A Model of the British Fishing Industry" 1976. M.Sc. project report. University of Birmingham.

13. J.A. Farrell, "Development of a container transportation system for the Herring Industry". 1971. M.Sc. project report University of Birmingham.

14. J.A. Kitchen, "A Distribution Model of the British Fishing Industry" 1978, M.Sc. project report, University of Hull.

15. "World Bank Loan for 14 integrated harbours in India", Fishing News International, v.14, n.8, 1975.

16. "Fishing Port Programmes for the Third World", Fishing News International, v.14, 1975.

THE STRUCTURE OF

FISH INDUSTRIES

A MULTI-CRITERIA MODEL FOR ASSESSING INDUSTRIAL STRUCTURE IN THE NORWEGIAN FISH-MEAL INDUSTRY

Lars Mathiesen

Norwegian School of Economics and Business Administration

5000 Bergen,Norway

INTRODUCTION

The Norwegian fish-meal industry consists of some 210 purse seiners, 200 trawlers and 40 processing plants located all along the coast converting capelin, mackerel and some other species into fish meal and fish oil.   Considerable yearly fluctuations in the size of individual stocks and the fact that this industry represents almost the only way of employment in large coastal areas are major characteristics.   In a recent governmental "Long Range Plan of the Fisheries"[2].several public goals are spelled out. These regard among others the desirability of maintaining the present distribution of residents, of securing stable jobs and protecting resource endowments from overfishing. At the same time the plan recognizes the considerable overcapacity as compared to prospective yields from the different species, and the very serious financial position which necessitates large yearly transfers. Besides being

The future size and structure of the industry, by which we mean the number of vessels of different types and sizes and the number and location of processing plants, will have to be set to meet these conflicting goals. Obviously some reduction from the present level is called for.   The reductions of capacities will have to be decided upon in the presence of uncertainty regarding the yearly quotas of the different species.

A two-stage LP-model[1] with multiple criteria[2] is formulated.
The focus of the model is on the conflicting goals, i.e., the
possible attainments and trade-offs that the resource endowments
provide.   These goals are represented by linear functions of the
decision variables of the model.   Alternative and exogenous
specifications of the industry are first-stage variables.   Feasible
operations within these capacities, i.e., the harvesting and processing in different states of fish resources constitute the second
stage variables.

The model is static in the sense that it pictures a stationary
situation.   Capital equipment is decided upon once and for the
whole 5-year target period 1981-85.   Yearly fishing quotas are
represented by random variables with distributions which are assumed
constant over time and independent of the operations of the industry.
The dynamics of the natural renewable resources are thus kept outside the model.

The model is used in two ways.   First it is subjected to
several parametric analyses where levels of goal fulfilment and
stipulations of the random variables are varied systematically.
This phase provides both information for validation of the model
and information for the users in the next phase, decision makers
of the industry.   For this purpose an interactive routine[3] is
available.   Through this routine a decision maker is confronted
with a solution, i.e., an industrial structure with accompanying
attainments on the various goals.   With reference to such a vector
of scores he will have to approve or disapprove of specified
changes, resulting in higher scores on some goals and lower scores
on others.   Depending on his answers, a new solution with
corresponding consequences is calculated and the procedure is
repeated.   This interactive process converges in a finite number
of steps.

In the next section we present the model verbally, while the
mathematical details can be found in the appendix. So far our

----

(1) A special class of LP-models where coefficients are allowed to
    be undertain.   The term "two-stage" refers to the interpretation
    of the variables in the model.   First-stage variables have to
    be decided upon before uncertainty is resolved, while second-
    stage variables can be decided upon afterwards.   For some
    theoretical results see Wets[7].
(2) Decision problems where multiple and conflicting goals are
    recognized are usual in real life.   In recent years several
    methods have been proposed to help a decision maker analyze
    such problems.   Zeleny[8] and Zionts[9] give several examples of
    theory and applications.
(3) The routine is based on Wallenius and Zionts[6]. The computer-
    code is written by Kristian Lund.

computational experience is limited, but some insights have been gained. These are presented in the third section, where we also discuss what kind of information such a model can give.

MODELLING THE CAPACITY ADAPTATION PROBLEM

Any modelling exercise should begin with a clear recognition of the purpose for which the model is formulated. Our main concern is the feasible attainments of and trade-offs between the conflicting (public) goals as to the optimal size and structure of the industry. The model reflects this focus, as it is biased towards some factors of this socio-bio-economic system, while others are dealt with rudimentarily. The presentation of the model is based on the three mathematical programming concepts of decision variables, objectives and constraints.

By size and structure of the industry we mean the number of vessels of different types and sizes and the number and location of processing plants. Different combinations of such units will give different scores on the number and location of jobs, the profitability of activities, the volume of catches etc. We will specify exogenously to the model a number of alternative combinations, their operating levels being decision variables. This is illustrated in fig. 1, where P_1 and P_2 are two alternative combinations. Let P_1 represent the present configuration of the industry, and let $x_1 = 1$ denote the corresponding level. Assume that P_2 implies half the number of vessels, but as many processing plants as do P_1, and let $x_2 = 1$ denote this. The solution P^* where $x_1^* = 0.3$ and $x_2^* = 0.6$ would then call for a 10% reduction in processing capacity and 40% reduction in the number of vessels compared to present capacities.

These variables represent the capital equipment to be used over the planning period. The actual yearly utilization of capacities, however, will be represented by a second set of variables. When modelling at the industry or national level the question of control over decision variables becomes a critical one.[4] While the individual vessels and processing plants are privately and to a great extent individually owned and operated, the authorities are able to influence the industry through various means like licences, financial aid etc. Furthermore, there is a central office routing vessels to the plants. As a consequence, the implicit assumption of control can at least to a certain degree be regarded as valid.

(4) Morgenstern (10, p.1167) states: "... Where there is not complete central control, i.e., where the outcome depends on several decision makers, as in game theory, linear programming does not give the complete answer. It can, however, provide ceteris paribus answers".

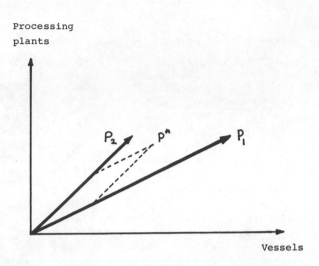

Figure 1. Illustration of the structural variables as vectors.

The capital equipment of the industry does already exist. The aggregate capacity is excessive and opportunity cost for the different units is only a fraction of their replacement cost. We therefore allow the cost function to differ with regard to scrapping and expanding capital equipment. This is done by separate activities for selling and buying equipment.

From the Long Range Plan 2 it is evident that the government and politicians in general attach special interest to certain consequences of the industrial structure and its activities. Among these are employment, the use of capital and natural resources, and profitability of operations. While the attainment of favourable scores on these criteria are viewed as goals, the relationship between them and the weights they are assigned are not at all clear, i.e.,the overall social welfare function where these are parameters is unknown. Our approach is then to model the way each separate goal depends on the decision variables and leave the unknown weighting to be resolved through the interactive process between a decision maker and the model. [5] [6]

---

(5)   This approach is inspired by the experiments of Wallenius et al.[5] with a short term macro-economic model of the Finnish economy.
(6)   In the industry there are many decision-makers. Our approach deals with one at a time.

The various public goals are summarized through three types of objectives concerning employment, profitability and catching ability. The employment goal is an obvious one: maximize the level of employment. When distinguishing between different locations (regions) for processing plants, it will be easy to expand this goal to include one for each region whereby the trade-offs between employment in different regions can be analyzed directly. The modifying element to obtaining the maximum employment is the requirement that the industry can be operated without too large subsidies[7]. We associate profitability with stable and well-payed jobs, which is another declared goal.

The social profitability goal can be described as:

maximize (yearly revenues from sales of products - yearly operating costs)+ β{contribution from the sale of equipment - expenditure on new equipment + present value of capital at the end of the planning period},

where β denote an amortization factor.   The input to this yearly income is the initially given capital equipment, the value of which is a constant and thus irrelevant to maximization.   It is also worth noting that costs in this context mean social opportunity costs.

The social profitability goal will in several ways differ from private profitability. Companies have to pay the market wage which is higher than the social opportunity cost of labour. They also have to repay loans irrespective of the actual value of the capital equipment. Finally, they may not be able to live through years of large deficits. Consequently we add a private profitability goal which will show more closely the financial position of the aggregate of private enterprises.

The final goal represents the catching capacity of the industry. In public discussion it is often voiced that under the new regime of the sea Norway should be able to catch the whole allotted quota, or else we might end up with a smaller quota in succeeding years. Inability to catch is clearly reduced by maximing (the value of) catches, since quotas are (assumed) constant with respect to capacities and fishing.

---

(7)   Given the widely agreed upon goal for maintaining the present geographical distribution of residents, social opportunity cost of labour will be low in some areas along the coast. Hence, the market price (wage) is not proper for evaluating the social profitability of activities, and some subsidies are warranted.

The constraints of the model will be of three types. Capacities and resources make up two sets of constraints on activities. In addition there are equations keeping track of the various types of capital equipment. These latter constraints amount to mere accounting:

capacity used in the target period + capacity sold out
- capacity added = present capacity.

The traditional analysis of fisheries economics is based upon a differential (or difference) equation explaining the dynamics of the stock. Catches are determined without regard to capacity of the exploiting industry. In fact, the capacity is assumed to be completely adaptable. Within the context of high variability of the stocksize, this approach would presumably require greatly varying capacity. We observe, however, that capital equipment is malleable only to a very limited degree and furthermore, that it will be uneconomical to maintain the maximally required capacity. The optimal management policy will therefore involve the minimum of two state variables, the allowable catch and the catching capacity. This fact makes an analytical analysis more complex[8], while in our numerical and static analysis it presents no problem. As an illustration of how capacities and stocks are handled in our model, let us consider a one species model and let

x denote the capacity of the industry (to be decided upon),

y denote the total catches (in one year),

ω denote the random stocksize,

$f(x;\omega)$ denote the catching and processing potential of capacity x given that the stocksize is ω,

$g(\omega)$ denote the quota given that the stock size is ω.

With these symbols maximum feasible catches will be

(1) $y = \text{minimum } \{f(x;\omega) , g(\omega)\}$

or equivalently

(2a) $y \leq f(x;\omega)$

and

(2b) $y \leq g(\omega)$

(2a) states that the total catch cannot exceed the catching and processing potential of the industry, i.e., merely a technical constraint. In (2a) we allow the potential catch to be a function of

---

(8)   An interesting contribution here is made by Clark, Clarke and Munro[1].

the stock-size.   A larger stock could for example be easier to
locate and might be available for profitable exploitation over a
longer period of time than does a smaller stock.   From existing
data some effect of this kind seems to be present, although it is
far from proportional.

(2b) implies that the total catch is not allowed to exceed the
exogenously specified quota.   Because the model contains no explicit
link between consecutive years this admittedly ad-hoc constraint has
to play the role.   In a dynamic model an equation would explain the
dynamics of the resource.   Exploitation in one year would then be
done with explicit regard to availability of fish in succeeding
years.   Hence a dynamic model can allow for "trade-offs" in catches
between consecutive years, while our static model cannot.   We
shall rather have to assume that catches in one year do not
influence upon what can be caught in succeeding years.   We therefore assume that quotas can be described by probability distributions
which are stationary over time.   This assumption does not presuppose
any special resource management policy.

Similarly we shall also not be able to incorporate in a
meaningful way the obvious uncertainty pertaining to the measurement of stock-sizes upon which quotas are stipulated.   Hence, this
probably major source of uncertainty will be neglected.[9]

Uncertainty regarding the outcome of a certain seasonal fishery,
for example because of bad weather preventing operations,can be
incorporated.   Technically this is done by expanding the state space.

The mathematical model is presented in the appendix.   There is
one further comment to be made here.   So far we have described a
stochastic multi-criteria model.   For computational purposes we
derive an equivalent deterministic multicriteria model, where goals
and constraints are approximated by linear functions of the decision
variables.   In this model we only use first moments of the random
criteria, i.e., the goals express expected values only and no higher
moments.   Consequently,the variations in goal attainment as such do
not enter explicitly, which implies that we assume risk neutrality.

INFORMATION COMING FROM THE MODEL

The project is now in the midst of generating numerical results
and will then continue with the interactive part.   As every
practitioner should know,the purpose of a mathematical programming
model in strategic applications is to develop insights into system

---

(9)   Within a Markov decision process this feature could be incor-
      porated, see Satia and Lave[4].   Technically it is done by
      expanding the state space.

behaviour which in turn can be used to guide the development of
effective plans and decisions.   Such insights are seldom evident
from the output of an optimization run.   So far our computational
experience does not seem to warrant definite conclusions although
some insights have been gained.

One approach to solving a multi-criteria model is to convert
all except for one to constraints simply by stipulating lower levels
of goal attainment.   Then one can study the score on the remaining
objective as a function of parametric changes in such stipulated levels.
This is parametric linear programming which is a standard routine.
The results can most conveniently be exhibited in a diagram. Figure 2
shows the social profitability of the industry for varying levels of
total employment.

We shall concentrate on two features of this frontier.   One is
its intersection with the abscissa, i.e., the employment level at

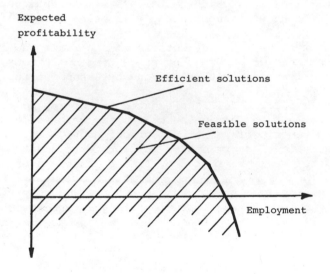

Figure 2. Trade-off between expected social
          profitability and total employment

which the industry breaks even.   The other is the slope of the
frontier, i.e., the "cost" of marginal employment in the neighbour-
hood of this intersection. (10)   These figures characterise the
performance of the industry as an employer. The maximum profit solution
so prominent in theory, would imply too large reductions to be
realistic.   Even break-even will cause considerable structural
change and resistance. Results so far indicate that the break-
even level is in the range of 60-70% of today's employment of
about 6000 workers, and that the "cost" of marginal employment at
break-even is in the order of N.kr 60.000,-. (11)   The corresponding
cost at the present level of employment is N.kr 80 .000,-. (12)

There are lots of qualifers to go with these numerical results.
Data, especially on quotas and their probability distributions, are
rough.   As explained above, we have also ignored major aspects, e.g.
the dynamics of the resources and the uncertainty regarding the
yearly measurement of the stock.   These two features, however,
cannot significantly lessen the quite pessimistic impression and
the dramatic conclusions to be drawn, unless they involve rather
unsymmetric relations.

To any point on the frontier in fig. 2 there corresponds an
optimal structure of the industry. One conclusion is that even if
capacities in total ought to be reduced, processing capacity in
Northern Norway should be maintained. This is rather obvious
because these plants are located closest to the capelin fishing
grounds and save fishing vessels long transportation time along the
coast. (13)   Next, the optimal structure seems to depend on the
target level for employment. Capacities are excessive and have
to be reduced. If employment is not permitted to fall, an alternative
is to operate the vessels with two crews. (14)

(10)   By the "cost" of marginal employment we here mean the decrease
       in profitability of a unit increase in employment.
(11)   The average remuneration to fishermen in this industry has been
       estimated to N.kr 90-160.000,- in 1977. It is likely to be some-
       what lower in 1978 and 1979. The social opportunity cost is
       stipulated to N.kr 40.000,- (1977-value) in this analysis.
(12)   It is interesting to note that Hansen[3], in the cost study of
       purse-seiners, estimated the marginal cost of employment in this
       fleet to be N.kr 115.000,- (1975 value). The main reason for
       the difference is that we allow for a change in structure and
       maintain employment, while he also preserves structure. He also
       uses replacement cost which is higher than opportunity cost.
(13)   As an illustration, a vessel unloading in Finnmark could make
       2-3 trips per week, while one trip to the south-west coast and
       back would require 8-10 days.
(14)   Hansen[3] advocated this solution. Taking into consideration
       reports that fishermen on average work close to 70% more hours
       than the average worker in Norway, such an arrangement seems
       reasonable. The merchant fleet already has some experience here.

Another kind of information from an optimizing model is the so-called shadow prices associated with the constraint   In our model these will be of three different types.   The shadow price corresponding to any quota tells the prospective increase in the objective resulting from a unit increase in the quota.   Such information would be important when negotiating with other nations on increases/decreases in quotas of the various species.   Our results show that these marginal values can deviate quite considerably from the net market value of the fish.   Restricted capacity in certain periods account for this fact.   The two other types of shadow prices correspond to capacities.

When interpreting results from a model with multiple criteria some caution should be exercised.   An optimal solution will usually imply a trade-off between these criteria, which implies that the shadow price bears the same kind of trade-off interpretation.   This means that the marginal value of an increased quota of a certain species can differ with the weight assigned to the criteria. Increases in the mackerel-quota for example, have less value to an industry that is mainly concentrated in Northern Norway than if it was located on the south-west coast.

CONCLUSIONS

The Norwegian fish-meal industry has considerable excess capacity as compared to prospective yields from the resources.   The financial position is quite serious and future prospects are even worse. Hence reductions of capacities and employment from the present level will be necessary.   This will, however, meet resistance in coastal areas where the fisheries constitute almost the only economic activity, and will also to some extent be in conflict with widely agreed upon public goals.

The aim of the project is to provide decision makers within the industry with a tool for making explicit and assessing trade-offs between different desired properties of the industry.   In this way it is hoped that the outcome of the ongoing debate on this topic will be enhanced.   For this purpose a mathematical programming model is formulated.

The project is now in the midst of generating numerical results and will then continue with the interactive part.   So far our computational experience does not seem to warrant definite conclusions although some insights have been gained.

REFERENCES

1. C.W. Clark, F.H. Clarke and G.R. Munro, The optimal exploitation
    of renewable resource stocks: Problems of irreversible invest-
    ment, Econometrica, vol. 47, no. 1, 1979.
2. Fiskeridepartementet: Langtidsplanen for fiskenaeringen, Oslo
    1977, (in Norwegian).
3. T. Hansen, The relationship between aggregate costs, employment
    and cargo capacity of the Norwegian purse seiner fleet,
    Scandinavian Journal of Economics, vol. 81, no. 1, 1979.
4. J.K. Satia and R.E. Lave, Markovian decision processes with
    probabilistic observations of States, Management Science,
    vol. 20, no. 1, 1973.
5. H. Wallenius, J. Wallenius and P. Vartia, An approach to solving
    multiple criteria macroeconomic policy problems and an
    application, Management Science, vol. 24, no. 10, 1978.
6. J. Wallenius and S. Zionts, An interactive programming method
    for solving multiple criteria problem, Management Science,
    vol. 22, no.6, 1976.
7. R.J.B. Wets, Stochastic programs with fixed reoource: The
    equivalent deterministic program, SIAM Review, vol. 16, no.3,
    1974.
8. M. Zeleny, ed., Multiple criteria decision making, Proceedings
    Kyoto 1975, Springer Verlag, 1976.
9. S. Zionts, ed., Multiple criteria problem solving, Proceedings
    Buffalo N.Y. 1977, Springer Verlag , 1978 .
10. O. Morgenstern, Thirteen critical points in contemporary
    economic theory: An interpretation, Journal of Economic
    Literature, vol. 10, no.4, 1972.

APPENDIX

Let:

$U(\cdot)$ denote the overall (but unknown) objective,

x_j denote the "activity" level of structure j, $x_j=1$ refers to the definition of the structure, i.e. how many units of the different types of capital equipment this structure comprises,

c_j denote total fixed non-capital (social) costs associated with $x_j=1$,

ℓ_j denote total employment when $x_j=1$,

Y_i denote total catch from the i'th resource stock (in mill.tons),

d_i denote the contribution (selling price net of variable social costs) from the sale of fishmeal and oil from one mill. ton of resource i,

$\tilde{\omega}_i$ denote the size of resource stock i (random parameter),

$\tilde{a}_{ij} = a_{ij}(\tilde{\omega}_i)$ denote the catching and processing potential of structure j with regard to the i'th resource (in mill. tons),

$\tilde{q}_i = q_i(\tilde{\omega}_i)$ denote the available quote of the i'th resource,

b_{kj} denote the number of units of capital equipment of type k when $x_j=1$,

b_k denote the number of units of capital equipment of type k in industry today,

$u_k^+ \; (u_k^-)$ denote the number of units of capital equipment of type k added (sold out),

$s_k^+(s_k^-)$ denote the investment cost (sales value) of one unit capital equipment k,

t_k denote the value per unit of capital equipment k at the end of the planning period,

α denote the discount factor, $\alpha=(1+r)^{-1}$ where r is the discount rate,

β denote amortization factor, $= \dfrac{r(1+r)^5}{(1+r)^5-1}$,

E denote the expectation operator.

With this notation the model can be presented. The objective is
(1)   maximize $E\{U(M_1,M_2,M_3,M_4)\}$,

where the arguments are defined as follows:

the social profitability goal

(2) $M_1 = -\sum_j c_j x_j + \beta\left[\sum_k s_k^- u_k^- - \sum_k s_k^+ u_k^+ + \alpha^5 \sum_k t_k \sum_j b_{kj} x_j\right] + \max_i \{\sum_i d_i y_i\}$

the private profitability goal

(3) $M_2 = -\sum_j \bar{c}_j x_j + \max_i \{\sum_i \bar{d}_i y_i\}$

the employment goal

(4) $M_3 = \sum_j \ell_j x_j$,

and the catching ability goal

(5) $M_4 = \max_i \{\sum_i d_i y_i\}$

The constraints on the maximization are the following:

change in capacities

(6) $\sum_j b_{kj} x_j + u_k^- - u_k^+ = b_k$, $\forall$ k ,

constraints relating catches to capacity

(7) $-\sum_j \tilde{a}_{ij} x_j + y_i \leq 0$, $\forall$ i , almost surely,

constraints relating catches to quotas

(8) $y_i \leq \tilde{q}_i$, $\forall$ i, almost surely,

nonnegativity constraints on variables

       $x_j \geq 0$, $y_i \geq 0$, $\forall j$, i .

The model (1) - (8) contains random coefficients $\tilde{a}_{ij}$, $\tilde{q}_i$.   The
interpretation of variables and random events is the following: at
first capacities, i.e., the x, u^+ and u^-, are decided upon in the
presence of uncertainty regarding yearly quotas.   Secondly quotas
and catch coefficients will become known and then the operations, i.e.,
the y can be performed.   Observe that the model does not allow
any revision of capacities other than at the very start.   With
this interpretation the model is of the two-stage type of a
stochastic program with recourse.   Note that (7) and (8)   are
stochastic constraints which are required to hold with probability

Because $\tilde{a}_{ij} > 0$ and $\tilde{q}_i \geq 0$, $\forall i$, j, there are always feasible
second stage solutions, i.e. y > 0, for any x > 0.   It is well
known, see Wets[7], that a two-stage type LP-model (one objective)
with random coefficients having discrete, finite probability
distributions has an equivalent deterministic linear formulation.
The extension to our case with multiple criteria is straight forward
when $U(\cdot) = \sum_h \lambda_h M_h$, which we are forced to assume anyway in order
to use the interactive routine.   Thus we have the following equivalent

deterministic multi-criteria linear model (in matrix and vector notation)

(9) $\max U(\overline{M}_1, \overline{M}_2, \overline{M}_3, \overline{M}_4) = \max \Sigma_h \lambda_h \overline{M}_h$

where the arguments now are defined by:

(10) $\overline{M}_1 = -cx + \beta \left[s^- u^- - s^+ u^+ + \alpha tBx \right] + E\{dy\}$

(11) $\overline{M}_2 = -\overline{cx}$ $+ E\{dy\}$

(12) $\overline{M}_3 = \ell x$

(13) $\overline{M}_4 =$ $E\{dy\}$

        Subject to
(14) $Bx + u^- - u^+$ $= b$

(15) $-A_n x$ $+y^n$ ≤ 0 $\forall n,$

(16) $\leq q_n$ $\forall n,$

        $x \geq 0,\ u^- \geq 0,\ u^+ \geq 0,\ y^n \geq 0$, $\forall n.$

(15) and (16) are the counterparts of (7) and (8) respectively for the n'th state of the discrete probability distribution. When there are N states, these constraints will be replicated N times, where A_n and q_n in general will differ from $A_{\bar{n}}$ and $q_{\bar{n}}$, for $n \neq \bar{n}$.

A MULTI-OBJECTIVE SIMULATION MODEL FOR THE NORWEGIAN

FISH-MEAL INDUSTRY

Trond Bjørndal

Norwegian School of Economics and Business Administration

Hellevn. 30, N-5000 Bergen, Norway

INTRODUCTION

The Norwegian fish-meal industry consists of some 210 purse
seiners and 200 trawlers with an aggregate cargo capacity of about
1.5 million hectolitres and 40 processing plants producing fish
meal and fish oil, with a daily production capacity of about
32,500 tonnes. Important fish species include capelin, blue
whiting, mackerel, North Sea herring, Norway pout, sandeel and
sprat. As pointed out by several studies (e.g. Hansen et al.[1]),
the major characteristics of the industry are: great fluctuations
in seasonal and yearly catches, considerable importance for employment in large fringe areas and great overcapacity. In addition,
the processing sector is geographically widely spread.

The governmental long-range plan for the fisheries[2] spells
out the major public goals for the industry. These include: maintaining employment in the fringe areas, operating profitably, and
securing the natural base of the resources.

The seasonal pattern of the Norwegian industrial fisheries is
indicated in figure 1. As catching takes place in a relatively
concentrated area while processing is carried out in various centres
spread along the coast, the catches have to be distributed. This
is illustrated in figure 2 which shows important fishing grounds
and indicates the distribution of catches.

In the near future decisions will have to be made concerning
the future structure of the industry and due consideration will
have to be given to the various public goals set for the industry.
This paper will describe a multi-objective simulation model for

295

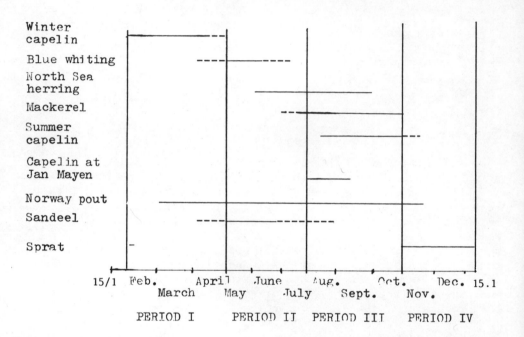

Figure 1. The seasonal pattern of the Norwegian industrial
          fisheries.

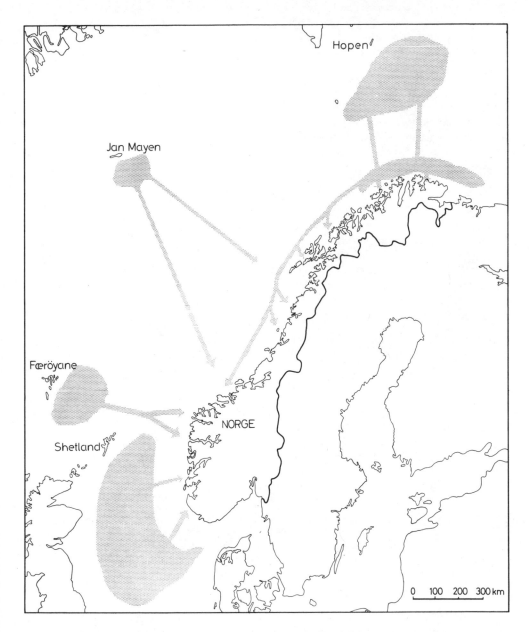

Figure 2. Important fishing grounds and the distribution
         of catches.

analyzing capacity adaptation in the Norwegian fish-meal industry.
The focus of the model is on industry structure. Alternative
fleet/processing sector combinations may be specified, the feasible
catching and processing operations simulated and the consequences
of conflicting goals estimated.

The model depicts a "steady-state" situation. Capital investments are undertaken once and for the whole five-year target
period 1985-90.   Yearly allowable fishing quotas are represented
by random variables with distributions which are constant over time
and independent of the operations of the industry in each individual
year.   Therefore, there is no explicit link between fishing in
consecutive years.

The model is used to simulate a typical year over the target
period. This may be done for any industry structure, facilitating
a comparison between various alternatives with regard to goal
fulfilment. It is the purpose of this study to present a flexible
tool for analyzing long-run capacity adaptation, where tradeoffs
between different public goals of the industry may be studied.

MODELLING OUTLINE

Simulation is an alternative to performing experiments in
reality.   Lundin[3] defines simulation as
        "the use of a model for representing over time
         important characteristics by the system being
         studied..." (p.3)
In our context, the definition implies that, by means of a model,
we will describe how the industry "functions" over time. The time
aspect in our analysis is, on the one hand, long-run capacity
adaptation of the industry and, on the other hand, yearly catching
and processing operations.

Experimenting with a model requires that we divide the input
variables into model parameters and decision variables. Moreover,
forecasts of the values of the model parameters have to be established.   The model is exposed to the same influences that are
expected to occur in reality and, in this way, we may obtain an
understanding of what might result if we choose one or the other
decision alternative.

Our purpose is to analyze capacity adaptation in the fish-meal
industry.   This is done in a situation with conflicting public
goals and several different fish species, represented by stochastic
variables.

With regard to the size and structure of the industry, the fleet
is defined by the number of vessels of different types and their

aggregate cargo capacity. The processing sector is defined according to location: the number of plants per region, their size, and
their aggregate production and storage capacity. Different
structures will give different costs. Moreover, the number and
location of jobs and the catching and processing prospects will
vary.   These aspects are captured by the model through its focus
on industry structure.

As previously stated the target period for the analysis is
1985-90 and the model will simulate a _typical year_ and not individual
years of the period. However, by estimating the variance of the
output-variables, we may indicate the variations around the average
year in the given period of time. This implies that our model
represents a "steady-state" system.

The following objectives are considered in the analysis:

1)   Maximizing catches
2)   Maximizing profits
3)   Maximizing employment

These objectives, based on the governmental long-range plan, have
also been considered in other analyses (e.g. Mathiesen[4]). The first
objective is established with the restriction that catches may not
exceed what is biologically feasible.   The rationale for this
objective is that quotas are divided between different countries,
partly on the basis of previous years' catches.   If Norway does
not fish her quota, this might eventually result in a smaller
allocation.   Employment and profits are estimated for the industry
as a whole. However, with regard to the former, it is perfectly
possible to establish one objective for each region. This would
make it possible to analyze directly tradeoffs between employment
in different regions. Profits are defined as sales-revenue less
operating costs and annual fixed costs.

In the model, the availability of each species is described by
a probability distribution, representing the basis for establishing
biological quotas in the individual years. The biologically established catch-quotas are distributed between Norway and other
countries, usually according to a percentage distribution.   The
fish are harvested by the fleet and delivered to the processors.
Due to their geographical spread, rules have to be laid down with
regard to the distribution of the catches among processing centres.
Some of the catches are used for human consumption but the major
part is processed into fish meal and oil.

Figure 3 gives a simple representation of the model.

In the model, the values of the long-run decision variables,
 i.e., the industrial capacities, are determined exogenously.

<u>Input:</u>   <u>Output:</u>

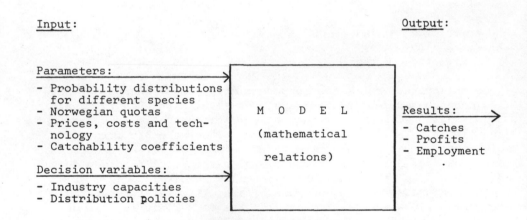

Figure 3. Model Outline.

Probability distributions for the fish species and the values of
other model parameters, including routines describing the distribution of catches among processing centres, are inputs to the model.

Prices of fish meal, fish oil and deliveries for human
consumption are based on forecasts by Hansen et al.[1] So far, the
effects of technological development have not been incorporated in
the analysis.

When the long-run capacity adaptation is undertaken (i.e. the
values of the long-run decision variables are determined), fishing
and processing operations may be simulated for a specified number
of years (times). In an individual year, Norwegian quotas are
determined as described above. With given capacities and chosen
distribution policy(-ies), yearly fishing and processing operations
are estimated. After simulating yearly catches for the given
number of years, the consequences, with respect to catches, profits
and employment, for the industry structure in question may be
estimated. This process is repeated for each industry structure
that is analyzed.

DECISION VARIABLES

Long Run

Investments in the industry are of a long run character – when
investments have been made, the industry structure is given for a
certain period of time. With regard to the level of aggregation,
the fleet is defined by the number of purse seiners and trawlers
and by their respective aggregate cargo capacity. When considering
the participation in the different fisheries, however, the purse
seiners and the trawlers are divided into subgroups. For the
processing sector, the capacities are defined by three regions:
Northern, Middle, and Southern Norway.

It is the main purpose of this model to present a tool for
analyzing the level of investments in the industry. Therefore,
when the results are presented for some tentative industry
structures at the end of the paper, profitability will be calculated
only for the industry as a whole. The division of the fleet into
subgroups and the processing sector into regions is necessary to
give a closer approximation to reality. Although these assumptions
facilitate the internal distribution of income in the industry,
these matters are of less interest in our context.

The costs of the industry have been estimated according to two
principles – private and social costs. The former are based on
studies of the costs of the industry for the years 1975-77. With
regard to the latter, it is based on a social opportunity cost for

labour of N.kr. 55,000 (1977-value). Capital costs are estimated
according to the replacement cost principle.   While it may be valid
to use replacement cost when considering an expansion of the industry, this may not be so when considering capacity reduction.
In today's situation, when the latter is being discussed, our costs
are overestimated because replacement costs are clearly higher than
opportunity costs. Here we only want to point out this discrepancy.
In further work on this project, we shall attempt to modify the
cost functions in order to study the possible reduction of industry
capacity.

Short Run

       The distribution of catches among processing centres was
roughly illustration in figure 2.   The choice of distribution
policy for a given fishery must be based on the size of the quota,
a given participation of the fleet and available processing capacity.
Therefore, it will be different from fishery to fishery and may
vary from year to year. This decision variable is, therefore, of
a short run nature.

       In laying down the distribution policies, it is necessary
to make assumptions with regard to "catchability" coefficients.
These express how many trips a boat may undertake a week between
the fishing grounds and the various production centres of a given
fishery. The coefficients are meant to cover time for going to
and from the fishing grounds, for searching, for the actual fishing
operations and for delivery ashore. Several factors will influence
the value of these coefficients, such as the number of boats
participating, the size of the quota and the weather conditions. The
values will also vary over the fishing season. Although very
exact relationships may be established for such functions, this
has not been done within the context of our analysis.   For some
fisheries, empirical studies have been made on catching efficiency
and catchability coefficients have been established which are
meant to represent average efficiency over the season.   These
results have been utilized by us while, for other fisheries, rough
estimates have been made. Sensitivity analyses (varying the values
of the catchability coefficients) imply that these simplifying
assumptions will affect the results only negligibly.

       The choice of distribution policy depends on the capacities in
relation to the size of the quota. If the capacities are small,
this calls for the most effective employment of the fleet and the
processing sector. On the other hand, if there is overcapacity,
a more lenient distribution policy may be used, e.g. giving a
geographically wider distribution of the catches. The distribution
policy also depends on whether the catches have to be processed
relatively soon, i.e., within a week or less, or whether they may

be stored for some time before processing (winter capelin may be
stored for up to eight weeks). Different routines describing
the distribution of catches are defined in the model.

STOCHASTIC VARIABLES

     The availability of the different species varies from year to
year. These variations are caused by natural factors and by
fishing effort.

     For the period 1985-90, we have established forecasts over the
total allowable yearly catches for the different species. This is
the quantity which the species on average may yield over time,
provided overfishing is not permitted. The yearly yield from the
species will, however, vary around the average yield because of
natural factors. The stochastic variables, i.e., the yearly yield
from the different species, will be described by probability
distributions. These have been established in cooperation with
research workers at the Marine Research Institute of Norway.
However, it should be stressed that the responsibility for the
assessments is ours.

     In the model, there is no explicit link between fishing in
consecutive years. Actual catches in a given year have no influence
on the stock size in subsequent years. This aspect would be very
important in a dynamic model. Valid objections may be raised
against these assumptions. However, we will stress that our model
only simulates a typical year in the target period, and not the
development over a number of years.   Moreover, in the model, actual
catches for the different species may not exceed allowable
catches, generated by the probability distributions. This
simplified constraint allows for probability distributions which
are stationary over time. However, a dynamic relationship may be
incorporated in our model with relative ease. The reason why this
has not been done so far is insufficient knowledge of these
relationships.

     When a random number is generated from a probability distribut-
ion, giving the allowable catches of a given species in a given year,
it is assumed that this will be distributed among Norway and other
countries according to a percentage distribution. Estimates of
deliveries for consumption are also given.

     The probability distributions used are relatively simple. To
some extent this reflects the uncertainty caused by the long time
horizon. However, more refined distributions may easily be
incorporated in the model.

THE SIMULATION MODEL

The model consists of a main model and a simulation model.
The model structure is indicated in figure 4.

The main model is a model for long-run capacity adaptation and
is run through only once for each industry structure. The simulation model, however, is a model for yearly catches, consisting of
submodels for the different fish species. This part is run through
a specified number of times (years). In the simulation of yearly
catches, the year is divided into four periods, as indicated in
figure 1.   In each period submodels are called for to simulate the
production taking place in the period. The choice of distribution
policy is undertaken in the submodels.   When moving from one period
to another, possible overlapping consequences are taken into
consideration.

For most fish species, assumptions have been made concerning
the duration of the fishing season. If the quota has not been
caught by the end of the season, fishing is halted in the model.
These assumptions have not yet been made for Norway pout, sandeel
and sprat. Fishing for these species may therefore be overestimated
in the model.

When all four periods have been run through, the consequences
for the year in question are calculated before the simulation starts
again for the next year. Having simulated the catches for the
specified number of years, the main model calculates the consequences for the industry structure in question.   This is based on
the realized operations in all years which the simulation model
encompasses. The model is written in FORTRAN.

To run the model, the values of the long decision variables,
i.e., industry capacities, have to be determined.   In addition,
the number of times (years) which the simulation model should be
run through must be stipulated.

RESULTS FROM THE ANALYSIS

Although an operating model has been developed, some important
work remains to be done. This relates partly to the data input
side, especially some of the assumptions about the duration of the
fishing seasons and the cost functions, but also the validation of
the model. So far, the emphasis has been put on model development
and obtaining relevant input data while less has been done for the
validation of the model on the basis of historical data. Moreover,
we will again point out the simplified assumptions concerning
technological development and the stochastic variables. For these
reasons, the results must be considered as preliminary.

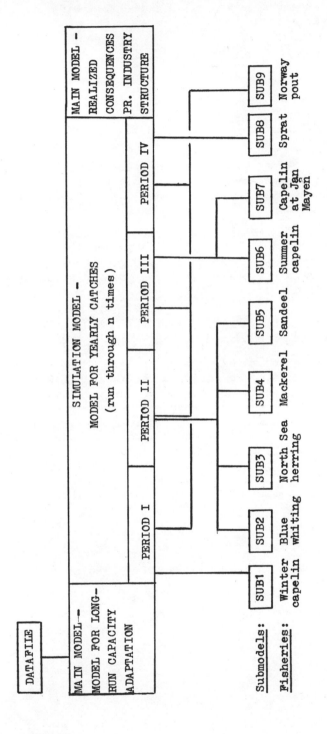

Figure 4. Model Structure

We will present some results for the following three alternative industry structures:

Alternative	Fleet	Processing Sector
I	Present capacities	Present capacities
II	25% capacity reduction	Northern Norway: Present capacities Middle & Southern Norway: 25% capacity reduction
III	50% capacity reduction	Northern Norway: Present capacities Middle & Southern Norway: 50% capacity reduction

The objective of maximizing catches will be expressed by showing
catches as a percentage of total available supply of fish. Social
costs have been used when estimating the profitability of the
industry. The results, based on 50 runs of the simulation model,
are given in table 1.

Each objective is defined in the way that a larger (positive)
value is preferred to a smaller. Hence, table 1 shows that no
alternative dominates another in the sense that one alternative
is better than the others with regard to all three objectives.
Therefore, when choosing between alternatives, all three have to
be taken into consideration. The choice will depend on what
weights are assigned to the different objectives.

The results illustrate the overcapacity of the industry very
clearly - 25% capacity reduction only reduces expected catches
negligibly. Therefore, the choice between alternatives I and II
involves maintaining about 1,300 workers in the industry and
reducing the yearly deficit by about 300 mill. 1977-N.kr.

Table 1. Some preliminary results

ALTERNATIVE	Catches as percentage of total supply of fish % (expectation)	Profits mill. 1977-N.kr. (expectation)	Employ- ment
I	99.8	- 500	5,900
II	99.4	- 200	4,600
III	95.9	+ 80	3,400

When comparing alternatives II and III, however, a reduction of about 3.5% in expected catches as a percentage of the total supply of fish must be taken into consideration. Moreover, the difference involves about 1,200 fewer workers and an improvement in profits of about 280 mill. 1977-N.kr. This imporvement in profits is somewhat less than in the choice between alternatives I and II, mainly because alternative III causes reduced catches.

As previously indicated, the model may simulate any industrial structure. However, alternatives I and III may, for political reasons, be viewed as the extremes. The former, because the yearly industry deficit is unacceptable. The latter for two possible reasons: the implied reduction in employment is unacceptable, and it implies monopoly profits (in our cost functions we have used a 7% rate of interest on invested capital). Therefore, the set of feasible alternatives is restricted to lie between I and III.

The decision makers of the industry are faced with a tradeoff problem: balancing the achievements of the three objectives. If there is no uncertainty involved in the problem, the decision maker will, for example, have to decide about how many jobs he is willing to give up in order to improve profits by some specified amount. If uncertainty is involved, the tradeoff-issue remains the same but the problem will become more difficult because it may not be evident what the consequences of the various alternatives will be.

The choice will depend on the subjective judgement of the decision maker. Value functions, based on the value structure of the decision maker, may be assessed in order to formalize the choice between alternatives. It is, of course, meaningless to compare the magnitudes of two objectives such as profits and employment which are measured in money units and number of jobs respectively. However, a value function combines the achievements of different objectives in a scalar index of preferability or value. A given industry structure, with given achievements of the three objectives, may therefore be represented by a single score on the value function. This enables a comparison of different alternatives. The problem of the decision maker is then to choose the alternative that maximizes his value function.

By means of the simulation model, the consequences of the conflicting goals may be estimated for a number of alternative industry structures. Through assessing a value function, the decision maker's choice between alternatives may be formalized.

SUMMARY

The multi-objective simulation model for analyzing capacity

adaptation in the Norwegian fish-meal industry presented in this
paper is mathematically very simple. However, it hopefully gives
a relatively good representation of reality, although this reamins
to be verified through a future phase of validation. The strength
and the value of the model lie in its great flexibility: the
analyst may specify any industry structure he wants, and the values
of the model parameters may easily be varied. Moreover, computer
time required for the simulation runs is relatively modest.

REFERENCES

1.   T. Hansen et.al., En strukturanalyse av sildeolje- og sildemel-
     industrien. Bergen, 1976 (in Norwegian).
2.   Fiskeridepartementet (the Ministry of Fisheries): Langtidsplanen
     for fiskerinaeringen. Oslo, 1977 (in Norwegian).
3.   R. Lundin, Metodproblem vid simulering. Företagsekonomiska
     studier i Göteborg (Gothenburg Studies in Business Adminis-
     tration) (in Swedish)
4.   L. Mathiesen, A Multi-Criteria Model for Assessing Industrial
     Structure in the Norwegian Fish-Meal Industry. Paper
     presented at the NATO Symposium on Applied Operations Research
     in Fishing, Trondheim, August 14th-17th, 1979.

MULTIOBJECTIVE OPTIMIZATION OF A LOCAL FISHING FLEET -

A GOAL PROGRAMMING APPROACH

Arnt Amble

Fisheries Division
Nordland Regional College
P.O. Box 309, N-8001 BODØ, Norway.

ACKNOWLEDGEMENTS

This paper is based on the author's Dr.ing. thesis, submitted to the Norwegian Institute of Technology in 1976 (Amble[1]), and on further research done at the Nordland Regional College. The work has been financially supported by the Norwegian University of Fisheries, the Nordland Regional College and the Norwegian Fisheries Research Council.

Several persons have given valuable advice, in particular Dr. Emil Aall Dahle of the Norwegian University of Fisheries and Mr. Åge Danielsen and other colleagues at the Nordland Regional College.

PROBLEM

General

Fishing is one of the most important industries in Northern Norway and the basis for the settlement in the region. How should the structure of the local fishing fleet be, to be optimal for the region?

The structure of the fishing fleet has a great influence on the possibilities of achieving objectives of many kinds (relating to employment of fishermen in the area, consumption of energy, stabilizing of fish deliveries through the year, capital needs for financing fishing vessels, and so on). Such objectives may be in conflict with each other, and it may be a hard political task to decide what the optimum fishing fleet should be.

The problem described may seem rather academic. Potential
vessel owners decide for themselves what kind of fishing vessels
they want to operate in order to maximize their own profit, not
thinking too much of what may be optimal for the society. But
governmental institutions of many kinds and on different levels do
have strong influence when it comes to deciding what vessels that
actually are to be financed and built, and how these shall be
allowed to operate. So far have the decisions of these different
governmental institutions only been coordinated to a modest extent.
If increased coordination is wanted, the model developed in this
project could be valuable.

Case area

In recent years the trend in the development of the fishing
fleet in Norway has been towards greater numbers both of small
and large vessels, and fewer of the ones in between. In spite of
this general trend, the local conditions vary quite a lot from
region to region, and so does the structure of the regional fleets.
In order to study the optimization of the fishing fleet in some
detail, it is necessary to pick some specific case area. In this
case I have studied the municipality of Øksnes in the Vesterålen
region of Northern Norway. This municipality is to a great extent
based on fishing and fish processing. 44 percent of the employed
people are fishermen, and another 22 percent are working in the
fishing industry on shore. The fishing fleet consists of a broad
variety of vessels, ranging from 20 feet one-man boats to 150 feet
stern trawlers.

METHOD OF APPROACH

Multiobjective optimization

When it comes to deciding the optimum solution to a problem
with many objectives involved, a traditional one-objective
optimization model is only suitable if all but one of the objectives
may be looked upon as absolute restrictions to the solution. If
this is not the case, for instance if the objectives are in conflict
with each other, so that it is impossible to achieve one goal without spoiling the attainment of another one, or if more than one
parameter is wanted to be maximized or minimized, a multiobjctive
decision model is needed.

It was decided to study the optimization of the fishing fleet
by means of a multiobjective optimization method. The early work
of Johnsen[2] and more recently Cochrane and Zeleny[3] gave surveys of

available multiobjective decision models. A later state of the art
is given by Hwang and Masud[4].   From these studies it is evident that
multiobjective optimization models represent a range of tools with
rapidly increasing application.

In order to keep the model as simple as possible, the Goal
Programming technique has been used in this study, as this is
basically a Linear Programming technique with some small but principal modifications.

Goal Programming

A thorough introduction to the Goal Programming technique and
various applications is given by Lee[5]

The mathematical formulation of a Goal Programming problem is
shown in the Appendix.   A flow chart for the Simplex procedure
of Goal Programming is shown in Fig. 1. A special FORTRAN program
for this algorithm was developed for this study and implemented on
a NORD-10 computer.

CASE STUDY

Objective function

Public plans for the development of the fishing industry in the
area imply that the fishing fleet shall contribute to the attainment of various goals that are wanted to be achieved.   These goals
are of many kinds.   In Norway, a policy for the fishing industry
has to be something more than just economics. By studying relevant
public plans and interviewing relevant persons, a list of 30 goals
were derived and included in the objective function of the model.
The goals are of the following 8 different categories:

A: Total yearly catch   ⩽ 18000 tons
    (divided on species)

B: Fish deliveries each month   ⩾ 1500 tons
    (slightly less in June, July, August)

C: Labour needed in fishing pr. year   ⩾ 2400 man months

D: Number of fishermen needed each month   ⩽ 400 men

E: Total yearly income to fishermen   ⩾ 20 mill. kr.

F: Total tax inflow from fishing   ⩾ 6 mill. kr.

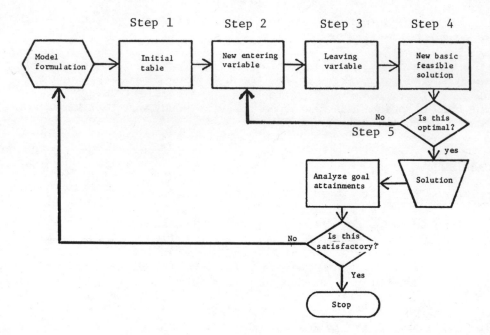

Fig. 1.   Flow chart of the simplex procedure of goal programming.

G: Minimize capital cost of fleet

H: Minimize fuel consumption of fleet.

     The complete list of goals in the objective function is in-
cluded in the Appendix.

     As some of the goals in the objective function may be in con-
flict with each other, an order of priority (ordinal ranking of goals)
has to be established. The ranking is represented by the priority
factors in the objective function.

Variation of objective function

     It is impossible to find a fishing fleet that will achieve all
the goals mentioned above completely. The Goal Programming model
will find a solution that achieves some of the goals completely
and makes the deviations from the other goals as small as possible.
The ranking of the goals will determine which goals that are to be
attained and the degree of fulfilment of each of the other ones.

     In this study, the optimum fishing fleet under various policies
were found. The various policies were given by varying the ranking
of the goals in the objective function. Effects of alterings in
each goal were also investigated.

Choice variables

     The choice variables in the model are the vessel types that
the model fleet may consist of. In this study six different vessel
types were included in the model:

     X_1 - 30 feet boat

     X_2 - 30 feet boat

     X_3 - 45 feet vessel

     X_4 - 45 feet vessel

     X_5 - 60 feet vessel

     X_6 - 150 feet stern trawler

The operation patterns for these six vessel types are shown in Fig.
2. These vessel types represent quite normal vessels in the existing fleet, and their operation patterns correspond to common
practice in the area in reality.

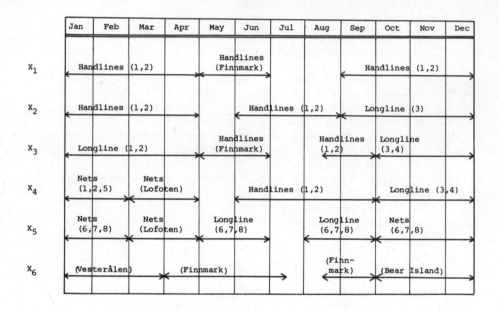

Fig. 2.   Operation patterns for fishing vessel types in model.

(Names in brackets refer to distant fisheries, numbers

in brackets refer to local fishing grounds).

Goal contributions from vessel types

Using data from the real fleet of the area, it was possible to calculate the contributions from each of the six different vessel types in the model to each of the 30 different goals in the objective function. These values represent what may be expected under "normal" circumstances and were entered to the model as technological coefficients.

RESULTS

By varying the ranking of goals, a great number of different problems were set up. The model found an optimum fleet in each case. Two examples will be presented here as illustrations.

Example I

In this example the objective function represented the following ranking order of goals:

	GOALS			DEVIATIONS IN OPTIMUM SOLUTION
1. priority:	Total yearly catch	$\leqslant$	18000 tons	0 (Achieved)
2. "	: Fish deliveries each month	$\geqslant$	1500 tons	- 310 tons
3. "	: Fuel consumption	=	0	+ 5570 tons
4. "	: Labour needed pr. year	$\geqslant$	2400 man months	- 623 man months
5. "	: Number of fishermen needed	$\leqslant$	400	0 (Achieved)
6. "	: Income to fishermen	>	20 mill. kr.	- 5.4 mill. kr.
7. "	: Tax inflow from fishing	$\geqslant$	6 mill. kr.	- 1.6 mill. kr.
8. "	: Capital cost of fleet	=	0	+ 6509 000 kr.

The optimum fleet composition being:

$X_1 = 4.17$

$X_2 = 13.03$

$X_3 = 0$

$X_4 = 0$

$X_5 = 4.59$

$X_6 = 7.06$

The model has found a fleet composition that gives the best solution to the fish delivery problem under the given catch constraints. If there exists more than one fleet composition that solves the delivery problem just as good, the model has found the fleet which has the least fuel cosumption. If there still are more than one possible optimum fleet, the goals of priority 4-8 have come into consideration.

Example II

In this example the fish delivery problem has been regarded as even more important than keeping within the catch restrictions. The ranking of the other goals has also been changed:

GOALS			DEVIATIONS IN OPTIMUM SOLUTION
1. priority:	Fish deliveries each month	≥ 1500 tons	0 (Achieved)
2. " :	Total yearly catch	≤ 1800 tons	+ 1077 tons
3. " :	Capital cost of fleet	= 0	+ 6583 000 kr.
4. " :	Tax inflow from fishing	≥ 6 mill.kr.	- 0.9 mill.kr.
5. " :	Income to fishermen	≥ 20 mill.kr	- 3.0 mill.kr
6. " :	Number of fishermen needed	≤ 400	0 (Achieved)
7. " :	Labour needed pr. year	≥ 2400 man months	- 200 man months
8. " :	Fuel consumption	= 0	+ 5595 tons

The optimum fleet composition being:

$X_1 = 0$

$X_2 = 32.10$

$X_3 = 0$

$X_4 = 0$

$X_5 = 6.01$

$X_6 = 6.75$

CONCLUSION

From these two examples and the other ones that have been examined, it is concluded that it is sensible to have a fleet in the case district consisting of several different vessel types. The optimum composition of the fleet is heavily dependent on a precise ranking of relevant goals. This ranking is in a high degree a political question. The model used in this study may help to illucidate this question.

The problem illustrated may seem a little academic, becoming real only if for instance the existing fleet in the area was completely destroyed instantly, and a whole new and optimal fleet had to be built.

In my current work with this model, I am collaborating with Mr. Erling Runde. We intend to use the model to illustrate the following problem:

If we want to keep most of the existing fleet, what smaller modifications of the fleet will be optimal?

When looking into this matter, we will also investigate the problem of finding integer numbers of vessels in the fleet. A more thorough examination of the sensitivity of the solutions will also be done.

REFERENCES

1. A. Amble, Optimalisering av fiskeflåten i en kommune belyst ved hjelp av en flerdimensjonal operasjonsanalytisk modell (Dr.ing. thesis, Norwegian University of Fisheries/Norwegian Institute of Technology, Trondheim, 1976) (in Norwegian).
2. E. Johnsen, Studies in Multiobjective Decision Models. (Student-litteratur, Lund, 1968).
3. J.L. Cochrane and M. Zeleny, editors, Multiple Criteria Decision Making (University of South Carolina Press, Columbia, 1973).
4. C.L. Hwang and A.S.M. Masud, Multiple Objective Decision Making - Methods and Applications (Springer-Verlag, Berlin, 1979).
5. S.M. Lee, Goal Programming for Decision Analysis (Auerbach, Philadelphia, 1972).

APPENDIX

Mathematical formulation of Goal Programming problem

$$\text{Minimize} \quad Z = \sum_{k=1}^{p} \sum_{i=1}^{m} (c_{ki} P_k d_i^- + c_{ki_2} P_k d_i^+), \quad (i_2 = i + m)$$

subject to

$$\sum_{j=1}^{n} a_{ij} x_j + d_i^- - d_i^+ = b_i \qquad (i = 1, 2, \ldots, m)$$

$$x_j \geqslant 0 \qquad (j = 1, 2, \ldots, n)$$

$$d_i^-, \, d_i^+ \geqslant 0 \qquad (i = 1, 2 \ldots, m)$$

$$P_k \ggg P_{k+1} \qquad (k = 1, 2 \ldots, p-1)$$

Here

a_{ij} = coefficient for the contribution to goal b_i from variable x_j

P_k = priority factor for goals of priority k

c_{ki} = weight factor for negative deviation from goal b_i

c_{ki_2} = weight factor for positive deviation from goal b_i

d_i^- = negative deviation from goal b_i

d_i^+ = positive deviation from goal b_i

Complete list of goals in objective function

A1:	Total yearly catch of cod	$\leqslant$ 10000 tons
A2:	" " " " haddock	$\leqslant$ 1000 "
A3:	" " " " saithe	$\leqslant$ 7000 "
B1:	Fish deliveries in January	$\geqslant$ 1500 "
B2:	" " " February	$\geqslant$ 1500 "
B3:	" " " March	$\geqslant$ 1500 "
B4:	" " " April	$\geqslant$ 1500 "
B5:	" " " May	$\geqslant$ 1500 "
B6	" " " June	$\geqslant$ 1000 "
B7:	" " " July	$\geqslant$ 500 "
B8:	" " " August	$\geqslant$ 1000 "
B9:	" " " September	$\geqslant$ 1500 "
B10:	" " " October	$\geqslant$ 1500 "
B11:	" " " November	$\geqslant$ 1500 "
B12:	" " " December	$\geqslant$ 1500 "
C1:	Labour needed in fishing pr. year	$\geqslant$ 2400 man months
D1:	Number of fishermen needed in January	$\leqslant$ 400
D2:	" " " " " February	$\leqslant$ 400
D3:	" " " " " March	$\leqslant$ 400
D4:	" " " " " April	$\leqslant$ 400
D5:	" " " " " May	$\leqslant$ 400
D6:	" " " " " June	$\leqslant$ 400
D7:	" " " " " July	$\leqslant$ 400
D8:	" " " " " August	$\leqslant$ 400
D9:	" " " " " September	$\leqslant$ 400
D10:	" " " " " October	$\leqslant$ 400
D11:	" " " " " November	$\leqslant$ 400
D12:	" " " " " December	$\leqslant$ 400
E1:	Total yearly income to fishermen	$\geqslant$ 20 mill. N. kr.
F1:	Total tax inflow from fishing	$\geqslant$ 6 mill. N. kr.
G1:	Minimize capital cost of fleet	
H1:	Minimize fuel consumption of fleet	

ON THE RELATIONSHIP BETWEEN FISHING CAPACITY AND

RESOURCE ALLOCATIONS

D.J. Garrod and J.G. Shepherd

Ministry of Agriculture, Fisheries and Food
Fisheries Laboratory,
Lowestoft, Suffolk NR33 OHT, U.K.

INTRODUCTION

It is now well established both in theory and in practice that
unrestricted access to a 'common property' fishery resource leads
to potential economic benefits being dissipated in the cost of
fishing. It is perhaps less widely recognized that this has been
aggravated by countries seeking to maintain some historic share in
a fishery and thereby being forced to maintain and even augment
their investment. The normal economic 'regulators' are less effective when a fishery is maintained by Government support policies with
the result that the resources themselves become more exposed to
overexploitation and its consequences, e.g. the widespread collapse
of Atlantic herring fisheries. Recognition of this risk finally
led to the attempt to control exploitation by extended limits of
national jurisdiction and by the regulation of the total allowable
catch (TAC).

The establishment of the TAC system contributes to the resource
management problem but neither it, nor extended limits, directly
address the economic problem. So far as national fishermen are
concerned the national allocation of a TAC represents to them a
common property resource and in the absence of additional controls
will still allow investment to continue towards overcapitalization.

In the UK this problem has been presented in a particularly
acute form through the loss of access to traditional distant-water
resources which had supplied up to one-third of the catch of demersal
species. This has necessitated a review of the national fishing
capacity in relation to possible future resource allocation, and
this paper describes the work being carried out in this area by a

321

UK research team drawn from the MAFF Fisheries Laboratory (Shepherd, Garrod), the White Fish Authority (Haywood, Curr, McKellar) and the DAFS Marine Laboratory, Aberdeen (Pope).

The methodology developed is described below. We note at the outset however that the overall problem is dynamic over time. Our present models are static, considering only the fleet requirements in response to possible catch allocations and stock equilibria under different conservation regimes. There is no feedback from a proposed pattern of allocation and exploitation to the resources in some subsequent year. Nor have we considered any long-term economic optimization within the model. The restriction is caused primarily by the complexity of the problem and the core space of computing facilities. But it is also true that we have deliberately tackled only a limited part of the total problem in order to under-stand at least one aspect of it. We also have reservations on the wisdom of planning over long time horizons in both the resource and economic fields. It seems to us very possible that given a start point, and a desired endpoint in terms of equilibrium situations, the direction, rate and scale of change can be interpolated outside a model as effectively as they can be deduced by complex modelling, given the confidence limits imposed by biological and economic imponderables.

THE ESTIMATION OF FISHING CAPACITY REQUIREMENTS FOR THE UK FLEET: THE LINEAR PROGRAMMING (LP) APPROACH

The structure of the UK fleet can be categorized according to the type of resources the vessels are designed to exploit, i.e. demersal pelagic or shellfish and crustacean species. There is overlap between them but our preliminary study was confined to the demersal fishery sector to restrict the problem. Even so it incorporates six vessel categories and four gear types; these are based on sixteen fishery districts which should be preserved in any model to examine regional effects, and these vessels have the opportunity to fish up to eight commercially valuable demersal species in any one of fifteen resource areas where they may be subject to quota restrictions. It is a large problem which is best summarized in the LP matrix and definitions (Appendix 1A).

In essence the fleet is composed of a number of vessels each capable of generating a specified number of hours fishing per season (quarter) according to the range of the resource involved from their base district. The catch generated is governed by the expected catch rate per resource of each species, subject to quota constraints. The prospective catch is then evaluated at market prices and compared with the annual fixed (investment) and operating costs to estimate a 'profit' margin in the objective function.

We found it extremely difficult to define 'profit' in a manner which could be appropriate to all aspects of the catching industry, bearing in mind the varied size and financial structure of the individual economic units and in practice we reverted to a simplistic definition based on the direct margin of earnings over operating and fixed costs.   We encountered additional difficulty in defining an objective function appropriate to the problem.   A strict maximization of 'profit' in these terms led to a very sharp reduction in fleet size and catches fell well short of the available allocations.

Alternatively with the catch allocations written as equality constraints the problem became infeasible because of incompatibilities created by the mixed fishery problem.   A workable procedure was established by specifying particular values of the catch as equality constraints in successive runs; this in effect minimizes costs for a prescribed level of earnings and enabled us to examine the implied size and structure of the fleet and its pattern of fishing associated with a value of the catch incremented within the over-riding catch constraint.

Fig. 1 summarizes a typical result and shows also the result of initial runs maximizing the profit margin in the objective function.   The development in fleet size is apparent, as is the progressive reduction in the profit margin to the break-even point close to the performance in 1974 on which this analysis was based. However detailed examination of the optimized solution close to the break-even point showed that of 260 cells in which fishing took place in 1974, 111 were deleted by the LP solution leaving 149 common entries and of these 86 were deemed 'profitable' and attracted increased fishing.   The solution also suggested 93 vessel x gear categories where profitable fishing might have been conducted.

Clearly the problem can be formulated in this way, and a solution found: the difficulty lies in seeing this as a real solution. Apart from the complexity of interpretation in the UK context, and the difficulty of expanding it to include other sectors of the total fishery, we became aware of a number of disadvantages in the LP approach:

(a) The mathematical rigour of the optimization procedure leads to solutions which may be unacceptable for a variety of reasons. The ruthless exploitation of marginal advantages leads to very sharp and unrealistic changes in fishing patterns which can only be overcome by the introduction of additional constraints – yet one is concerned to preserve flexibility in order to examine alternative solutions.

(b) The optimization procedure provided only a single solution to each set of constraints with no information of alternatives

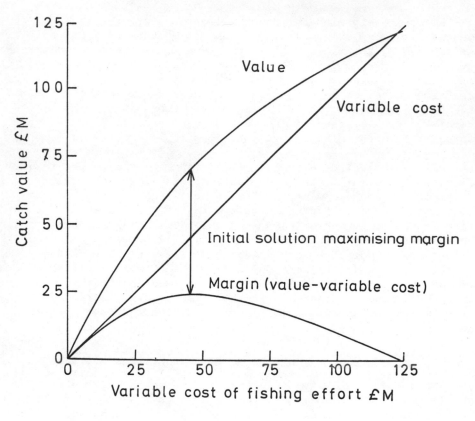

Figure 1.

close to the optimum without recourse to complex analysis of
shadow costs etc. The particular example of this is the
extremely high cost or infeasibility of meeting catch allocations very exactly because of the mixed fishery problem.
(c) The procedure of compiling the solution by selecting vessels
    on a hierarchy of relative profitability is equally unrealistic
    because vessels fish simultaneously: the solution is biassed
    by the 'wisdom of hindsight' in being provided with a set of
    prospective catch rates on which to base the optimal strategy.
    This could never be known to fishermen.

We also noted the solution was unavoidably restricted to vessel
categories and resources for which catch rate data were available
and the optimization took no account of non-linear variation in
value caused by changes in supply. In principle these can be
overcome by interpolation of dummy variables (e.g. for new resources)
or by segmented linearization of the price/supply function (if it
were known) but the Group was more concerned by the technical
limitations of the approach, particularly the substantial changes
in response either to small (but real) advantages of certain types
of fishing, or to random variability in the data used. This led
us to consider an alternative approach.

THE MARK II MODEL

(i) Rationale

Industrial change on the scale envisaged by the optimized LP
solution would have enormous and unacceptable social, political
and economic consequences outside the model area. Administration
and the Industry are innately conservative and concerned to maintain the status quo if possible Scientists must recognize the
gravity of decisions to fundamentally alter this balance and the
implication that a Government is unlikely to optimize any given
situation because of the scale of change that it would imply. It
is more appropriate to identify the direction of change that would
be beneficial and at least initiate some movement.

In the UK the essential problem is to determine the amount of
fishing and hence the number of vessels required to meet catch
allocations in a total fishery which is extremely heterogeneous
with respect to the vessels and gears concerned and the mix of
species each of them might take both by area and by season. This
should be coupled with the economic implications of any proposed
degree of change. Note also that for the UK the problem has two
stages, first to determine the amount of fishing appropriate to
the reduced number of resources available, and second to explore
further modifications that may become desirable within each of the
resources in the development of conservation policies.

Our second approach therefore explores the potential benefits of any adjustment in fishing effort regardless of an optimal solution to the problem as a whole. This has been done by incorporating into the objective function penalties associated with change. These might be expected to be positive and increasing for any direction of change and are therefore fundamentally non-linear. This led to the formulation of a non-linear optimization problem in about 1500 variables which, if posed in an unconstrained form, can be solved using recent developments (Powell[2]) of the conjugate gradient algorithm (Fletcher and Reeves[1]).

The LP formulation contained three sets of constraints, the quotas, the number of vessels and, internally, the annual effort potential of an individual vessel. We chose to formulate the Mark II model in terms of effort directly leaving its interpretation in terms of vessels outside the minimization procedure. The limit on the number of vessels (potential effort) was also made redundant by introducing a penalty for departure from some historic (baseline) pattern to ensure effort was not channelled excessively into only the most profitable effort categories. The quota constraints were avoided in the same way, by imposing a penalty for deviation from the prescribed quota levels. The penalty functions thus provide a set of flexible constraints based on a desirable objective (the status quo) but which can be relaxed within the model if the potential benefit of doing so is sufficiently large. It will be seen that this, when added to the economic objective, provides a flexible composite objective where each component can be weighted according to some judgement of its relative importance in developing a strategy. For example, if the fulfilment of a TAC is more important than retaining a historic pattern of fishing then the 'quota match' penalty is given a higher weighting to give it less flexibility. Similarly the economic component can be weighted heavily to place a premium on economic efficiency and, at the limit, this transforms into a problem analogous to the LP optimization.

The Mark II model then has the potential to move in any direction depending on the weight attached to different policy objectives and these are related to some baseline situations (e.g. the most recent year) to evaluate the trade-offs between different factors in a progressive rather than an optimized way. The decision variables are the weighting factors attached to each sub-objective, the prospective catch allocations, and ultimately the catch rates at which the allocations may be caught.

(ii) Choice of penalty function and the model

The use of the conjugate gradient algorithm requires that the objective function can be differentiated to provide the gradient in terms of penalty function. We considered three options for

this penalty function, a chi-square analogy, a logarithmic and a
maximum entropy form.   These are illustrated in Fig. 2.   The chi-
square analogy was chosen for the quota matching (to discourage
overshoot) and the maximum entropy form was chosen for the fleet
disruption penalty as being most neutral to the sign of any inequality.   Given this penalty function the objective function as a
whole can be formulated as described in Appendix 1B.   The operation
of the model requires catch and effort statistics by vessel x gear
category for the base year, and estimates of stock size in that
year, plus the anticipated stock size and catch allocation under
any prospective resource scenario.   The routine establishes the
base year catches, fishing pattern and catch per day absent from
the fishery statistics, varies the prospective catch per day in
proportion to the change in stock size, where appropriate, and then
adjusts the pattern of fishing at the new catch rate to meet the
new catch allocation in accordance with the minimization of the
objective function.   Note however that the minimization is terminated when it has achieved a 90 per cent reduction in gradient: we
found the modification incurred by further pursuit of the optimum
did not justify the computing time involved.

(iii) Validation and illustrative results

Table 1 presents a validation of the model.   The actual amount
of fishing by vessel length category in 1977 is first compared with a
'prediction' of 1977 effort based on the 1977 catch data.   The close
correspondence simply shows that the model is arithmetically
satisfactory.   The second comparison is a prediction of fishing
effort in 1977 to meet the catches in 1977 but based on the level
and pattern of fishing in 1976, i.e. a short-term predictive mode.
The Mark II Model solution suggests a preferential use of smaller
vessels at the expense of the 140 ft+ fresh fish, distant-water
trawlers in particular.   In practice the effort of that larger
category was not reduced as rapidly as had been expected by the
industry itself, but the trend generated in the model was correct.
Similar trials from different starting points and following the
minimization further indicate the reliability of the procedure:
it does appear to be approaching a true minimum rather than locating
local minima.

Table 2 summarizes three variants of possible fishing effort
in 1978 (based on fishing in 1977), the variants being generated
by attaching different weighting to the economic efficiency
component of the objective function.   These are a non-optimized
set with zero economic weighting where the solution is determined
only by the 'quota match' and 'fleet disruption' components.   The
'cautious' and 'ruthless' optimizations are generated by a
progressive increase in the importance (weighting) attached to
economic efficiency as defined by the objective function. This
indicates a progressive concentration of fishing within the

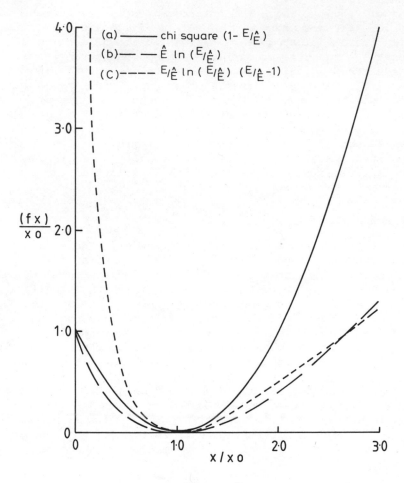

Figure 2.

Table 1. Fishing effort in the UK fleet: a comparison of
          modelled estimates of the fishing effort require-
          ments in 1977 with recorded effort in that year.
          (i)   1977 effort estimated from catches, catch
                rates and pattern of fishing in 1977;
          (ii) 1977 effort estimated from catches in 1977
                but catch rates and pattern of fishing in
                1976

Vessel length category (feet)	Number of days absent (000)		
	1977 actual	Model estimates	
		(i) 1977 on 1977	(ii) 1977 on 1976
40- 64	155.2	151.5	175.2
65- 79	66.6	67.0	70.8
80-109	20.5	20.7	22.2
110-139	29.6	29.6	27.3
140+ freshfish	14.3	14.3	7.3
140+ freezer	9.6	9.9	10.1
	295.8	293.0	312.9

smaller vessel categories followed ultimately by some preference
for freezer trawlers to exploit particular fisheries.   It serves
to show the characteristics and flexibility of the model and the
possible opportunity for improving economic performance, as
measured here, to the extent that it could make a significant
contribution to offset the fixed costs of fishing as well as the
operating costs - if economic efficiency were the only criterion.

    This model embraces the entire UK fishery for all species and
can be interrogated at different levels of aggregation for regional
effects and their local consequences for fleet structure and
employment etc. outside the model.   The chief advantage over the
LP optimization lies in the flexibility of the method and the way
in which the solution can be held to represent a realistic situation
or controlled departures from it.   Further it can be applied to
judge fishing capacity requirements appropriate to any future
catch or resource scenario.

Table 2. The effect of variation in weighting attached to
          the economic efficiency component of the objective
          function on modelled estimates of fishing effort
          requirements in 1978.   Estimates for 1978 based
          upon hypothetical catch allocations in 1978,
          expected catch rates in 1978 coupled with the catch
          rates and fishing pattern in 1977.

          Non-opt.   non-optimized run where economic
                     weighting is zero;
          Cautious) two examples of the progressive increase
          Ruthless) in the relative importance attached to
                     economic efficiency

Vessel length category (feet)	Fishing effort (days absent, OOO)			
	1977	1978		
		Non-opt.	Cautious	Ruthless
40- 64	155.2	158.5	155.0	162.6
65- 79	66.6	66.7	64.5	64.8
80-109	20.5	20.5	18.0	10.5
110-139	29.6	26.0	25.2	19.0
140+ fresher	14.3	10.8	10.3	9.4
140+ freezer	9.6	6.4	7.5	10.4
Total catch (index		1.00	1.04	1.10
'Profit' margin (index)		1.00	1.56	2.52
Fixed cost (% of margin)		2.32	1.46	0.90

The disadvantages of the Mark II model are common also to the LP solution.   These concern the acquisition and reliability of fleet and stock (catch rate) data, the absence of a price/supply relationship, and that it is a static approach which, as it stands, cannot evaluate a strategy over time.   Some of these points can be judged subjectively in interpreting the model solution.   For example, the price/supply relationship may not be important in the invest‐ igation of alternative ways of landing the same quantity of fish. The lack of dynamics is important.   Mathematically the model could be easily extended given a sufficiently large computer but some aspects of the resource and economic data become more speculative. Nevertheless it is still informative to examine fishing capacity requirements under varying resource allocations and particularly the implication of varying resource management policies that will influence catch rates.

RESOURCE MANAGEMENT OBJECTIVES IN RELATION TO ECONOMIC CONSIDERATIONS

The first use of the Mark II Model described above and summar‐ ized in Table 2 indicates how the amount of fishing and fleet might be adjusted to make improved use of resources presently available. Equally it may be used to investigate the implications of any change in resources, and particularly changes in their abundance (catch rates) that could be achieved by resource management policies.

Fig. 3 shows the application of the static model described here to estimate profitability at different 'equilibrium' levels of exploitation of a mixed fishery, the catch rate being adjusted to the stock level associated with reduced exploitation outside this model.   Clearly the potential increase in profitability compared to the recent 'break-even' value (100) is very large, and would be larger still if the effect of demand had been included in assessing value.   Drastic reduction in exploitation to take advant‐ age of this may be one interpretation of the efficient utilization of resources but it raises many difficulties particularly in the internal pressures created by the improved return on investment and the implications to employment of any reduction in fishing activity.

Fig. 3 does suggest that a significant improvement in the margin of earnings over operating costs could still be achieved but with much less impact on employment etc. by reduction in fishing effort which are very much lower than those implied by the optimal solutions.

CONCLUSION

The Mark II Model described here offers a different approach

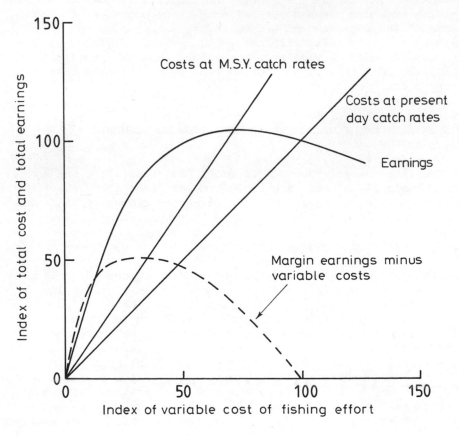

Figure 3

to the allocation of shares within a national allocation in a way which retains traditional fishing patterns and therefore takes account of the mixed fishery problem, which meets the catch allocations with a realistic degree of precision and which allows an evaluation of the economic impact of resource allocations on the catching industry.   We hope to use this as a basis for further models to indicate how the resource allocations themselves might be varied to further improve the economic effects without violating the resource management requirements, and it may be that this can be compounded with the international allocation of shared stocks by a similar 'minimum change' approach.

REFERENCES

1. R. Fletcher and C.M. Reeves, Function minimization by conjugate gradients, Comput. J. 7 (1964) 149-154.
2. M.J.D. Powell, Restart procedures for the conjugate gradient method, Rep. Comput. Sci. Syst. Div. AERE CSS 24 (1975) 24.

Appendix 1A   Linear programming matrix for UK demersal fishery

		E $adlgp$	N dig	VALUE	COSTS	PROFIT (FREE)	LIMIT
OBJECTIVE	(Z)					1	
PROFIT	(E)			1	-1	-1	O
LAND	(G)			1			X
VAL	(E)	Earn_{adlgp}		-1			O
COS	(E)	Run_{adlgp}	Fix_1		-1		O
Q_{bs}	(L)	Exp_{adlgps}					Quota_{bs}
R_{dlgp}	(L)	Rec_{adlgp}	-1				O
V_{dlg}	(E)		1				Num_{dlg}

1.   Categorization, coding

a –   for fishing area (1 to 15)   g – for gear (1 to 4)
d –   for fishery district (1 to 16)   p – for period of the year (1 to 4)
l –   for vessel length (1 to 6)   s – for species (1 to 8)

2.   Variables

E_{adlgp} – for fishing effort, expressed as hours fishing per quarter (by area, district, length, gear and period)
N_{dlg} – for numbers of vessels in the fleet (by district, length and gear). For convenience we also define the following subsidiary variables by means of internal equality constraints.
VALUE – for total value at first sale of fish landed in the year
COSTS – for total annual costs (including fixed costs) of the whole fleet.
PROFIT – for VALUE–COSTS, i.e. the national total annual profit. This is a free variable (i.e. allowed to be negative).

3.   Objective

Formally the objective is PROFIT (to be maximized) but with

a constraint on VALUE this is exactly equivalent to minimizing COSTS.

4. Constraints

We use three equality constraints to define our subsidiary variables:

$$PROF = VALUE - COSTS - PROFIT = O$$

$$VAL_{adlgp} = \sum_{adlgp}\sum\sum\sum\sum Earn_{adlgp} \; E_{adlgp} - VALUE = O$$

$$COS = \sum_{adlgp}\sum\sum\sum\sum Run_{adlgp} \; E_{adlgp}$$

$$+ \sum_{dlg}\sum\sum Fix_1 \; N_{dlg} - COSTS = O$$

The true constraints are those on the value of landings, or the 'quotas', on the effort available from the vessels in operation, and on the number of vessels available.

$$LAND = VALUE > X \text{ (a specified upper limit to the value landed)}$$

$$Q_{bs} = partial \sum_a \quad \sum_d \sum\sum_{lg} \sum_p Exp_{adlgps} \; E_{adlgp} \leqslant Quota_{bs}$$

$$R_{dlgp} = \sum_a Rec_{adlgp} \; E_{adlgp} - N_{dlg} \leqslant O$$

$$V_{dlg} = N_{dlg} \qquad\qquad = Num_{dlg}$$

5. LP matrix coefficients

(a) $Earn_{adlgp}$ is the earning rate (£ per hour fished)

(b) Run_{adlgp} is the running cost (£ per hour fished)

(c) Exp_{adlgps} is the expected catch rate (cwt/hour fished).

(d) Rec_{adlgp} is the reciprocal of the maximum feasible effort (in hours) available from a single vessel.

(e) Fix_1 is the annual fixed charge for a vessel of length 1.

(f) $Quota_{bs}$ are the 'quotas' for each species in each resource area

(g) Num_{dlg} is the number of vessels in each length and gear category in each district.

Appendix 1B   Mark II Model: Formulation of objective function

$$F = W_1 F_1 + W_2 F_2 + W_3 F_3 \qquad \text{where}$$

F_1, quota mismatch penalty

$$F_1 = \bar{P} \sum_{bs} (C_{bs} - Q_{bs})^2 / Q'_{bs}$$

F_2, fleet disruption penalty

$$F_2 = \sum_{advp} Z_v \{ E_{advp} \ln(E_{advp}/\hat{E}_{advp}) - (E_{advp} - \hat{E}_{advp}) \}$$

F_3, economic efficiency objective

$$F_3 = E_{advp}(Z_v - H_{advp}) \hat{E}_{advp} \exp(X_{advp})$$

and W_1, W_2, W_3 are arbitrary weighting factors of the order unity and selected by trial. Subscripts as in Appendix 1A but note

a	=	fishing area which is not necessarily equivalent to
b	=	resource area by which quotas are established
v	=	1 x g of Appendix 1A
$\bar{P}$	=	average price
C	=	actual catches in a base year
Q	=	quota allocation
Q'_{bs}	=	$0.9 Q_{bs} + 0.1 \hat{Q}_{bs}$ where $\hat{Q}_{bs}$ are actual catches in the base year to avoid zero divisor
Z	=	cost of fishing per day absent
E	=	fishing effort in the target year
$\hat{E}$	=	fishing effort in the base year
H	=	earnings per day absent
X	=	$\ln(E_{advp}/\hat{E}_{advp})$

Note inclusion of $\bar{P}$ in F_1 and Z in F_2 are required to bring each component of the objective to comparable (monetary) terms.

A FRONTIER PRODUCTION FUNCTION FOR THE NORWEGIAN

COD FISHERIES

Røgnvaldur Hannesson, Olav R-Hansen and Svein Age Dale

Department of Economics,
University of Bergen,
Norway.

THE QUESTIONS WE ASK

The problems we shall be dealing with concern economic efficiency.   More precisely we shall examine whether the proportion in which labour and capital are combined in the design and operations of individual fishing vessels is efficient when related to the social opportunity cost of these factors.   Furthermore we shall consider returns to scale; that is, whether the design and operations of the fishing vessels is such that economies of scale, to the extent they are present, are fully utilized.

As is well known from economic theory, a socially optimal combination of factors which are substitutes in production requires that the ratio of their marginal products is equal to the ratio of their social opportunity costs.   That is,

$$\frac{MP_L}{MP_K} = \frac{w^o}{r^o} \qquad\qquad (1)$$

where MP denotes marginal product, L and K are subscripts denoting labour and captial, and w^o and r^o are the social opportunity costs of a unit of labour and capital respectively.   With regard to scale efficiency this requires that the size of all production units be so determined that the scale elasticity of output is equal to one for all units, as it is then impossible to increase the output obtained by a given set of inputs by altering the number of units of production.

Is there, then, a presumption that the efficiency condition (1)

will or will not be satisfied in the fishery?   Again economic
theory tells us that it will be satisfied in a competitive market
economy without market failures of any sort[1]   There is one
particular market failure of interest in this context which we
suspect to prevail in the fishery.   Norwegian fishermen have
access to cheap credit from a public bank set up for the purpose of
financing the construction of fishing vessels, so one would expect
that the cost of capital,as it appears to the fishermen, will be
lower than the social opportunity cost of capital - that is, the
value of the marginal product of capital in its best alternative
use.   As to the private opportunity cost of labour, we must take
into account that the remuneration of Norwegian fishermen is
determined not by a wage rate but by a share in the value of the
vessel's catch.   This implies a certain resemblance with the labour-
managed firm.   In that case, the relevant private opportunity cost of
labour to consider in a long run equilibrium analysis is the one at
which there are no incentives either to enter or leave the fishery.
As a first approximation it is not unreasonable to set this
opportunity cost equal to "the" wage rate outside the fishery,
disregarding the details of different individual preferences, etc.
Since private opportunity costs rather than social opportunity
costs determine which particular combination of factors will be
chosen by individual entrepreneurs, the right hand side of Equation
(1) must be replaced by the ratio of private opportunity costs (w/r).
Taking the wage rate outside the fishery as representing the social
opportunity cost of labour (w=w^o) and considering our argument that
$r < r^o$, we obtain the following characterization of the long term
equilibrium in the fishery:

$$\frac{MP_L}{MP_K} = \frac{w}{r} > \frac{w^o}{r^o} .$$

(2)

According to the law of diminishing returns, which says that the
marginal productivity of one factor relative to that of another
will increase if its use is decreased relative to the other, we
may therefore expect that less labour tends to be used in the
fishery relative to capital equipment than is warranted from the
point of view of economic efficiency. We will refer to this later

_____

(1)   That this will occur as a result of the maximization of profits
      by capitalistic firms taking the prices of all commodities
      and inputs as fixed parameters is a theorem presented in every
      introductory textbook on economics.   What is less widely known
      is that this will be the result as well in an economy consist-
      ing of labour-managed firms maximizing the remuneration per
      employee, provided entry is free and that no industry
      operates under decreasing returns.   See[5], pp. 27-41.

as the hypothesis of overinvestment.

THE SPECIFICATION OF THE PRODUCTION FUNCTION

Before deriving quantitative estimates of marginal products or elasticity of scale, it is necessary to specify the form of the production function relating the use of factors of production to the resultant output and estimate its parameters.   It is only appropriate to acknowledge that the conclusions may already be prejudiced to a degree at this stage.   Having chosen a specific form of the production function, the question becomes how well does it fit the facts.   There is, however, reason to believe that the relationship between factors of production and output one is trying to capture by the production function will obey certain rules.   This, together with sheer computational convenience, limits the class of functions to be considered.   To investigate the hypothesis of over-investment and the elasticity of scale we shall estimate the parameters of the following production function, first proposed by Zellner and Revankar[6]:

$$X^{\alpha} e^{\beta X} = AL^{\delta} K^{1-\delta},$$ (3)

where X is output, L is labour, K is capital, e is the base of the natural logarithms, and α, β, A and δ are parameters to be estimated. We have chosen this particular specification for the sake of computational convenience, but it also has the proprties that we find essential, namely (i) it permits the returns to scale to vary with the size of the production unit, measured in terms of out-put, and (ii) implies isoquants convex to the origin in L,K space.

As the right hand side of (3) is in fact a Cobb-Douglas prod-uction function, the marginal rate of substitution between K and L is

$$-\left.\frac{dK}{dL}\right|_{dX=0} = \frac{MP_L}{MP_K} = \frac{\delta}{1-\delta} \cdot \frac{K}{L}$$ (4)

This expression is a monotonically increasing function of the capital/labour ratio, and so the isoquants are convex to the origin.

Taking the logarithms of (3) we get:

$$\alpha \ln X + \beta X = \ln A + \delta \ln L + (1-\delta) \ln K$$ (5)

Differentiating this gives

$$\alpha \frac{dX}{X} + \beta dX = \delta \frac{dL}{L} + (1-\delta) \frac{dK}{K} .$$

For a proportional change in L and K, dL/L = dK/K = λ , we get

$$\alpha\frac{dX}{X} + \beta dX = \delta\lambda + (1-\delta)\lambda = \lambda$$

which gives us the following expression for the elasticity of
scale:

$$\varepsilon = \frac{dX/X}{\lambda} = \frac{1}{\alpha + \beta X} \qquad\qquad (6)$$

Thus, if economies of scale are to diminish monotonically as the
size of the production unit - X that is - increases, it is
necessary that $\beta > 0$. So, for $0 < \alpha < 1$, there are economies of
scale for "small" production units, but these will be exhausted
when some critical size has been reached. The possibility that
$\beta < 0$ seems unreasonable, as this would imply that, beyond some
critical size of the production unit, economies of scale would
increase without limit. As a limiting case we shall allow for
$\beta = 0$. Since negative scale elasticity makes no sense, we shall
require that $\alpha > 0$.

DATA AND DEFINITIONS

Beginning in 1968, the Norwegian Directorate of Fisheries
annually surveys the profitability of a sample of fishing vessels
larger than 40 feet, "efficiently operated throughout the whole
year". We have used a series of these data for the period
1971-1977. The data refer to single fishing vessels, without
it being possible to identify the individual vessel. For each
vessel there is a set of data for one or more "seasons" each
year, where a change of season occurs if the mode of operations
changes so radically as to necessitate separate accounts to be
kept. This happens, for example, if there is a switch to a
different type of fishing gear and a change in the number of
fishermen employed, as the rules for dividing the catch between
the vessel owner and the crew may be different in each case.
This makes it possible for us to divide the vessels into groups
that are characterized by a higher degree of homogeneity with
respect to fishing methods and fish stock(s) exploited than
would otherwise be possible. This is important, because we want
to isolate the effects of labour and capital on output.

Among the vessels represented in our sample we have retained
nine groups which we expect to be reasonably homogeneous with
respect to the natural resource factor. These groups are listed
in Table 1. The fisheries that are represented exploit two main
species, cod and saithe, the latter only represented by one group
(group 2). Among the cod fisheries the natural resource factor

Table 1

The vessel groups for which a production function was estimated.

Group no.   Definition

1   Bottom trawl fishery, Barents Sea and Bear
               Island/Spitzbergen area.

2   Saithe fishery, gill nets.

3   Spring cod fishery in Finnmark.

4   Winter cod fishery ("skrei"), long line.

5   Winter cod fishery, gill nets.

6   Winter cod fishery in Finnmark.

7   Winter cod fishery in Troms.

8   Winter cod fishery in Vesteralen.

9   Winter cod fishery in Lofoten.

may be expected to vary because differing groups of vessels
exploit different year classes of fish, which variations in
abundance are not at all synchronized in time. The bottom trawl
fisheries in the Barents Sea and Bear Island/Spitzbergen area
exploit mainly the young year classes (2-5 years old); these are
represented by group 1. These very same year classes approach
the coast of Finnmark in spring chasing the capelin; a spring
cod fishery cashing in on this migration has long traditions
and is represented by group 3. The reason why we have considered
groups 3 and 1 separately, despite their exploiting, in part at
least, the same year classes of cod, is that the availability of
the migrating fish may be quite different from what it is when
these same year classes are dispersed in the Barents Sea and the
Bear Island/Spitzbergen area. Lastly there is the fishery
exploiting the mature year classes of cod migrating to the
spawning grounds concentrated mainly in the Lofoten area. The
vessels participating in this fishery have been aggregated into
groups according to two alternative criteria, that is, type of
fishing gear used and area of operations. Groups 6-9, which
represent the geographical grouping, therefore overlap to a large
extent with groups 4 and 5. There are good reasons for
discriminating among the vessels according to both of these
criteria. Age-composition analyses of catches have shown that
gill nets and long line do not select the year classes in quite
the same way, so that the natural resource base will not be quite
the same for both these types of fishing gear. As to the regional
criterion, the concentration of the migrating fish, and thereby
its availability, may be different in different areas, according
to oceanographic conditions.

     Ideally, one would wish to measure output in physical units,
whereas the available data only inform us of revenue in terms of
money. There is reason to believe that the composition of catches
in terms of species caught and quality classes sold is similar
for all vessels in each group. Because the length of the fishing
season may vary considerably among the vessels in each group,
we have defined output as revenue per day within the season,
after deducting the costs of intermediate inputs such as fuel,
ice, salt and containers. Two variables are available for measuring
the length of the season: days at sea and days of operation.
The former appears more attractive, as one would expect it to be
more closely related to the effective operation of the vessel.
This we have used in all cases, except with respect to bottom
trawl because of incomplete records.

     We measure labour simply as the number of fishermen employed
on each vessel. With regard to capital this is recorded as
estimated repurchase value of the vessel hull, motor, and other

equipment (mostly hydraulic and electronic equipment), sometimes
also fishing gear). The capital stock thus defined as the
aggregate value of the various objects that constitute the
physical capital, valued at current prices. This implies that
we regard capital as yielding an even flow of services over
its entire length of life, and the suddenly depreciating to
zero.[2]

ESTIMATION PROCEDURES

There are, roughly speaking, two classes of production
functions which it may be possible to estimate from a set of
data such as the one at hand. These are known as "average"
and "frontier"functions respectively. In estimating average
functions one assumes a symmetric error term, so that
deviations from the "normal" conditions one is trying to get
an idea of may go either way. Ordinary statistical methods
do not permit us to carry out such estimations in this case,
as variations in the use of labour and capital were found to
explain extremely little of the variations in output. There

---

(2) Our way of aggregating the different capital objects is,
     strictly speaking, permissible only when all of them
     depreciate at the same rate, (in our case, last for the
     same length of time), or else are always employed in
     the same proportion within each group of vessels.
     For if the composition of capital varies among vessels
     and the different capital objects depreciate at
     different rates, different gross yield rates on aggregate
     capital will be required to give the same net yeild for
     all vessels. Neither requirement is fully satisfied
     in our case. An alternative approach would be to use the
     flow of capital services as a factor of production,
     instead of the aggregate stock of capital, the flow of
     services being defined as the rate of depreciation times
     the stock of capital, summed across all capital objects
     (cf. Forsund and Hjalmarsson [2]). When trying this out
     it was found that the estimated coefficients of the
     production function were only slightly affected. We will,
     instead, take into account the variable composition of
     capital across vessel groups by calculating the aggregate
     rate of depreciation and adjust the required gross rate
     of return to give a similar net social rate of return
     for all groups of vessels.

are two explanations for this anomalous result. First, the
division of the data into groups of vessels fishing under
sufficiently similar conditions to permit a meaningful
estimation leaves us with rather similar vessels within
each group. Secondly, there is undoubtedly a third factor
of production on which we have no information, and that is the
difference in skills among captains and crews.

By contrast, the idea behind the concept of a frontier
production function is that there can only be a one-sided
deviation between the actual output of a production unit and
that which an efficient use of the resources at its disposal
would have yielded. For the case of constant returns to
scale this may be illustrated as in Figure 1. The unit isoquant
if the curve X_o, showing all the different combinations of
capital and labour which, if used efficiently, would yield one
unit of the product. All observations of actual use of labour
and capital will lie either on or to the Northeast of the
isoquant, like the points marked by x's in the figure. In our
view, the frontier is a more appealing concept than the average
function, as it would show the scope of substitution between
the factors of production when using, in each case, the most
efficient methods of production. There are, however, two
sets of complications one may expect will arise when trying
to estimate a frontier function. The first is easily dectected
and puts an effective end to the whole exercise: if the
production units embody an economically efficient design and
relative factor prices have stayed constant for a long period
of time, they would, in the absence of technological progress,
be identical. Stochastic variations in efficiency would then
produce input coefficients distributed along a ray through
the origin showing the optimal factor combination. No
production function implying possibilities of substitution
between the factors of production could in that case be
estimated. So, if it is to be possible at all to estimate
a function with these characteristics, the input coefficients
observed must show variations in factor intensity among the
production units. Such variations may, in the absence of
technological progress, be due to either of two causes:
relative factor prices have changed over time, or some units of
production have not been designed for the correct factor price ratio.

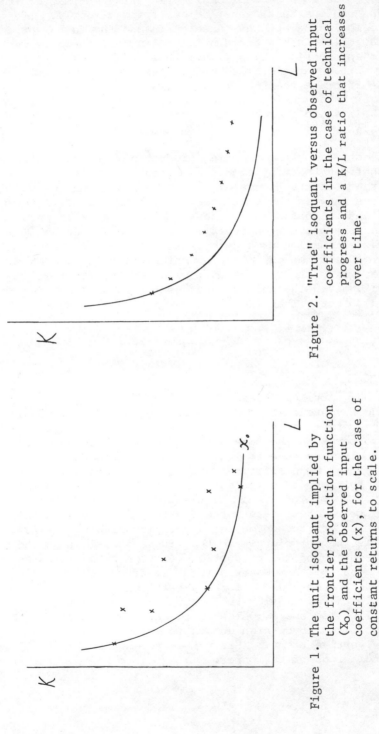

Figure 2. "True" isoquant versus observed input coefficients in the case of technical progress and a K/L ratio that increases over time.

Figure 1. The unit isoquant implied by the frontier production function (X_0) and the observed input coefficients (x), for the case of constant returns to scale.

This alerts us to a second complication with estimating frontier functions and one that cannot be dectected directly. If technological progress has taken place and the factor price ratio has drifted in a certain direction over the same period, the "old" production units (i.e., those that were designed for a factor price ratio different from the one now prevailing) will be less efficient than newer units that might have been built if the factor price ratio had remained constant. This will distort the isoquants and thereby the parameters of the production function.(3)

In estimating the frontier production function we follow an approach first taken by Aigner and Chou.[1]   This, for the case of constant returns to scale, involves minimizing the sum of deviations between the unit isoquant and the observed input coefficients. In the more general case of variable returns to scale and the production function specification above (Equation (5)), the estimation procedure is as follows. For each production unit the maximum production which is theoretically possible (i.e., the production implied by the frontier production function), given the unit's input vector, cannot be less than the observed output. That is:

$$\alpha \ln X + \beta X - \ln A - \delta \ln L - (1-\delta) \ln K = U; \quad U \leqq 0.$$

---

(3)   In our particular case it is likely that the more modern vessels are characterized by a higher capital/labour ratio than the older ones. If new vessels incorporating the same capital/labour ratio as older vessels were to be built, they would probably be a good deal more efficient than the older ones. Our observations of efficient vessels will therefore be as shown by the x's in Figure 2, whereas the relevant isoquant, the one that incorporates the most modern technology at any capital/labour ratio, would be as the curve drawn in the figure. Estimating the parameters of the production function by adapting an isoquant to the observed input coefficients (the x's) will therefore be biased; for any particular capital/labour ratio the marginal rate of substitution (Equation (4)) will be too small, which is to say that δ turn out in most cases to be high enough to support the hypothesis of overinvestment. This rather strengthens the conclusions of the paper.

The estimation, as indicated above, involves maximizing the sum
of the negative U's; that is, making the sum of deviations
between observed and theoretically maximum production in each
production unit as small as possible. This is accomplished by
solving the following LP-problem:

$$\max \alpha \sum_{i=1}^{n} \ln X_i + \beta \sum_{i=1}^{n} X_i - n.\ln A - a\sum_{i=1}^{n} \ln L_i - b\sum_{i=1}^{n} \ln K_i,$$

S.T. $\alpha \ln X_i + \beta X_i - \ln A - a.\ln L_i - b.\ln K_i \leq 0,$

$a + b = 1$

$\alpha > 0, \quad \beta \geq 0, \quad a \geq 0, \quad b \geq 0,$

where the side constraints follow directly from the specification
of the production function and the conditions that we impose upon
it. Note that n stands for the number of efficient production
units.

During the process of estimation it was discovered that in
a number of cases an L-shaped isoquant was implied, as δ assumed
either the value of one or zero. This is not an unknown problem
among those who have tried similar procedures and is due to
"outliers"; that is, exceptionally efficient or inefficient units
which force the LP-program to produce L-shaped isoquants. This
is well illustrated by what may be said to be our "typical"
case. The observed input coefficients form a cone-like pattern, as
in Figure 3, so that the isoquant that produces the best fit will be
L-shaped (disregarding for the moment the possibility that variable
returns to scale may account for some of the more extreme
input coefficients). Excluding the inefficient production units
(that is, units using more of at least one factor of production
than another unit while producing less) is often enough to
eliminate the L-shaped isoquants. This is quite acceptable
theoretically, because the frontier function is, by definition,
a concept which only applies to technically efficient units of
production, and we have therefore generally used this procedure.
In those cases where the frontier function estimated for the
efficient units implies L-shaped isoquants, this is due to some
vessel(s) being exceptionally efficient. In terms of Figure 3,
an exceptionally efficient vessel is characterized by an input
coefficient that is much closer to the origin than the rest,
thus pulling the middle part of the isoquant toward the origin.
Timmer's [4]   way of getting around this problem was by eliminating
successively the most efficient vessel, until the estimated value

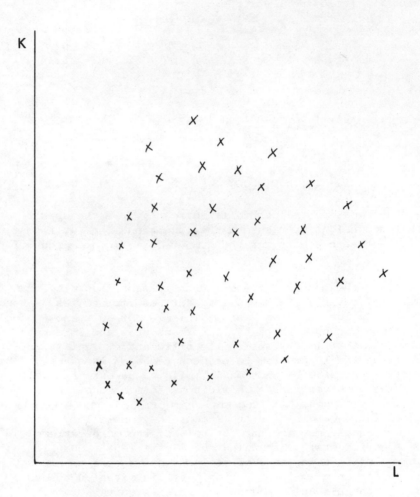

Figure 3. A "typical" case of observed input coefficents (x)
a group of vessels.

of the parameters of the production function had become stabilized.
In our case this is a little more complicated, since variable
returns to scale may account for an input coefficient being closer
than others to the origin, rather than extreme efficiency per se.
We have therefore amended the Timmer approach slightly and
eliminated the vessel which produces the largest change in the
estimated value of δ. We have not carried this further than to
eliminating one or two vessels, because of the essentially
ad hoc character of this procedure.   The defence for this procedure
lies in the possibility that the recordings of output per day need
not always be quite correct, due mainly to inexact reporting of
the number of days at sea.   Some particular vessels might therefore
appear much more efficient (or inefficient) than they in fact are.

THE RESULTS I: RETURNS TO SCALE

        Table 2 shows the estimated values of the parameters α and
β.   Equation (6), it will be recalled, expressed the elasticity of
scale (ε) as a function of these parameters.   As is clear from
Equation (6) the optimum size of the production unit is determined
only when β>0.   In a number of cases our estimation procedure
failed to yield a positive value of β, but the estimated values
of α are consistently smaller than one, except for the trawlers
(group 1) in 1973, so indicating increasing returns to scale.

        For the cases in which the estimated value of β is greater
than zero it is impossible to claim that the optimum size of the
production unit has been determined with a high degree of accuracy.
The optimum vessel size, measured as the amount of output produced
efficiently, varies wildly from one year to another within each
group of vessels.   There appears to be some difference nevertheless
between the groups with respect to whether the largest vessels in
each group are large enough to utilize fully the economies of
scale.   The three estimates of scale-optimal output levels
(X; ε=1) obtained for the trawlers (group 1) all indicate that
the optimal vessel is larger than the existing ones. Underutilization
of economies of scale also is a pervasive tendency in the results
obtained for the saithe fishery (group 2) and the winter cod
fishery except the long-line vessels (groups 5 - 9); the scale-
optimal output level is higher than the highest one realized
in a clear majority of cases.   For the long-line winter cod
fishery (group 4) and the spring cod fishery (group 3) the
opposite is true; there is indeed not a single result showing
a scale-optimal output level exceeding the highest one realized
for the long line winter cod fishery.   For these two groups, then,
economies of scale would appear to be fully utilized by the largest
existing vessels.

        So, to sum up, it seems that there are economies of scale for

Table 2

Estimated values of α and β. $X(\varepsilon=1)$ shows, where applicable, the optimal size of the production unit, measured as efficiently produced output. X_L and X_H show the highest and lowest recorded value of output.

	1971	1972	1973	1974	1975	1976	1977
Group 1							
α	.0	.20857	1.02912	.81275	.0	.45776	.3008
β	.00007	.0	.0	.0	.00001	.0	.284-05
$X(\varepsilon=1)$	14286				100000		246197
X_H	9117	10947	11544	15740	17609	30554	25256
X_L	499	3286	6551	6307	9146	9160	13132
Group 2							
α	.0	.45782	.52566	.06664	.2025	.12644	.32231
β	.0001	.00003	.0	.00016	.00015	.00003	.00004
$X(\varepsilon=1)$	10000	18073		5834	5317	29119	16942
X_H	8804	7347	5963	8917	7639	16848	12000
X_L	581	458	296	228	823	162	1023
Group 3							
α	.0	.42466	.08685	.0	.31624	.13732	.0
β	.0002	.0	.00018	.00013	.00004	.00004	.0002
$X(\varepsilon=1)$	5000		5073	7692	17094	21567	5000
X_H	6965	7214	6295	8634	5330	12322	12698
X_L	868	506	32	1434	481	1421	1409

Table 2 (continued)

	1971	1972	1973	1974	1975	1976	1977

Group 4

	1971	1972	1973	1974	1975	1976	1977
α	.0	.0	.0	.0	.5991	.49258	.34277
β	.00027	.00028	.00037	.00018	.0	.0	.00005
$X(\varepsilon=1)$	3704	3571	2703	5556			13145
X_H	7538	6885	4159	11072	7095	9526	27051
X_L	1023	760	1001	530	360	1085	3441

Group 5

	1971	1972	1973	1974	1975	1976	1977
α	.17087		.45004	.00460	.2495	.0	.0
β	.00004		.0	.00011	.00002	.00005	.00006
$X(\varepsilon=1)$	20728			9049	37525	20000	16667
X_H	9727		22737	9954	5332	16660	22739
X_L	1841		1184	719	616	1233	758

Group 6

	1971	1972	1973	1974	1975	1976	1977
α	.46384	.0	.44895	.0	.43253	.17072	.40617
β	.0	.00014	.0	.00011	.0	.00002	.66 −05
$X(\varepsilon=1)$		7143		9091		41464	89974
X_H	6049	4286	4159	7441	3679	9526	13443
X_L	531	1713	1001	1102	616	129	3357

Group 7

	1971	1972	1973	1974	1975	1976	1977
α	.3567	.1499	.0	.0	.0	.6975	.0
β	.00014	.0	.00001	.00011	.00008	.0	.00006
$X(\varepsilon=1)$	12500		100000	9090	12500		16667
X_H	9727	9347	10190	11072	5332	16660	12338
X_L	3850	2497	1308	381	1969	3333	6227

Table 2(continued)

	1971	1972	1973	1974	1975	1976	1977
Group 8							
α	.2692	.0	.6286			.0	.6977
β	.0	.00013	.0			.00022	.0
$X(\varepsilon=1)$		7692				4545	
X_H	5005	4098	4890			3845	10090
X_L	2538	760	1723			1685	6122
Group 9							
α	.2522	.0	.02315	.20021	.47452	.25211	.0
β	.00007	.00012	.00026	.00002	.77-05	.00002	.00006
$X(\varepsilon=1)$	10683	8333	3757	39990	68244	37395	16667
X_H	6919	11035	5340	7900	4537	5412	10158
X_L	741	113	104	719	360	1085	3592

all groups of vessels over quite a wide range, but there is some
uncertainty as to whether the largest existing vessels are
big enough, or perhaps bigger than needed, for fully exploiting
the economies of scale. Our estimates of the scale parameters
do not appear to be accurate enough to answer that question
with an acceptable degree of certainty.

THE RESULTS II: THE OPTIMUM CAPITAL/LABOUR RATIO

   Combining Equations (1) and (4) we obtain

$$\frac{\delta}{1-\delta}\frac{K}{L} = \frac{w^o}{r^o} \, , \qquad\qquad (7)$$

which allows us to calculate the optimal capital/labour ratio once
we have estimated δ and found the appropriate relative opportunity
cost of labour (w^o/r^o). The quality of our estimates of δ is,
however, hardly sufficient to warrant such a straightforward
procedure. Looking at Table 3 we see that the estimated δ-values
vary a great deal without any clear trend within each group of
vessels, and so it would be impertinent to claim that δ has been
estimated with any high degree of accuracy. We shall, therefore,
deal with the question of overinvestment by calculating the δ-
values which would make the average capital/labour ratios
identical to the optimum ratio, given an appropriate relative
opportunity cost of labour. Comparing the δ-values so calculated
with the estimated values will help us to judge whether the
average capital/labour ratio observed for any group of vessels
is likely to be the optimal one.

   Before we proceed to consider the results of this analysis, it
may be appropriate to alert the reader to the fundamental
relationship between δ, the capital/labour ratio, and the relative
opportunity cost of labour which holds for any optimal combination
of the factors of production. Consider Equation (7). For any given
relative cost of labour (the right hand side of Equation (7)),
the lower the optimal capital/labour ratio, the higher δ must be.
Different groups of vessels can thus be characterized by different
optimal capital/labour ratios even though a common factor price
ratio applies to both, if the value of the parameter δ is not the
same for all groups. But if the δ-value is the same for all groups
and the optimal capital/labour ratio nevertheless is different,
the same factor price ratio cannot possibly be applicable to both.
We can see this by dividing Equation (7) by K/L; then, for any
given δ, the higher the capital/labour ratio, the higher the
relative cost of labour must be, if the equality is to be
preserved.

What, then, are the appropriate prices of labour and capital
(w^o and r^o) to be used for expressing the relative opportunity cost
of labour? The appropriate price of capital in this context is
the rate of return on capital in its best alternative use. Any
thorough investigation of this question is beyond the scope of
this paper, but a net alternative rate of return of 5-10% rate
of real depreciation (cf. Table 3) for all vessel groups except
the trawlers (group 1),for which we assume a 10% rate of
depreciation (cf. Table 3), gives corresponding gross rates
of return of 20-25% and 15-20% respectively.(4)

Similarly, the relevant price of labour to be used is the value
of the commodities those who are employed in the fishery would
produce in their most productive alternative occupation. We
have assumed that this value is equal to the average gross labour
cost in manufacturing industry (the gross labour cost pr. man-
year in manufacturing industry is shown in Table 4). This is,
needless to say, only a rough guess. The most tempting alternative
assumption would be to set a lower opportunity cost of labour,
because of restricted employment opportunities. If we had done
this, the relative opportunity cost of labour would be lower
than we have assumed, and so would the optimum capital/labour
ratio, in which case any possible conclusions to the effect
that overinvestment has occurred would be strengthened.

The δ-values calculated on the assumption that the average
capital/labour ratio observed each year was in fact the optimal
one are shown in Table 3, together with the estimated δ-values
(we shall henceforth refer to these as the calculated and the
estimated values respectively, for short). These calculations
have been carried out on the basis of the two alternative gross
rates of return on capital mentioned above. Looking at these
results, the most striking difference is between the trawlers
(group 1) and all other groups taken together. The average
capital/labour ratio for the trawlers is much higher than it is
for any other vessel group. So, if all ratios were in fact
optimal, it would be necessary that the δ-value characterizing
the trawlers were lower than it is for the other vessel groups,

---

(4)   The average rate of depreciation, as shown in
      Table 3, is a bit higher than 15% for groups 2-9,
      and 11-12% for the trawlers. Subtracting these from
      the assumed gross rates in fact gives us a bit less
      than 5 and 10% net rate of return respectively. The
      values of r^o that we use are, given that the capital
      stock is expressed in 1000 kroner, 200 and 250 for
      groups 2-9, and 150 and 200 for group 1.

Table 3

Estimated values of δ ($\hat{\delta}$), average capital/labour ratio ($\overline{K/L}$) in 1000 kroner per man, rate of depreciation (D) including maintenance of capital equipment, and the values of δ which would make the average capital/labour ratio optimal for a given relative opportunity cost of labour, with subscripts to δ indicating the gross rate of return on capital that has been used when setting the relative opportunity cost of labour. The line NE shows the number of efficient vessels eliminated in each particular case in order to obtain a smooth isoquant, while the line N shows the number of (efficient) vessels in the sample that were used for estimating the parameters of the production function. Missing values of δ indicate that L-shaped isoquants were obtained even after eliminating one or two vessels.

	1971	1972	1973	1974	1975	1976	1977
Group 1							
$\hat{\delta}$	.6423		.8397	.7031	.5559	.6329	.5587
$\overline{K/L}$	386	412	500	622	621	724	756
$\delta 15\%$	.38	.39	.37	.36	.40	.39	.34
$\delta 20\%$	.32	.33	.31	.29	.33	.32	.28
D	12%	12%	12%	12%		10%	12%
NE	1		1	1	0	0	1
N	6		6	5	9	9	11
Group 2							
$\hat{\delta}$	.7753	.7791	.8106	.7039	.8253	.5762	.6538
$\overline{K/L}$	113	118	145	202	184	196	201
$\delta 20\%$	.62	.63	.61	.56	.63	.63	.65
$\delta 25\%$	.56	.58	.55	.51	.58	.58	.60
D	15%	16%	16%	15%	17%	17%	17%
NE	0	1	0	0	0	1	1
N	7	12	7	9	14	17	6

Table 3 (continued)

	1971	1972	1973	1974	1975	1976	1977

Group 3

$\hat{\delta}$	.8730	.6509	.7541	.8574	.5735		
$\overline{K/L}$	108	132	156	176	162	157	283
$\delta 20\%$	.63	.60	.59	.60	.66	.68	.57
$\delta 25\%$	.57	.54	.54	.54	.61	.64	.51
D	16%	14%	15%	15%	16%	17%	16%
NE	0	1	2	2	2		
N	7	6	17	11	8		

Group 4

$\hat{\delta}$	.7407	.7166	.8247	.7843	.5951	.6049	.7255
$\overline{K/L}$	92	116	130	142	148	167	265
$\delta 20\%$	.66	.63	.63	.65	.68	.67	.59
$\delta 25\%$	.61	.58	.58	.60	.63	.62	.53
D	16%	14%	15%	15%	16%	15%	15%
NE	2	0	2	0	1	1	2
N	8	8	7	8	10	10	10

Group 5

$\hat{\delta}$	.5426		.7145	.6368	.3696	.5710	.5578
$\overline{K/L}$	83	97	118	145	157	177	210
$\delta 20\%$	.69	.67	.66	.64	.66	.66	.64
$\delta 25\%$	.64	.62	.62	.59	.61	.61	.59
D	17%	15%	15.%	14%	15%	16%	17%
NE	1		0	0	0	1	1
N	11		15	15	16	11	19

Table 3 (continued)

	1971	1972	1973	1974	1975	1976	1977
Group 6							
$\hat{\delta}$		.8575	.8041	.8113	.8660	.6321	.6882
K/L	83	101	141	143	127	162	175
δ20%	.69	.66	.61	.64	.71	.68	.68
δ25%	.64	.61	.56	.59	.66	.63	.63
D	17%	16%	14%	18%	18%	16%	19%
NE			2	1	2	2	2
N			8	7	4	7	9
Group 7							
$\hat{\delta}$		.4577	.8752	.6698			
K/L	115	99	144	192	166	165	231
δ20%	.61	.67	.61	.57	.65	.67	.62
δ25%	.56	.62	.56	.52	.60	.63	.56
D	19%	18%	18%	16%	19%	20%	20%
NE		0	1	0			
N		3	4	10			
Group 8							
$\hat{\delta}$	.7625	.5133	.5842			.4103	.5964
K/L	87	72	115	110	118	119	141
δ20%	.68	.74	.66	.70	.72	.74	.73
δ25%	.63	.69	.61	.65	.68	.70	.68
D	14%	15%	16%	14%	15%	14%	17%
NE	0	1	1			1	1
N	4	3	5			5	5

Group 9

$\hat{\delta}$	.8424	.7159	.3059	.6479	.5445	.4263	.5478
$\overline{K/L}$	81	97	109	138	153	165	153
$\delta 20\%$	.69	.67	.67	.65	.67	.67	.70
$\delta 25\%$	.64	.62	.62	.60	.62	.63	.66
D	14%	14%	13%	13%	13%	13%	14%
NE	2	2	2	1	2	2	1
N	9	9	9	7	7	9	14

Table 4

Gross labour costs per man-year in manufacturing industry in Norway 1971-1977, in kroner. Source: Wage Statistics. Central Bureau of Statistics, 1977.

1971	1972	1973	1974	1975	1976	1977
36,327	39,887	44,822	51,734	62,043,	68,354	74,874

for any common relative opportunity cost of labour. Since the
rate of depreciation is somewhat lower for the trawlers than for
the other vessels we have assumed a lower gross rate of return
on capital when calculating the relative opportunity cost of
labour for the trawlers (cf. footnote (4)),but the calculated
δ-values are still lower than for any other vessel group. The
δ-values estimated for the trawlers are consistently higher than
the calculated ones however, indicating that too much capital is
invested on the averate relative to the labour used. There is,
moreover, a basis for a stronger statement than this. Setting δ
equal to 0.55, which is close to the lowest value that was estimated
for the trawler group, and using the "low" gross rate of return
on capital (15%), gives optimal capital/labour ratios that for
every single year are well below the lowest values observed for
the different vessels. The trawlers would thus appear to represent
a fishing technology that is too capital intensive to be
justified by the relative opportunity cost of labour characterizing
the Norwegian economy.

    For the other vessels there does seem to be an indication of
overinvestment for at least some groups, but a much less clear
one than for the trawlers. For the saithe fishery and the
spring cod fishery the estimated δ-values are in most cases higher
than the calculated ones, but there is one case for each group in
which the estimated δ-value is low enough not to indicate any
overinvestment on the average even at 25% gross rate of return on
capital. For the winter cod fishery the results are somewhat
different for the gill net and long line groups. Vessels using
long line are more capital-intensive than vessels using gill nets,
but nevertheless the δ-values estimated for the long line vessels
are higher than the ones estimated for the gill net vessels, so
the average capital/labour ratio of both groups cannot possibly
be efficient, given a common factor price ratio. For the long
line vessels the estimated δ-values are in most cases higher than
the calculated ones, while the opposite is true for the gill net
vessels.

    The conclusion is, therefore, that while overinvestment
seems to be characteristic of the long line vessels, this is
definitely not the case for the gill net fleet. If on the other
hand we group the vessels according to region (groups 6-9) we
arrive at a rather inconclusive result. Since there is no clear
ranking of regions according to the average capital intensity
of the vessels, there is no reason to expect the δ-value to be
markedly different. For the four groups taken together we
find that the estimated values of δ exceed the calculated ones
exactly as often as they fall below these (nine cases each),
while they are about equal in two cases.

The general conclusion to emerge with respect to the overinvestment hypothesis is that it is most clearly supported with respect to the most capital intensive vessels, the trawlers that is, and to a degree supported with respect to the vessels engaged in the saithe fishery, the spring cod fishery and the long line vessels participating in the winter cod fishery. On the other hand it is not supported with respect to the gill net vessels engaged in the winter cod fishery, which are the ones coming at the bottom of the league with respect to capital intensity.

References

(1) Aigner, D.J. and Chu, S.F.: On Estimating the Industry Production Function. American Economic Review, Vol. 58, 1978,pp.226-239.

(2) Forsund, F.R. and Hjalmarsson, L.: Frontier Production Functions and Technical Progress: A Study of General Milk Processing in Swedish Dairy Plants. Econometrica, Vol. 47, No. 4, July 1979, pp. 883-900.

(3) Johansen, L.: Production Functions. North-Holland, Amsterdam, 1972

(4) Timmer, C.P.: Using a Probabilistic Frontier Production Function to Measure Technical Efficiency. Journal of Political Economy, Vol. 79, 1971, pp. 776-794.

(5) Vanek, J.: The General Theory of Labour-Managed Economies Cornell University Press, Ithaca & London, 1970.

(6) Zellner, A. and Revankar, N.S.: Generalized Production Functions. The Review of Economic Studies, Vol. 36, 1969, pp. 241-250.

INTEGRATED ANALYSIS

A STRATEGIC PLANNING   MODEL FOR THE FISHERIES SECTOR

Thorkil Bøjlund, Lars Kolind, Rolf Aagaard-Svendsen

OAC, Operations Analysis Corporation A/S

Sofielundsvej 23, DK-2600 Glostrup, Denmark.

INTRODUCTION

During the late sixties and the early seventies a large amount of resources was spent on the development of the Greenland fisheries sector.   This development included the establishment of a fleet of modern fishing vessels designed for catching primarily cod in the open waters of West Greenland. Storage and production facilities were set up in a number of places along the west coast of Greenland.

Due to climatic and other reasons, however, the resource background for these investments later made it necessary to limit the cod quota considerably.

Later the national fishing limits for the Greenland fishing industry were expanded and shrimp quotas rose considerably.

On this background an economic and social evaluation of possible development trends for the Greenland fisheries sector was necessary. Many possible alternative solutions existed and it was necessary to evaluate these possible solutions in an easily comparable way.

The great uncertainties (fish resources, energy and economic conditions) - and the amount of money involved in adding new vessels and necessary production and storage facilities - made it worthwhile to apply a computer based economic model to support the decision making process.

The purpose of the model was to develop a planning tool for the Greenland fisheries sector with the aim to evaluate the

363

- employment
- private economic
- public economic

consequences of various

- resource forecasts
- degrees of exploitation of the resources
- investment in fleet and plants on shore
- forecasts of cost- and price development

for the Greenland sector as a whole.

It was moreover a wish that the model should be able to reflect
different solutions desirable from different political standpoints
and thus leave the balancing of the political factors to the
politicians.   Thus the model was designed for consequence evaluation without any optimization taking place in the model itself.

BUILDING THE MODEL

Systems analysis

The physical and economic system underlying the catching,
storage and production operations was analyzed before construction
of the model.   The process can be described as a causal chain
system as shown in figure 1.

Each block in the chain is affected by the results from the
proceeding operations and some exogenous factors.   In each block
the costs, employment and energy requirements are calculated.

The catching operation includes a transportation from harbour
to fishing ground, the catching operation and the transport back
to the processing place.   Processing on sea plays an important
role.

The activities on shore includes the landing, storage,
processing and transport.

Model description

The model structure is briefly shown below (figure 1).
The model design reflects the structure of the physical system.

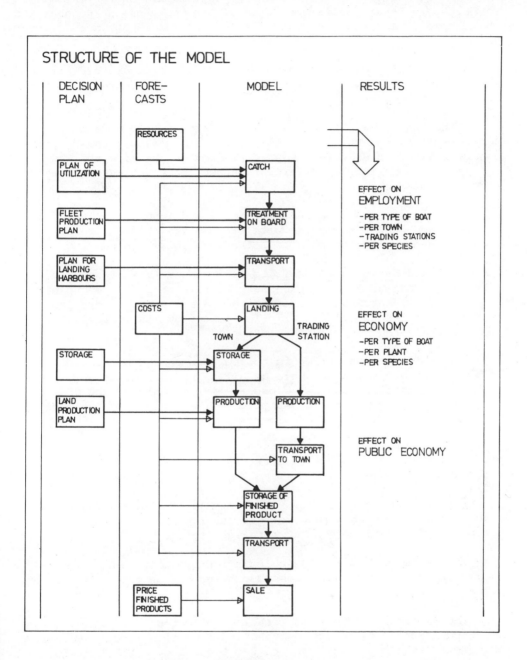

Figure 1. Model Structure

The quantity of shrimp caught depends mainly on the shrimp resources, weather conditions, type of fishing vessels and gear and the skipper. A map of the main fishing grounds is shown on figure 2.

Due to ice, the best fishing grounds are inaccessible part of the year, and moreover the ice conditions can vary greatly from year to year.

After the trawl has been hauled and the shrimp have been separated from the catch of other fish, i.e. halibut, the shrimp can be treated in different ways depending on the size and the equipment of the trawler. Three treatments are reflected in the model:

Raw icing This treatment is simple and involves only a storage in boxes covered with crushed ice. Limits are set on the number of days allowed between catching and landing.

On board freezing If this is applied the trawler can remain on grounds as long as necessary to fill the storage capacity. The daily catch is limited by the capacity of the freezer and the length of each trip is limited by the storage capacity and the oil consumption.

Sea boiling The shrimp is boiled, frozen and packed on board and not processed on shore at all. The trawler can remain on the grounds as long as necessary limited only by the storage capacity and the oil consumption. The daily catch is limited by the capacity of the boiler and the freezer.

The activities on shore include landing, raw material storage, production, product storage and transport. Each activity is reflected in a model block.

The landing operation is fairly simple and includes mainly calculations on the time and cost necessary in order to bring catch from the ship to the raw material storage.

The raw material storage calculation block includes economic, energy and capacity calculations depending on the input and output patterns.

The production module is rather complicated. It allows for a spectrum of production methods reaching from traditional manual processing to a highly industrialized technology.   It includes various defrosting, boiling, peeling, freezing and package methods. At each point, energy, economic and employment consequences are calculated. Capacity demands are calculated.

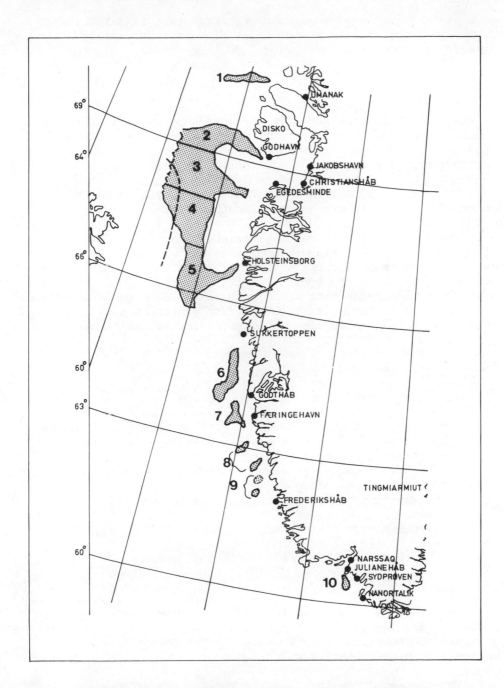

Figure 2. Main fishing grounds at West Greenland

The cold storage demand and cost structure is calculated
on the basis of the input and output patterns. A variety of
diminishing rules can be applied.

DATA

At the moment the model handles seven towns, ten fishing
grounds, three types of fishing vessels, three types of landed
products, a large number of final products and a variety of
combinations of production facilities on shore.

The main results from the model are on a yearly basis, but
intermediate results are given in monthly figures.

The input variables can appropriately be divided into two
categories, control variables and exogenous variables. The control
variables are those factors which - at least in the model - are
assumed to be under the control of the decision maker. In figure
1 they are associated with the blocks in the decision plan column.
The exogenous variables are forecasts or empirical data on
resources, costs, capacities, employment ratios and technical
parameters.

Control variables

For each town it is possible to specify

- the amount of shrimp demanded to each town in each month,
- the number of fishing vessels of each type available
- which grounds they are going to
- the percentage of the catch to be landed as raw iced,raw
  frozen and sea boiled
- the quantity of frozen shrimp to be transported to other
  towns
- the percentage of the shrimp going to different production
  lines

Exogenous variables

Resource data appear in the model as average daily catch per
month per fishing ground per type of fishing vessel. These figures
are based on empirical data provided by the national and international fisheries statistics.

The distance from the towns to the fishing grounds have been
calculated as the distance from the towns to the catch centre of
gravity of each ground. Since the concentrations of shrimp move
during the year, a distance has been calculated for each month.

MODEL IMPLEMENTATION

    The model has been programmed in the computer language APL, which permits the model to be run interactively from a typewriter terminal.

    When necessary input data has been specified, the results follow immediately, and if the results are not satisfactory to the use or if one wants to try another alternative, the input can be modified accordingly to obtain alternative results.

    The model is run for one town at a time. By running a number of towns a national plan can be made.

RESULTS

    The model has been used to analyse a number of alternative development trends for the fisheries sector.

    The flexibility of the model is illustrated by showing briefly some of the main results from four basic alternatives run to compare possible catching and processing strategies for one town using only one type of trawler.

    The four alternatives compared here are the following:

A.   Sea boiled shrimp is landed all year except February, March and April where catching is prevented by ice, and May, June and July, raw frozen shrimp for export is landed.

B.   Raw frozen shrimp landed all year except for the months February, March and April where fishing is impossible due to ice. The shrimp is processed on shore by machine and IQF freezing. 80% of the IQF frozen shrimp is sold in bulk packaging while 20% is sold in retain packages.

C.   Raw iced shrimp is landed all year except for the months February, March and April where fishing is impossible due to ice. Processing is similar to alternative B.

D.   Raw iced shrimp is landed all year except for the months February, March and April where fishing is impossible due to ice. The catch is peeled by hand and frozen in blocks.

    These main alternatives all include the supply by three trawlers of one specified type. Catching takes place on the best fishing grounds not far away from the processing plant and fishing vessels are assumed to be fully utilized for this purpose. (A sample model

```
REJEMODEL          Ⓐ                                    OAC   12. MAJ 1978
BYOVERSIGT:                    560 HOLSTEINSBORG        FANGST - PRODUKTION


              FORUDSAETNING OM TRAWLERKAPACITETEN I 1978
ANTAL TRAWLERE TIL RAADIGHED HVER MAANED:
                                      MAANED
KATEGORI    1    2    3    4    5    6    7    8    9   10   11   12
     1      3    0    0    0    3    3    3    3    3    3    3    3
     2      0    0    0    0    0    0    0    0    0    0    0    0
     3      0    0    0    0    0    0    0    0    0    0    0    0

BEREGNET BEHOV FOR TRAWLERE:
                                      MAANED
KATEGORI    1    2    3    4    5    6    7    8    9   10   11   12
     1    3.0   .0   .0   .0  3.0  3.0  3.0  3.0  3.0  3.0  3.0  3.0
     2     .0   .0   .0   .0   .0   .0   .0   .0   .0   .0   .0   .0
     3     .0   .0   .0   .0   .0   .0   .0   .0   .0   .0   .0   .0

              TILFQRTE REJEMAENGDER I 1978 (TON):
        PLAN      TILFQRT     HERAF:     SQKOGTE  RAA FROSNE  RAA ISEDE
        6315       6291                   3931      2360         0

              TRAWLERQKONOMI (TUSIND KR)                   ARBEJDSKRAFT
        DRIFT   ENERGI   LQN   OMK IALT  INDTAEGT            MAND-AAR
        15941    4127   9557    29625     27639                91.8

              FORUDSAETNINGER OM PRODUKTION OG LAGRE (PCT):
R FR  ISEDE MPIL. HPIL. MP.FR   IQF     LAGER OMS.HAST. PR MAANED
EXP   MPIL. HERM. HERM. BLOK   BULK      RAA   SQK   EXP  R FR   FR
100    100    0     0     0     80       4.0   1.0   1.0  1.0    1.0

                     FAERDIGVAREMAENGDER (TON):
        SQKOGTE   R.FRZ.EXP.  IQF SALG   IQF BULK   BLOKKE    HERMETIK
         3931       2360         0          0          0          0

                    OMKOSTNINGER              ARBEJDS-        KAPACITET
                     TUSIND KR                  KRAFT
                 DRIFT ENERGI  LQN  IALT    MAND AAR    NUV   BEHOV ENHED
FANGSTBESK.                                   91.8
LANDINGSPRIS                        27639
LANDING          259   100    332   691       7.5
RAAVARELAGER      0                   0                * 203     0   M2
PRODUKTION:
 FABRIK  **      6185    50   1330  7565                *3680    0   M2
 PILLEANLAEG       0      0     0     0        .0        8      .0 ANTAL
 IQF  ANLAEG       0      0     0     0        .0        2      .0 ANTAL
 BLOK ANLAEG       0      0     0     0        .0        0      .0 ANTAL
 HERMETIK          0      0     0     0        .0
 HAANDPILN.                    0     0        .0        0       0 HPPL.
 MELLEMTRSP.                         0
 FRYSELAGER      625                625                * 188   888   M2
 UDTRANSPORT                        5127

TOTAL           7069   150   1662 41648       99.3
SALGSINDTAEGT                       68802

(*)  TOTALT AREAL TIL LAGRING OG PRODUKTION AF REJER OG FISK
(**) EXCL. AARLIGE OMK. TIL EVT. UDVIDET FABRIKSAREAL
```

Figure 3. Sample model output

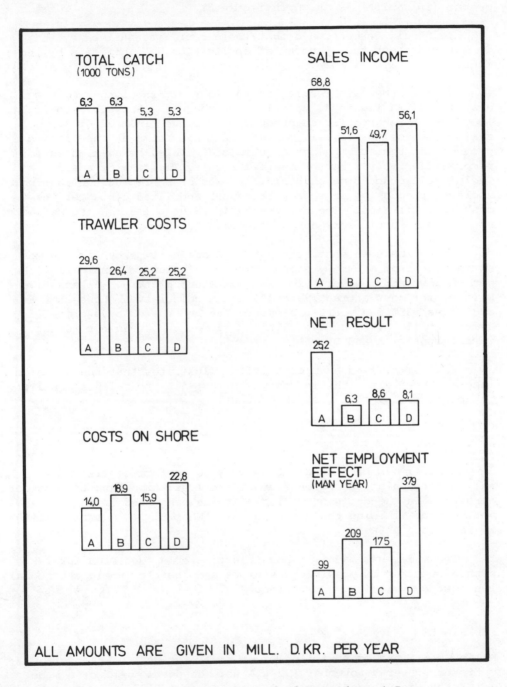

Figure 4. Comparison of alternatives A–D

output (in Danish) is shown in figure 3).

It should be mentioned that these four alternatives by far
do not represent the spectrum of application possibilities by the
model.

Some of the main results of the four alternatives are shown
graphically in figure 4. It is seen that it is possible to catch
1000 tons more when frozen products are landed instead of iced
shrimp because the vessels can stay a longer time at the fishing
grounds and thus can be better utilized. This, however, does not
necessarily imply a better overall result. The trawler costs
include all processing on board.   Costs on shore include landing,
storage, processing on shore, transport and fixed costs for the
onshore facilities. The difference in sales price is also reflected
and the net result for each alternative is shown.

It can be seen that alternative A seems to be by far the most
attractive from a purely economic point of view.   Considering the
net employment, however, yields a different picture. Alternative
D is by far most attractive and a rough calculation shows that an
additional 280 man years employment can be created by a price of
some 17 mill. D.kr. per year (the difference between net result
in alternative A and D).

This simplified example briefly illustrates that the decision
maker is now left with his most important job: to balance conflicting political aims against each other.

CONCLUSION

The model has been used extensively to perform sensitivity
analysis to point out which factors are most important to consider
careful   in the decision making. It has been used in an interactive mode allowing the users to ask: "What .. if" questions with
an immediate answer.

The model has made it possible to provide decision makers
with a spectrum of possible decisions and their economic and social
consequences within a short time.

REFERENCES

Kolind, L: Notat om udvikling af en model for det grønlandske
    fiskerierhverv, november 1977 (A note on development of a model
    for the Greenland Fisheries Sector, November 1977), in Danish.

Bojlund, Th., Kolind, L., Aagaard-Svendsen, R.: Rapport om

udvikling af en model for det grønlandske fiskerierhverv, maj 1978
(A report on the development of a model for the Greenland
Fisheries Sector, May 1978), in Danish.

BIO-ECONOMIC SIMULATION OF THE ATLANTIC SEA SCALLOP FISHERY:

A PRELIMINARY REPORT

Guy Marchesseault[1], Joseph Mueller[2], Lars Vidaeus[3] and
W. Gail Willette[4]

1. New England Fishery Management Council, Peabody,
   MA 01960, U.S.A.
2. National Marine Fisheries Service, Gloucester,
   MA 01930, U.S.A.
3. World Bank, Washington, D.C., 20854, U.S.A.
4. U.S.D.A., Economics Division, Washington, D.C.,U.S.A.

INTRODUCTION

The issue of optimum time path of resource extraction has been
extensively examined in the fisheries management literature for
more than two decades. Initially, the decision problem was commonly viewed as one of finding a steady state solution, e.g. the
maximum biological or economic sustainable yield. More recently,
however, it has been recognized that the choice between stationary
states is significantly affected by the discount rate, and that
maximum economic sustainable yield would be optimizing economic
rent from the fishery over time under a zero discount rate (Clark[1]).
Furthermore, it is increasingly recognized that by restricting the
analysis to stationary states some essential characteristics of
the fisheries may be hidden. For example, random fluctuations in
recruitment are observed for many fish stocks suggesting that short-
term variations in the rate of exploitation is a most relevant
problem to consider (Hanneson[2]).

The prediction of annual recruitment to fish stocks, sensitive
to natural periodicity and variability, and the associated determination of appropriate multiple year harvesting strategies are the
major challenges facing the planning for management of the fisheries
off the U.S. Northeast Coast. A major U.S. and Canadian fishery
in this area has traditionally been directed towards the Atlantic
sea scallop resources on Georges Bank and off the Mid-Atlantic Coast.
Biological evidence presented by MacKenzie, et al.[3] suggests that

375

annual fluctuations in recruitment to these stocks are substantial. The New England and Mid-Atlantic Fishery Management Councils, established pursuant to the enactment of the U.S. Fisheries Conservation and Management Act of 1976[4], are currently planning for the management of this fishery. Objectives for management have been established, and policy issues including the allocation of harvests between domestic and foreign interests, the entry of additional vessels to the fishery, and allocation of benefits among domestic user groups have been identified. Industry pressures for continued high levels of exploitation stem from recent dramatic rises in consumer and ex-vessel prices and peak catch rates. High levels of profitability in the harvesting sector have prompted considerable new investment in vessels.

This paper presents a bio-economic framework for evaluation of alternative multiple year harvesting strategies for Atlantic sea scallops (Figure 1). Predictions of stock dynamics are made as stock abundances are simulated over a 10 year plan period in response to allowable levels of harvests. These predictions are incorporated into a financial model simulating present values of expected net returns to capital and labour in the U.S. harvesting sector. This simulation model is capable of evaluating the policy decisions in the areas noted above.

BIOLOGICAL SIMULATION MODEL

The ability to project stock dynamics over time is a prerequisite for identifying and evaluating multiple year harvest strategies for the sea scallop fishery. As with most efforts to model the dynamics of exploited species, a technique must be found for predicting recruitment to the stock and evaluating the impact of harvests on stock abundance. Several aspects of the biology of the sea scallop populations inhabiting the shelf off New England and the Mid-Atlantic are unknown. Specifically, the relationship between parental stock and subsequent recruitment is unclear as is the extent to which the geographical components of the resource are reproductively independent. Historically, the recruitment patterns of the various resource components have differed. The periodic occurrence of unusually abundant year classes, sustaining intensive fishing of limited duration, has characterized the southern, Mid-Atlantic component. The more northern, New England components have been characterized by more stable patterns of recruitment, sustaining fisheries at relatively higher levels of annual yield.

In the absence of data supporting a stock/recruitment relationship, a major task of the biological simulation is to capture the periodic and variable nature of recruitment to each of the resource components. The pattern of recruitment to the Mid-Atlantic resource component appears to be the most easily deciphered despite a limited

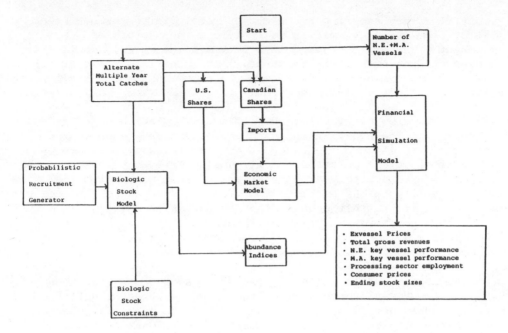

Figure 1

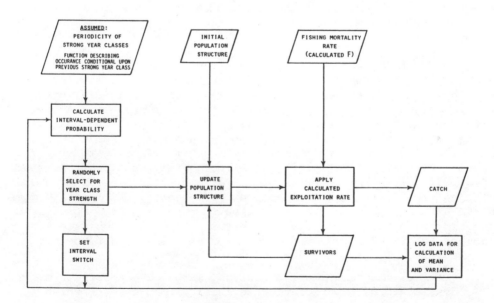

Figure 2

amount of survey and fishery data. Current efforts to model sea
scallop population dynamics have therefore focused on this resource
component. The approach used is believed to be equally applicable
to other components of the resource. As specific details of the
model are discussed fully elsewhere (Marchesseault and Russell[5])
major aspects are summarized below.

A logic flow diagram for the biologic simulation model is shown
in Figure 2. The modelling approach incorporates trends in the
estimated strength and frequency of past recruiting year classes
into a stochastic simulation model. Characteristic patterns in the
strength of recruiting year classes are treated probabilistically,
and variability in the estimates of year class abundance is treated
as a random input. Although in the simulation recruitment is
stochastic in nature, natural and fishing mortalities are handled
deterministically using standard population dynamics equations to
advance cohorts through succeeding age classes.

The absence of an extensive time series of stock assessment
data for the Mid-Atlantic sea scallop resource makes it difficult
to empirically derive a function describing the probability associated with the characteristic recruitment of occasional strong year
classes to the fishery. However, using a Bayesian approach it is
possible to assign a biologically meaningful function to initially
describe the recruitment process.[1] In the case of the sea
scallop resource, the probablistic occurrence of exceptionally
strong year classes may be related to both natural cycles in key
environmental parameters as well as to the temporal proximity of
previous abundant year classes which have pre-empted available
resource space.

The function shown in Figure 3 attempts to capture the
probablistic occurrence of strong recruiting year classes. This
a priori function suggests that the presence of an initial strong
year class will inhibit the close succession of another. After
a few years, however, the probability of a second strong year class
increases as the environment becomes increasingly capable of
accommodating it and cycles in various environmental parameters
become increasingly influential. Although the function falls short
of directly relating recruitment to various environmental parameters
(the nature of any such interactions is unknown), the function does
serve as a useful first approximation to the recruitment process by
incorporating what is observed in nature. The function is clearly
subject to verification or modification as relevant data become
available.

_____

(1)   Previous authors (Clark & Lackey[6], Lamb[7])have adopted similar
      approaches to incorporating decision variables, for which
      there is considerable associated uncertainty, into the manage-
      ment analysis process.

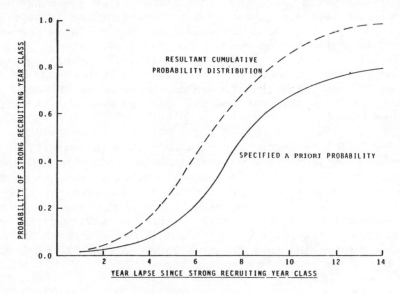

Figure 3

SEA SCALLOP SIMULATION PROGRAM OUTPUT

7.00 ⎫
6.00 ⎪
5.00 ⎬
4.00 ⎬ input data, catch in thousands of metric tons.
3.00 ⎪
3.00 ⎭

Simulation Year	Fishing Year	Avg. Initial Stock Level 1000 MT	95% C-I	Average Catch 1000 MT	95% C-I	F
1	1976/1977	19.692 +/- 0.016		7.000 +/- 0.001		0.35
2	1977/1978	19.645 +/- 0.030		6.000 +/- 0.000		0.34
3	1978/1979	17.228 +/- 0.095		5.000 +/- 0.001		0.35
4	1979/1980	14.331 +/- 0.226		4.000 +/- 0.003		0.37
5	1980/1981	12.286 +/- 0.391		3.000 +/- 0.003		0.36
6	1981/1982	11.726 +/- 0.559		3.000 +/- 0.003		0.49

Figure 4

The simulation process (Figure 2) starts with an initial
population size and age structure and calculates numerous time
paths of change in stock abundance. These time paths are sensitive
to various selected patterns of future recruitment. The recruitment probability function (Figure 3) is used to influence an otherwise random selection of recruiting year class strength by controlling the relative probabilities associated with selecting from one
"strength class" or another. Once the "strength class" of the
recruiting year class is determined, the absolute abundance is
selected randomly according to the expected mean and variance of
the selected class.

The impact of incorporating uncertainty in recruitment on
projected stock abundance is evaluated by conducting 300 stochastic
experiments. In each of these experiments the sequence of allowable
catch values describing an individual harvest strategy is held
constant. The model calculates (1) mean stock sizes with associated
confidence intervals and (2) mean fishing mortality rates over time.
An output sample is given in Figure 4. Absolute stock abundance
values are converted to a relative index as input to the economic
model discussed in the subsequent section.

THE ECONOMIC MODEL

The objectives adopted for management of the Atlantic Sea
Scallop resource (N.E.R.F.M.C.[8]) suggest that an appropriate
criterion for the selection of a harvesting strategy would be the
present value of the sum of net earnings in the harvesting sector
and consumer surplus. Measurement of consumer benefits faces
empirical problems relating to prevailing consumption patterns and
a limited data base. As a consequence the objective of the economic
model presented below is specified as the maximization of the
present value of expected net vessel and crew income. The economic
framework for evaluation of harvesting strategies relative to the
objective is comprised of an ex-vessel price forecasting model, a
sector catch model, a financial simulator, and a bookkeeping routine
for strategy comparisons. Each of these components is discussed
briefly below.

The Ex-Vessel Price Forecasting Model

A price forecasting model was developed using results of recent
scallop demand studies (Storey and Willis[9]). Appropriate changes
were made in the formulation of the ex-vessel price relation with
a view to enhancing the predictive ability. Thus, the price
equation contained in this model specifies domestic sea scallop
landings, imports, inventories, the price of substitute goods (king
crab) and consumer income as determining prices.

Sector Catch Model

The focus of the analysis is on a key group of 25 vessels landing in New Bedford, Massachusetts. During the period 1975-1977 these 25 vessels averaged approximately 70 percent of the total New England sea scallop catch and demonstrated that a close statistical relationship exists between their scallop catch and the scallop catch by all vessels. This relationship is used in the model to predict the catch by key group vessels given an overall fleet quota. The specific share that a typical vessel captures in the model is based on its relative fishing power. In the absence of new entrants, it is assumed that this share relationship will continue in the future. The model can easily be modified to capture the effects of new entrants on the share to key group vessels. Individual vessel catches are constrained by limits on the effort that a single vessel can generate. Accordingly, a vessel's original catch allocation is reduced if the number of fishing days required to harvest the share exceeds the limit imposed.

Financial Simulator

In order to properly evaluate impacts of alternative harvesting strategies on the net benefits to the harvesting sector it is necessary to specify the cost structure of representative vessels. Cost data were obtained from actual settlement sheets between the vessels and the crews. Settlements follow a standard lay system illustrated in Figure 5. An important aspect of the lay system is the deduction of fuel and food costs from the gross crew share. Thus, if the catch per unit effort declines over time, the resultant increase in the variable cost per pound of scallops landed is born in its entirety by the crew.

Various "expense" functions were specified and estimated in cross section analysis. Some expenses are thus linked to total vessel landings (quantity or value), whereas others, e.g. fuel and food expenses, are related to catch per unit of effort which, in turn, is directly associated with the stock size determined from the biological model. Relative factor costs and the exogenous variables influencing the ex-vessel price are predicted to change annually by a constant percentage over the plan period. Consequently, all calculations are made in current dollars throughout the simulation period.

Bookkeeping Routine

Individual management strategies are identified in terms of total annual allowable levels of harvests and U.S. and Canadian shares in overall harvests in the New England and Mid-Atlantic

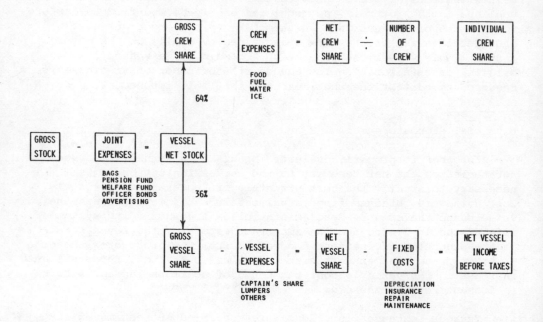

Figure 5. Determination of net vessel and crew income in the
New Bedford sea scallop fishery.

areas.   U.S. imports of scallops from Canada are predicted based
on an estimated linear relationship between total Canadian production (most of which originates from U.S. waters) and exports to the
U.S.   U.S. imports and the specified U.S. catches are fed into
the ex-vessel price forecasting model in which other explanatory
exogenous variables have been collapsed into the intercept term.
Given the predicted ex-vessel price, the annual shares to the key
vessel group (quantity and value) are calculated. Net income to
vessels and crew are generated through the financial simulator.
The bookkeeping routine discounts the value of net income streams
and selects the strategy with the highest present value through
simple numerical comparisons.

APPLICATION

     The development of a biological simulation model for the New
England sea scallop resource is at this time incomplete. Prediction
of stock abundance changes for all resource components under alternative annual harvesting strategies are, therefore, currently not
possible. Rather, strategies had to be initially developed for
the Mid-Atlantic resource component. Total Atlantic scallop catches
are subsequently projected under the assumption that historic average relative catches from New England and Mid-Atlantic scallop
stocks are maintained over the plan period. Thus, under all
strategies, 74 percent of the total Atlantic scallop catch is derived
from the New England resource. Similarly, the annual stock abundance
index associated with individual strategies reflects expected
changes in the Mid-Atlantic resource component given the simulation
of stock dynamics in that resource area under the particular harvesting regime considered. These stock abundance changes were generalized to the entire Atlantic sea scallop resource.

     In a preliminary application of the presented bio-economic
framework six individual harvesting strategies were examined. These
strategies (Table 1) were identified with reference to the adopted
management objectives and the current debate between management
agencies and industry regarding policy issues[2]. Thus, an individual strategy reflects a goal of either a high or a low stock
abundance level at the end of the plan period. The high goal requires that after a ten year period the stock should be at the 1976-
77 average level. The low goal allows for a 50 percent reduction
in that stock size. Relative stock abundance indices are given in
Table 3. Each strategy further describes a typical time-path of

_____
(2) Under each strategy the Canadian share of the total allowable
     sea scallop catch in New England waters (i.e. on Georges Bank)
     is assumed to equal the percent allocation specified in the
     U.S.-Canada Agreement on East Coast Fishery Resources currently
     under U.S. Senate review.

Table 1. Harvesting Strategies Used in Preliminary
          Application

| Plan Time-path Of Harvest | End of Planning Period Stock Goal | |
	High	Low
Low - High	S_1	S_4
Steady - At Intermediate Level	S_2	S_5
High - Low	S_3	S_6

The detailed harvesting strategies are shown in Appendix 1.

harvest over the plan period, i.e. annual catch levels either
describe increasing, steady (at an intermediate level), or decreasing trends.

The predetermined terminal stock sizes constrain the feasible
set of time-paths of resource extraction.   Comparative evaluation
of strategies within each of the two sets $\{S_1, S_2, S_3\}$ and $\{S_4, S_5, S_6\}$ can be made without reference to what is left of the stock at
the end of the plan period. Comparisons between sets have to
acknowledge the difference in terminal stock values. Such comparisons reflect one of the trade offs to be evaluated by the policy
maker, i.e. what present value of expected industry net benefits is
gained or lost by constraining the set of harvesting strategies to
satisfy one particular terminal stock size goal as opposed to
another.

The above six strategies were examined under alternative
assumptions regarding (a) the number of vessels in the key group,
(b) factor prices, and (c) the appropriate discount rate.   In
particular the sensitivity of the solution to new entry of vessels,
increased fuel cost, and a higher discount rate was tested.

RESULTS AND POLICY IMPLICATIONS

Base Case Analysis

The results of the simulations in this preliminary application
of the model are presented in terms of the present value of the
expected net returns in Table 2.   Expected stock abundance changes
(Table 3) partially reflect the annual pattern of recruitment
generated from the biological simulation.   This pattern is

Table 2.
Results of Preliminary Simulations of the
Atlantic Sea Scallop Fishery.

	Base Case [1]			New Entry [2]			Increased Fuel Costs [3]			Increase In Discount Rate [4]		
		Expected Net Return to Vessels	to Crew		Expected Net Return to Vessels	to Crew		Expected Net Return to Vessels	to Crew		Expected Net Return to Vessels	to Crew
	Total			Total			Total			Total		
High End of Period Stock Goal												
S_1	109.8	22.2	87.6	105.9	18.3	87.6	108.6	22.2	86.4	94.8	19.0	75.8
S_2	108.0	21.7	86.3	104.2	17.9	86.3	106.9	21.9	85.0	94.7	19.1	75.6
S_3	98.0	19.7	78.3	98.1	17.3	80.8	96.5	19.8	76.7	87.8	17.9	69.9
Low End of Period Stock Goal												
S_4	115.3	24.5	90.7	113.9	21.6	92.3	113.9	24.7	89.2	99.8	21.1	78.7
S_5	115.2	25.7	89.5	116.6	21.8	92.9	113.3	25.8	87.5	101.8	22.7	79.1
S_6	102.1	21.9	80.1	104.7	20.5	84.2	100.1	22.1	78.0	90.4	19.5	70.9

1/ The base case assumes 25 vessels in the key group.

2/ 5 new vessels are assumed to enter the fishery in year 1.

3/ The cost of fuel is increased by a factor of 20 percent.

4/ The discount factor is raised to 10 percent.

determined from the specified probability function for predicting the recruitment of strong year classes (Figure 3). Therefore, under each strategy it is expected that a strong year class would occur during the latter rather than the former half of the period.

Within each of the two strategy sets, strategies with an increasing allowable catch trend (i.e. S_1 and S_4) show rising stock abundance in the very beginning of the plan period. Thereafter stock abundances adjust downward to their pre-determined terminal levels. This general pattern results from relatively low allowable catch levels at the start of the plan period coupled with continued growth of the strong year class initially in the fishery. Strategies with the reverse trend (i.e. S_3 and S_6) are initially associated with falling stock abundance as allowable catches are kept at relatively high levels. By mid-period stock abundances have reached their lowest values. During the latter years they rise to their targeted end of plan period values. The harvesting strategies characterized by steady harvests at intermediate levels (S_2 and S_5) show relatively less fluctuation in expected stock abundance. These increases (decreases) in stock abundance are postulated to result in proportional increases (decreases) in catch per unit effort and decreases (increases) in effort required to harvest a given quantity of sea scallops. The stock abundance changes are thereby translated into changes in total annual variable costs of harvesting.

In interpreting the results, it becomes important to recognize the implications of a fixed fleet size and a specified individual vessel effort limit (expressed in terms of days absent). Specifically, during years of high allowable catch levels and/or low catch rates the fleet may not have the capability to harvest the entire annual catch allocation (Table 3).[3]   This phenomenon is best exemplified by actual catches predicted under both S_3 and S_6.   In the initial year of the simulation the allowable harvest is reached. However, because of the relatively high catch, stock abundance falls. As a result catch per unit effort drops, causing substantial increases in the variable cost of fishing. Therefore, the present value of the expected net returns of S_3 is lower than for S_1 and S_2.   The same holds true for S_6 in relation to S_4 and S_5.

As expected, strategies aimed at a "high" terminal stock abundance yield lower expected net returns than those targeted towards a 50 percent reduction in end of plan stock abundance. Thus, strategies S_4 and S_5 yielding a "low" terminal stock size dominate strategy S_1 by 5 million dollars in present value of expected net

---

(3) In the current formulation there is no feedback loop into the biologic model if the allowable harvest is not taken in its entirety. If in fact total quotas are not harvested, future stock sizes may be higher than shown.

Table 3.

Total Allowable U.S. Catches, Stock Abundance Indices, and Projected Effort Days per Vessel Under Six Alternative Atlantic Sea Scallop Harvesting Strategies.

	YEAR									
	1	2	3	4	5	6	7	8	9	10
S₁										
Total Allowable U.S. Catches	154.90	154.90	154.90	193.70	193.70	193.70	232.40	205.30	255.60	267.30
Actual U.S. Catches	154.90	154.90	154.90	193.70	193.70	193.70	232.40	205.30	255.60	267.30
Abundance Index	1.00	1.09	1.07	.99	.91	.89	.93	.98	.97	.86
Effort Days/Vessel	108.00	99.00	101.00	136.00	148.00	151.00	173.00	146.00	182.00	216.00
S₂										
Total Allowable U.S. Catches	193.70	193.70	193.70	193.70	193.70	193.70	182.00	166.60	185.90	189.80
Actual U.S. Catches	193.70	193.70	193.70	193.70	193.70	193.70	182.00	166.60	185.90	189.80
Abundance Index	1.00	1.03	.96	.85	.80	.77	.85	.97	1.03	1.03
Effort Days/Vessel	135.00	131.00	140.00	158.00	168.00	175.00	149.00	119.00	126.00	128.00
S₃										
Total Allowable U.S. Catches	309.90	309.90	309.90	232.39	104.50	108.50	108.50	135.60	147.20	154.90
Actual U.S. Catches	309.90	309.90	219.90	139.30	104.50	108.50	108.50	135.60	147.20	154.90
Abundance Index	1.00	.85	.60	.38	.35	.47	.66	.86	.98	1.06
Effort Days/Vessel	216.00	254.00	255.00	255.00	208.00	161.00	114.00	109.70	104.40	101.70
S₄										
Total Allowable U.S. Catches	154.90	154.90	154.90	232.40	232.40	232.40	251.80	263.40	294.40	298.23
Actual U.S. Catches	154.90	154.90	154.90	232.40	232.40	232.40	251.80	263.40	256.60	201.60
Abundance Index	1.00	1.09	1.07	.97	.84	.78	.78	.78	.70	.55
Effort Days/Vessel	108.00	99.00	101.00	167.00	193.00	207.00	225.00	235.00	255.00	255.00
S₅										
Total Allowable U.S. Catches	271.10	271.10	271.10	271.10	189.79	112.30	178.10	232.40	247.88	224.65
Actual U.S. Catches	271.10	271.10	263.90	190.60	146.60	112.30	178.10	232.40	230.90	201.60
Abundance Index	1.00	.92	.72	.52	.40	.47	.59	.66	.63	.55
Effort Days/Vessel	189.00	205.00	255.00	255.00	255.00	169.90	210.10	244.90	255.00	255.00
S₆										
Total Allowable U.S. Catches	348.60	348.60	348.60	147.20	77.50	127.80	174.30	220.80	228.50	220.80
Actual U.S. Catches	348.60	289.60	168.60	95.30	77.50	127.80	174.30	220.80	223.60	201.60
Abundance Index	1.00	.79	.46	.26	.30	.42	.54	.61	.61	.55
Effort Days/Vessel	242.00	255.00	255.00	255.00	180.00	211.00	225.00	252.00	255.00	255.00

returns. By embarking on a policy of matching beginning and end
of plan period stock size, rather than allowing for a 50 percent
reduction in stock size over the period, the 25 vessels in the key-
group are foregoing a present value of expected net returns (vessel
& crew) of 5 million dollars.

Additional Analysis

     The expected total net returns from an increase in the number
of participating vessels depends on whether the additional effort
will increase the total catch or simply spread the catch over a
larger number of operators. Since in the Base Case analysis of
S_1 and S_2 the actual catches equal the total allowable catches, new
entry simply implies a smaller average annual catch per vessel.[4]
The addition of associated "lumps" of fixed costs to the industry
decreases the net returns to vessel owners. Since the crew shares
are related proportionally to the gross stock, the total net return
to the labour sector is unchanged, but the individual crew share
is reduced. Under S_3, the increase in the number of vessels
results in a higher total catch compared to that realized in the
Base Case. This brings a higher return to the labour sector but
a lower return to capital. During periods of unchanged stock
abundance, the crew shares change proportionally to the total catch,
while the return to capital depends on the relative total changes
in gross boat shares and total fixed costs.

     Under S_4, S_5, and S_6 the effect of the increase in vessels is
an increase in the net return to labour but a reduction in the net
return to capital. (Only for S_4 does the decrease in return to
capital outweigh the gain to labour). The reasons are as stated
above relative to S_3. Within the two sets of strategies identified the optimal strategy under the new entry scenario is S_5.

     The impact of rising fuel costs (an increase of 20 percent) is
essentially to penalize those scenarios that result in relatively
greater declines in stock abundance. As noted previously, the
impact of increased fuel costs falls entirely on the labour sector.
The return to capital actually increases slightly due to the
reduction in Federal taxes associated with a reduction in crew
earnings (See Figure 5).

     Currently there are no property rights assigned in the U.S.
sea scallop fishery. Under such a situation, there is considerable

---

(4) In the current formulation it is assumed that new entrants are
     as productive as existing vessels in the fleet. In fact, it
     probably takes the vessels some extended period of time to
     develop comparable levels of production.

uncertainty as to the individual vessel owners' ability to harvest in the future those resources which he leaves in the ocean bank today. Thus, the vessel owner might assign a relatively high discount rate to future earning potentials. To examine the implication of such a situation, the discount rate was increased from 7 to 10 percent. As a consequence, the optimal strategy for selection is S_5 rather than S_4 as in the Base Case. This result is primarily due to a substantial decrease in the value of catches at the end of the period as evaluated under the increased discount rate. Clearly, under the current institutional framework the higher discount rate may be appropriate.

SUMMARY

The purpose of this report is to present the analytical framework for and some preliminary results of ongoing investigations into the determination of optimal intertemporal harvesting strategies for the management of the Atlantic sea scallop fishery. The strategies identified and results reviewed are solely for the purpose of demonstrating the analytical framework.

Several aspects of the presented bio-economic model and its associated simulation programs are being refined. First, a biological simulation model(s) is being developed for the more northern New England resource component(s). Simultaneously, the nature of the proposed probabalistic recruitment function for the Mid-Atlantic resource component is being reviewed. This will permit the differences in recruitment patterns between resource components to be incorporated into the analytical decision framework.Secondly, the economic model is being extended to cover additional fleet sectors. As a result, it should be possible to apply the model to policy issues relating to user group allocations. Finally, technical modifications are being made within the network of simulation programs. These include the development of appropriate routines to ensure that (a) stock abundance responds to actual catches as determined in the economic model and (b) strategy generation, evaluation and comparison become a continuous process internal to the program.

REFERENCES

1. Colin W. Clark, Mathematical Bioeconomics, John Wiley & Sons, New York (1976).
2. R. Hanneson, Economics of Fisheries: Some Problems of Efficiency, Studentlitteratur, Lund, Sweden (1974).
3. C.L. MacKenzie, A.S. Merrill and F.M. Serchuck, Sea Scallop Resources Off the Northeastern U.S. Coast. Marine Fisheries Review, 40:19-23 (1978).

4. U.S. Congress, Fishery Conservation and Management Act of 1976. P.L. 94-265, H.R. 200.

5. G.D. Marchesseault and H.J. Russell, Jr. A Simulation Approach to Identifying Multiple Year Harvest Strategies in the Sea Scallop Fishery. Northeast Fish and Wildlife Conference, April, 1979. New England Fishery Management Council Research Document 79 SC 4.1. (1979).

6. R.D. Clark, Jr. and R.T. Lackey, A Technique for Improving Decision Analysis in Fisheries and Wildlife Management. Virginia Journal of Science 27(4):199-201 (1976).

7. N.D. Lamb, A Technique for Probability Assignment in Decision Analysis. General Electric Technical Information Service. MAL Library, Appliance Park, Louisville, KY. 20 pp.(1976).

8. New England Regional Fishery Management Council, Atlantic Sea Scallop Fishery Management Plan; Statement of the Problem - Part 1 (Draft) Res. Doc. No. 79 SC 2.1. (1979).

9. David A. Storey and Cleve E. Willis, Econometric Analysis of Atlantic Sea Scallop Markets. A Report Submitted to the New England Regional Fishery Management Council (1978).

Appendix 1.
Allowable Catch Levels Under Six Atlantic Sea
Scallop Harvesting Strategies.

STRATEGY		1	2	3	4	5	6	7	8	9	10
						YEAR					
S_1	USN	66.74	66.74	66.74	83.43	83.43	83.43	100.11	88.43	110.13	115.13
	USM	88.18	88.18	88.18	110.23	110.23	110.23	132.28	116.34	145.50	152.12
	C	183.71	183.71	183.71	229.62	229.62	229.62	275.55	243.40	303.10	316.88
	TOTAL	338.63	338.63	338.63	423.28	423.28	423.28	507.94	448.68	558.73	584.13
S_2	USN	83.43	83.43	83.43	83.43	83.43	83.43	78.42	71.75	80.09	81.76
	USM	110.23	110.23	110.23	110.23	110.23	110.23	103.62	94.80	105.82	108.03
	C	229.62	229.62	229.62	229.62	229.62	229.62	215.85	197.48	220.44	225.03
	TOTAL	423.28	423.28	423.28	423.28	423.28	423.28	397.89	364.02	406.35	414.82
S_3	USN	133.49	133.49	133.49	100.11	45.05	46.72	46.72	58.40	63.41	66.74
	USM	176.37	176.37	176.37	132.28	59.52	61.73	61.73	77.16	83.77	88.18
	C	367.40	367.40	367.40	275.55	124.00	128.59	128.59	160.74	174.51	183.71
	TOTAL	677.25	677.28	677.25	507.94	228.57	237.04	237.04	296.30	321.70	338.63
S_4	USN	66.74	66.74	66.74	100.11	100.11	100.11	108.46	113.46	126.81	128.48
	USM	88.18	88.18	88.18	132.28	132.28	132.28	143.30	149.91	167.55	169.75
	C	183.71	183.71	183.71	275.55	275.55	275.55	298.51	312.29	349.03	353.62
	TOTAL	338.63	338.63	338.63	507.94	507.94	507.94	550.27	575.67	643.31	651.86
S_5	USN	116.80	116.80	116.80	116.80	81.76	48.39	76.75	100.11	106.79	96.78
	USM	154.32	154.32	154.32	154.32	108.03	63.93	101.41	132.28	141.09	127.87
	C	321.47	321.47	321.47	321.47	225.03	133.18	211.25	275.55	293.92	266.36
	TOTAL	592.60	592.60	592.60	592.60	414.82	245.50	389.42	507.94	541.80	491.01
S_6	USN	150.17	150.17	150.17	63.41	33.37	55.06	75.09	95.11	98.45	95.11
	USM	198.41	198.41	198.41	83.77	44.09	72.75	99.21	125.66	130.07	125.66
	C	413.32	413.32	413.32	174.51	91.85	151.55	206.66	261.77	270.96	261.77
	TOTAL	761.91	761.91	761.91	321.70	169.31	279.37	380.95	482.54	499.47	482.54

USN = Annual U.S. scallop catches in New England waters.
USM = Annual U.S. scallop catches in Mid-Atlantic waters.
C   = Annual Canadian scallop catch in New England waters.
TOTAL= Annual total scallop catch in New England and Mid-Atlantic waters.

Appendix 2.
Total Allowable U.S. Catches, Actual U.S. Catches, Stock Abundance Indices, and Projected Effort Days per Vessel Under Six Alternative Atlantic Sea Scallop Harvesting Strategies.

					Y E A R					
	1	2	3	4	5	6	7	8	9	10
S₁										
Total Allowable U.S. Catches	154.90	154.90	154.90	193.70	193.70	193.70	232.40	205.30	255.60	267.30
Actual U.S. Catches	154.90	154.90	154.90	193.70	193.70	193.70	232.40	205.30	255.60	267.30
Abundance Index	1.00	1.09	1.07	.99	.91	.89	.93	.98	.97	.86
Effort Days/Vessel	108.00	99.00	101.00	136.00	148.00	151.00	173.00	146.00	182.00	216.00
S₂										
Total Allowable U.S. Catches	193.70	193.70	193.70	193.70	193.70	193.70	182.00	166.60	185.90	189.80
Actual U.S. Catches	193.70	193.70	193.70	193.70	193.70	193.70	182.00	166.60	185.90	189.80
Abundance Index	1.00	1.03	.96	.85	.80	.77	.85	.97	1.03	1.03
Effort Days/Vessel	135.00	131.00	140.00	158.00	168.00	175.00	149.00	119.00	126.00	128.00
S₃										
Total Allowable U.S. Catches	309.90	309.90	309.90	232.39	104.50	108.50	108.50	135.60	147.20	154.90
Actual U.S. Catches	309.90	309.90	219.90	139.30	104.50	108.50	108.50	135.60	147.20	154.90
Abundance Index	1.00	.85	.60	.38	.35	.47	.66	.86	.98	1.06
Effort Days/Vessel	216.00	254.00	255.00	255.00	208.00	161.00	114.00	109.70	104.40	101.70
S₄										
Total Allowable U.S. Catches	154.90	154.90	154.90	232.40	232.40	232.40	251.80	263.40	294.40	298.23
Actual U.S. Catches	154.90	154.90	154.90	232.40	232.40	232.40	251.80	263.40	256.60	201.60
Abundance Index	1.00	1.09	1.07	.97	.84	.78	.78	.78	.70	.55
Effort Days/Vessel	108.00	99.00	101.00	167.00	193.00	207.00	225.00	235.00	255.00	255.00
S₅										
Total Allowable U.S. Catches	271.10	271.10	271.10	271.10	189.79	112.30	178.10	232.40	247.88	224.65
Actual U.S. Catches	271.10	271.10	263.90	190.60	146.60	112.30	178.10	232.40	230.90	201.60
Abundance Index	1.00	.92	.72	.52	.40	.47	.59	.66	.63	.55
Effort Days/Vessel	189.00	205.00	255.00	255.00	255.00	169.90	210.10	244.90	255.00	255.00
S₆										
Total Allowable U.S. Catches	348.60	348.60	348.60	147.20	77.50	127.80	174.30	220.80	228.50	220.80
Actual U.S. Catches	348.60	289.60	168.60	95.30	77.50	127.80	174.30	220.80	223.60	201.60
Abundance Index	1.00	.79	.46	.26	.30	.42	.54	.61	.61	.55
Effort Days/Vessel	242.00	255.00	255.00	255.00	180.00	211.00	225.00	252.00	255.00	255.00

PUBLIC EXPENDITURE AND COST-RECOVERY IN FISHERIES: MODELLING

THE B.C. SALMON INDUSTRY FOR POLICY ANALYSIS AND GOVERNMENT

INVESTMENT DECISIONS

D.A. Pepper[1] and H. Urion[2]

1. Economic Development Directorate, Department of
   Fisheries and Ocean, Ottawa, Ontario, Canada.
2. Consultant, Vancouver, B.C., Canada.

INTRODUCTION

The need for knowledge in the fisheries is ever-increasing.
This need is partly the result of the intrinsic features of
fisheries management whereby the analysis of the dynamics of fish
populations require a continual flow of information and also the
result of the growing complexity of fisheries management of those
species. This need for knowledge has increased the use of simulation
models as analysts seek to explain and measure more and more
phenomena. Unfortunately, the dialogue between modellers and
managers in the fisheries management field is not that extensive.
However, presented here will be a specific case in which a model was
developed, not only for the use by managers, but also for the larger
purpose of analyzing public investment in the fisheries. We will
describe the genesis of the Cost and Income model of the Salmon
Fishing Industry (CISFIN) and its relationship to some specific
needs in analyzing public expenditure and cost-recovery in the
Salmonid Enhancement Program (S.E.P.) in B.C. as well as the use of
modelling generally in the Canadian fisheries management context.
One of us (D.A. Pepper) required the model for the industry
consultations on cost-recovery and the other (H. Urion) developed
the technical aspects of the model.

We would be remiss if, however, we did not acknowledge our great
debt to the "back-room" experts of the Pacific Region who made it
all work which must include David Reid, Rob Morley and Jay Barclay,
to name a few. We thank them.

Traditional Operations Research Analysis

There seem to be three possible alternatives available for the evaluation of public expenditures such as those made in fisheries and the economic policies underlying them.[1]   First, through ex ante and ex post measurement (usually benefit-cost analysis) it is possible to measure the impact of public expenditure upon certain sectors of the economy. Second, some types of ex post analyses are enough for evaluation if extensive data exist.   It is then possible to measure the results of some expenditure or policy. However, both these methods suffer from the problem of large data-requirements and our inability to hold all variables constant in the real world.   These problems result in the appeal of the third approach of operations research, modelling and computer simulation. (We do not differentiate among these activities and shall use them interchangeably).

Models and simulations have existed in fisheries management for many years, especially for the biological aspects. Indeed, populat-ion dynamics analysis can be said to be one modelling exercise related to the real world.   The familiar models are those of Schaefer, Beverton-Holt, Ricker and others. Watt[2] has categorized them into four types.

These various types are not explored here other than to note that there is an extensive and growing literature of models for estimating fish populations. In general, stock assessments are now growing more sophisticated and demanding in their general need for prediction and reliable estimates.

Economic models also exist in fisheries although they may be considered not as well developed as the population-dynamics models. Perhaps this is because economic variables are much more difficult to quantify than biological ones. The Scott-Gordon model[3] of the common-property aspects of fisheries is perhaps the most widely known and certainly the simplest. Copes[4] has extended the analysis to analyze the distribution of the various economic benefits and their relationship with the industrial organization in the industry. Finally, Smith and Clarke[5], have developed models attempting to incorporate the population dynamics and economics into what they call bio-economic models.

Just as the biological models appear to focus on yield, the economic models focus on economic rent and there have been a number of recent studies that have modelled the fisheries to analyze the possible extraction of economic rent[6].

However, it seems apparent to those who are involved in fisheries management that there is a wide gap between the theory in the models and actual practice in fisheries management.   This may be the result of not having models that are suited to the specific purposes of management. The CISFIN case examined here will demonstrate the design of a model that is directly related to some fisheries management needs and investment decisions.

New Modelling Needs: The Canadian Case

While it is difficult to find explicit directions for researchers and managers to construct models there is one case, Canada. Canada has clearly outlined her general management principles and goals in what is called "The Policy Document"[7]. In particular, precise objectives have been drawn up (and are currently being reviewed). They number twenty in all and are related to three broad areas:

a) resource use and allocation,
b) economic development,
c) social/cultural development.

It is interesting that, within these three categories, there are specific objectives that directly relate to model-building. For example, in the area of resource-allocation, objective 3 requires:

"Incorporation in resource-meanagement models, not only of biological and environmental, but also of major social and economic components of the system."[8]

The necessary resource management models are not directly named or specified in any way but the Gordon-Scott model of over-fishing has been established for over 20 years and appears to be the cornerstone model from which more sophisticated and dynamic models are built, at least in the Canadian context[9]. In reviewing some of the various Canadian models, there appears to be no fully developed models that take into consideration the specific subobjectives in Objective 3. But social/cultural development models are scarce.

The traditional problem in fisheries management has been over-fishing and the dissipation of economic rent. With control over fishing effort the new problem has been the necessity for equity, however measured, in the allocation of the benefits from rational management. While these problems can be defined simply, it is nonetheless apparent that the demands of fisheries management in the Canadian context are complex and entail many trade-offs. The design of any large-scale investment project would certainly go beyond traditional benefit-cost analysis if it becomes necessary to measure for comparison the trade-offs outlined in the Policy Document. This problem is not addressed but it is worth noting here that these precise objectives in fisheries management have an

impact upon modelling exercises.   In other words, the Policy
Document has very specific requirements for the specifications and
outputs of models.   We now turn to our particular example which
considers some of these requirements and others.

THE B.C. SALMON INDUSTRY

Description of the Industry

     The B.C. salmon industry has been extensively studied with
prolific literature in both the biological and economic fields.
(We exclude the extensive recreational salmon fishery from this
analysis).   Fisheries management, per se, is quite well developed
in the salmon fishery. The principal feature of the commercial
fishery from a fishery management point of view has been the control
of fishing effort by license limitation place upon the vessels[10].
The   success or failure of the limitation scheme is not reviewed
here.

     The B.C. commercial salmon fishing is complex from an
industrial organization and fisheries management point of view.
There are five types of salmon: 1) sockeye, 2) coho, 3) chums,
4) pink and 5) springs.

     Each of these sub-species has a different life cycle and over
150 streams in B.C. are part of the total spawning habitat.   These
fish are exploited by three gear-types: 1) seines, 2) gillnetters,
and 3) trollers.   They vary in their dependence upon particular
species and river-system "runs".   Salmon then end up in a number
of product forms but are in three broad categories, fresh, frozen,
or canned.   The primary (catching) industry is also characterized
by a   number of organizations reflecting various interest groups,
unions, cooperations, vessel-owners, etc.

     The industrial organization of the secondary industry is
characterized by two large firms and a number of other relatively
smaller firms but vertical integration is a notable feature.   As
will be shown, all these intrinsic features of the industry had to
be incorporated into the CISFIN model because of the variable impact
of S.E.P. projects and cost-recovery upon each species and ultimately
each interest group.   The cyclical variations in catch and value
and among the gear-types is apparent and part of the increased
variation is biological and the variation of catch by gear is
affected as different gear-types "target" different species.   There
is also an obvious fluctuation in the area of catch but given the
mobility of the fleet this factor is not too significant.

THE SALMONID ENHANCEMENT PROGRAM

The Program

In May 1977, the federal government of Canada and the Province of British Columbia completed a $6 million two-year planning phase for the design of the Salmonid Enhancement Program (S.E.P.). The program had as its general objective to double the present level of catches (to historic levels) and had the specific objective of maximizing the net social and economic benefits subject to technical and budgetary restraints. Phase I of the program was for an expenditure of $150 million over 5 years to increase the production of salmon. A planning framework was developed to measure the impact of this major investment. The framework is multi-objective and utilizes what is called the "5-account" system to evaluate the impact in the following areas:

1) National Income,
2) Regional Development,
3) Native People (status Indians),
4) Employment,
5) Environmental and Resource Preservation.

Using the planning process in the Canadian government, Planning, Programming, Budgeting System (P.P.B.S.) and the five accounts, it was possible to rank a large variety of enhancement projects, both large and small, in terms of benefit-cost analysis (national income) and the other accounts[11]. Thus, decision-makers had a process by which rational investment decisions could be made subject to the budgetary and technical constraints. The planning process, project selection and public involvement aspects are well-documented in the S.E.P. Annual Report[12]. To conclude, the program was (and is) a major investment undertaking requiring tremendous amounts of technical and economic analysis. However, the investment decision by the federal government was subject to what was euphemistically called "cost-recovery mechanisms". In other words, there was to be a direct return on the investment by the major recipients of the benefits: the fishermen and the processing industry. This concept and the related role of the CISFIN mldel is outlined in the next section.

Cost-recovery

The principle of cost-recovery for public expenditures in general and investments in particular is accepted in a wide variety of forms but a specific cost-recovery plan for the S.E.P. investment is notable. This is because of the magnitude of the investment, $150 million, and the identification of the beneficiaries - the

Structure of model

The model is comprised of 219 endogenous variables grouped into 14 modules in three recursive blocks: (a) the primary sector; (b) the secondary sector; and (c) the cost-recovery mechanisms. The following five modules represent the primary sector of the industry:

1) Operating Costs,
2) Fixed Costs,
3) Vessel Maintenance Costs,
4) Gross Income,
5) Net Income.

The secondary sector is comprised of seven modules:

6) Operating Costs,
7) Fish Processing,
8) Fixed Costs,
9) Fish Product Inventory,
10) Inventory Costs,
11) Export/Domestic Fish Product Sales,
12) Net Income.

The cost-recovery methods under consideration for the primary and secondary sectors are respectively implemented in modules 13 and 14.

It should be noted that the increased production of salmon from S.E.P. is exogenous to the model. Two methods were used to estimate possible future production. One was a "fish production model" under development separate from the CISFIN model, and the other was called the "Worksheet 1A" method. This latter method was simply to take the expected fish production from the proposed list of enhancement projects and input the results into CISFIN. These exogenous values from either or both methods are inputted through module 4:"Fishermen's Gross Income Module".

Fig. 2 provides an overview of these modules and indicates the general flow of information between the modules. The landing prices paid to fishermen for their catch in Module 4 are passed on as an operating cost to the fish processors (Module 6). Information from the processor fixed-cost modules and wholesale values (modules 8 and 11, respectively) are used in calculating revenue obtained by the public sector via the processor-related costs recovery policies (module 14). The fish-processing module converts the supply of salmon bought from fishermen into ten fish products. Inventory levels and their costs are determined in modules 9 and 10. Gross wholesale values of fish products to domestic and foreign markets are calculated in module 11. All costs are subtracted from gross income levels to determine fishermens' and processor's net income (modules 6 and 12).

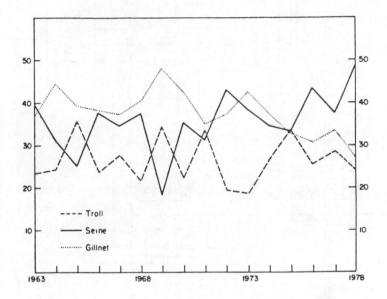

SOURCE: Canada Annual Statistical Review

Figure 1. Landings of Salmon Percentage
By Gear Type - (1963-1978)

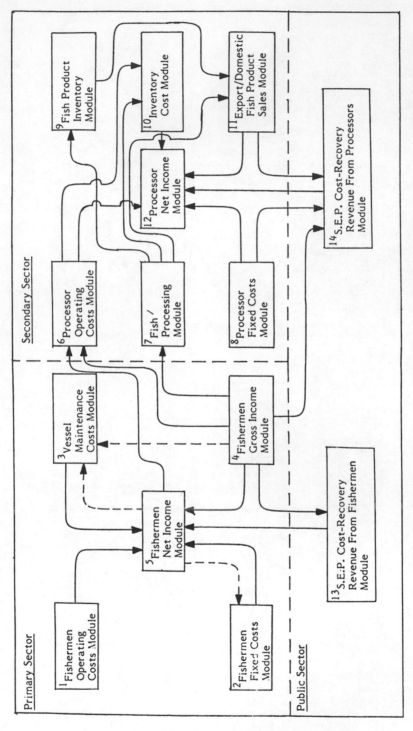

Note: A broken line connecting two modules indicates a lagged variable interdependency whereas a solid line indicates a current year variable interdependency.

Figure 2. Overview of Cost–Income Model of British Columbia Salmon Fishing Industry

Table 1. B.C. Landings of Salmon By Species and Gear - 1969-1978
(metric tons, round weight)

GILLNET

	Springs Tonne	Sockeye Tonne	Coho Tonne	Pinks Tonne	Chums Tonne	Total Tonne	Value $ 000	Deliveries '000	Catch per Delivery Kg	Value per Delivery $
1969	1 176	8 662	1 608	2 345	3 949	17 800	12 186	98	182	124
1970	1 495	6 941	3 664	8 102	10 601	30 856	16 780	130	237	129
1971	1 374	10 038	2 489	4 507	3 656	22,138	15 308	106	209	144
1972	1 516	7 342	2 301	2 614	14 777	28 633	18 345	109	263	168
1973	1 204	15 276	1 796	2 719	15 705	36 751	40 288	109	337	369
1974	982	11 026	1 689	2 685	7 943	24 367	26 672	89	274	300
1975	1 031	4 223	1 524	2 353	2 845	12 014	15 014	61	197	246
1976	1 027	7 655	1 177	3 013	4 841	17 746	25 825	78	228	331
1977	1 129	10 793	1 357	5 470	3 229	22 028	34 416	83	265	415
1978	997	7 819	1 281	2 319	6 641	19 102	38 687	69	277	561

SEINE

	Springs Tonne	Sockeye Tonne	Coho Tonne	Pinks Tonne	Chums Tonne	Total Tonne	Value $ 000	Deliveries 000	Catch per Delivery Kg	Value per Delivery $
1969	366	1 798	615	2 087	2 102	6 969	3,885	9	774	432
1970	567	3 281	2 126	13 510	6 012	25 498	11 492	16	1 593	718
1971	427	5 357	1 904	10 264	1 704	19 658	10 679	11	1 787	971
1972	459	2 059	1 026	14 139	15 362	33 053	14 587	15	2 204	972
1973	583	5 784	2 084	7 943	16 943	33 339	31 817	17	1 961	1 872
1974	531	7 882	1 611	7 547	4 460	22 034	20 319	13	1 695	1 563
1975	558	1 152	1 901	6 026	2 483	12 128	11 934	10	1 213	1 193
1976	492	4 395	1 197	12 790	6 044	24 925	26 712	14	1 780	1 908
1977	926	6 237	1 981	12 814	2 696	24 661	29 697	16	1 541	1 856
1978	886	11 107	1 108	12 326	8 901	34 337	57 796	13	2 641	4 446

TROLL

	Springs Tonne	Sockeye Tonne	Coho Tonne	Pinks Tonne	Chums Tonne	Total Tonne	Value $ 000	Deliveries 000	Catch per Delivery Kg	Value per Delivery $
1969	4 911	487	5 768	1 841	26	13 037	11 729	139	94	85
1970	4 503	1 202	7 859	2 410	158	16 135	16 804	155	104	108
1971	6 901	1 934	9 699	2 858	60	21 456	18 489	156	138	119
1972	6 377	116	7 206	1 389	54	15 146	17 409	141	107	123
1973	5 769	467	7 370	2 644	115	16 368	27 893	132	124	211
1974	6 123	2 786	7 078	974	76	17 043	27 007	126	135	214
1975	5 701	305	4 312	1 851	60	12 241	19 965	121	102	165
1976	6 257	288	6 948	1 254	37	14 790	39 405	134	110	294
1977	5 467	358	6 519	6 439	107	18 893	44 612	147	129	303
1978	6 004	3 395	6 763	686	313	17 165	61 681	150	114	411

SOURCE: CANADA ANNUAL STATISTICAL REVIEW

industry.   The arguments for and against cost-recovery are not
detailed here.   For a general review in the Canadian context,
Bird[13] is good and the specific discussion is in the S.E.P.
"Discussion Paper - Cost-Recovery Mechanisms in S.E.P."[14]
The principle of cost-recovery from the industry produced the
need for a model which would show the linkages between the investment in a particular project increasing a certain species and its
impact upon the income of the various gear-types and the final
impact upon the processing sector.   The model had to serve a real
purpose in the industry consultations that would help to design the
appropriate "cost-recovery mechanisms" and also gain acceptance of
the principle of cost-recovery, through demonstrating the individual
benefits of various investment scenarios.

     To summarize, the CISFIN model had to answer certain specific
policy questions and was additionally constrained by the necessity
of being realistic and credible.   However, we focused on one
question that would seem to come to grips with all the demands of
the policy-makers, fisheries management, and the industry.   It was:

          How much would net incomes of fishermen and
          processors change if various levels of S.E.P. investment
          took place and if various types of cost-recovery were
          implemented?

We now describe the model.

THE CISFIN MODEL

     The Cost-Income model of the Salmon Fishing Industry (CISFIN)
is a recursive deterministic model designed to provide economists
and policy-makers with quantitative estimates of the economic
impacts of the Salmonid Enhancement Program on the B.C. salmon
fishing industry.   The model is also designed to facilitate
evaluating the effect of cost-recovery methods such as license fees,
landing royalties and export taxes on the income level of fishermen and processors.   In addition, the model calculates estimates
of the annual revenues generated through cost-recovery methods and
determines the present-worth values of key economic variables used
to evaluate the economic feasibility of S.E.P.

     In order to evaluate the economic benefit of salmon enhance-
ment with the implementation of cost-recovery methods, it is
necessary to determine the net income levels of fishermen and
processors.   Consequently, operating (variable) and fixed costs
are explicitly incorporated into the model.

The total cost-recovery revenues collected from fishermen and fish processors (modules 13 and 14, respectively) are subtracted from the corresponding gross incomes in determining net income levels (modules 5 and 12). Various taxes, vessel license fees and other levies were explored in detail.

As noted, future S.E.P. salmon production for each of the five species of salmon was provided exogenously to the model (module 4). The amount of each species of salmon caught by gill-netters, seiners and trollers is determined by applying coefficients which reflect the percentage of each salmon species caught by each of the three types of gear over the past 10 years. The average annual amount of salmon caught over the past 5 years was considered the "base catch" and the average value of income from the sale of other fish are also provided exogenously to CISFIN. It is also assumed that the "base catch" and other income will remain constant in the future.

It should be noted that no price module exists in the model. After careful analysis of landed fish prices and domestic/export fish product prices, it was concluded that it would be impossible to project values from prices with any degree of confidence. Consequently, price information is provided exogenously to the model. Average costs of harvesting and processing each of the five principal species of salmon caught on the B.C. coast were determined using historical information and it was assumed that these costs would not change over the projection period.

Hence, most of the relationships between the variables in the model are essentially accounting type equations and are, therefore, very simplistic. The limitations of time and detailed historical data precluded the development of behaviorial equations. It is also worth noting that the CISFIN model does not "capture" the entire industry from the population dynamics to marketing of the products. However, it does model the impact of exogenously determined increased fish production on the primary and secondary industry and thus is almost a complete industry model.

Use of the Model

As part of the development process, the model was first used to produce estimates of the costs and incomes of fishermen and processors during the years 1968 and 1977. The model was found to track historical data fairly well. Annual estimates from the model for fuel and other operating costs of fishermen are in error by plus or minus five per cent and those for fixed costs are in error by plus or minus ten percent. Estimates of fishermen's annual gross amount vary from known values by usually no more than plus or minus ten per cent, although estimates of total incomes averaged over the 10-year period are within plus or minus two per cent.

Although lack of data prevented thorough verification of the operating and fixed costs of the secondary sector, these estimates can be assumed to be in error by no more than plus or minus ten per cent. The estimates of processors' gross income produced by CISFIN vary by plus or minus twenty per cent. However, the annual gross income estimates averaged over the 1968-1977 time period is underestimated by only eight per cent. The test of the model against historical data enabled users to gain a better understanding of the model itself and its use in producing projected income levels of fishermen and processors.

Numerous runs of the model are made to investigate the sensitivity of certain coefficients and exogenous variables on projected income levels. Initially, the model was executed to produce annual projected cost and income levels for the "Base Case Scenario" in which it was assumed that no additional salmon will be produced. This would correspond to a decision by the government not to undertake the Salmonid Enhancement Program and allow fishermen to continue to catch at existing rates.

A variety of alternative S.E.P. production levels and cost-recovery methods were then provided exogenously to the model. Annual license fees collected per vessel and landing royalty taxes applied to either the landed weight or value of salmon were the principal cost-recovery methods investigated. Projections of costs and income levels of the fishermen and processors under these S.E.P. policy scenarios were produced by the model. A separate report generation program produces reports on the projected values of any variable either provided exogenously or calculated by the model. The present-worth values of income levels and revenues generated through cost-recovery calculated by the model enabled economists to compare alternative levels of S.E.P. production and cost-recovery. In addition, the model provided a quantitative indication of any imbalance of benefits among the different fishing fleets which might occur due to different production levels of different species of salmon under S.E.P. Specifically, with the present planned S.E.P. production levels, gillnet and seine fishermen would realize the highest increase in net incomes. Given the relatively lower production levels under S.E.P. of coho and spring salmon, (which are caught primarily by trollers), the amount of revenue collected from trollers either through license fees or landing royalties would be lower (as would be the benefits). A variable (lower) tax could possibly be used to equalize benefits. Alternatively, the large S.E.P. productions of sockeye, pink and chum salmon might cause a shift in the traditional fishing patterns of trollers from coho and springs to those species.

Review of Results from the Model

Users of the CISFIN model were pleased with its applicability in modelling the total performance of the salmon industry. The flexibility of the model in producing outputs from a wide variety of different inputs related to changes in estimated fish production, catch-ratios by the various gear-types and various cost-recovery mechanisms was particularly noteworthy. The discussions with a particular industry sector (e.g. the trollers who catch spring and coho from "southern" rivers) were augmented by the very detailed data that could be produced from the model. In other words, the particular impact on income of a specific enhancement project with an assumed level of cost-recovery could be presented to that particular group. This was the desired result posed by the question in section 4.

In addition, the "internal" analysis and evaluation of specific projects could be supplemented by outputs from CISFIN and this was especially noted in the selection of projects that were not subject to traditional benefit-cost analysis (national income account). The impact of various "mixes" of projects that maximized one of the five accounts could be quantified. The trade-offs among them then could be measured.

While the users of CISFIN were generally pleased with the application of the model one hoped-for outcome did not materialize. This was for the model to be demonstrated to or used by one of the industry groups. It was felt that a "hands-on" use of the model by an industry group would give the model the added credibility of being "tested" by those who would ultimately be affected by some of the policy conclusions of the model. This did not come to pass as the industry consultations ended with only the general views of the various groups upon the principle of cost-recovery itself and various cost-recovery mechanisms. The detailed discussions of im- pact that the model could produce were not able to be fully explored by the industry. The federal election of May 1979 has intervened and at present the status of cost-recovery in S.E.P. is under review. To conclude, the model had both an "internal" and "external" use, for the time being it is only being used in the former mode.

CONCLUSION

The results and use of the CISFIN model must be seen against the larger needs of the S.E.P. program and the stated objectives in the Policy Document as well as compared to the characteristics of models generally noted in Section 2. We think there are three general conclusions that are apparent:

1) The CISFIN model is relatively simplistic given that

there are no non-linearities and behavioural equations.

2) The model satisfied most of the needs required in the analysis and evaluation of the cost-recovery aspects related to S.E.P. but did not get the chance to take full advantage of the detailed impact results feature in the model. On balance, the model had more internal policy analysis use rather than in the external consultation process to gain acceptance of a specific cost-recovery plan.

3) The model satisfied some of the criteria outlined by the Policy Document as necessary to the development of fisheries management policy. Especially, the allocation of benefits from S.E.P. to the various groups could be tracked very well by the model. The other aspects of fisheries management contained in the five-account system were not directly assessed by the model. However, the outputs from the model were inputs into the various analyses of the policy options related to the trade-offs among the various accounts.

There are, we hope, implications for other countries who might contemplate public investment in the fisheries sector and seek to recover some of the costs (students of economic rent might also be intrigued).   It might be useful to conclude with the words of Sinclair[15] which, paraphrased, state that any recommendations upon fisheries management have to meet three conditions. They need to be:

a) Administratively feasible,
b) Politically acceptable,
c) Publicly defensible.

As such, they are useful guidelines for any modelling exercises in fisheries management such as the CISFIN model presented here.

BIBLIOGRAPHY

1.   Thomas H. Naylor, "Computer Simulation Experiments with Models of Economic Systems", John Wiley and Sons,Toronto   (1971).
2.   Kenneth E. Watt, "Ecology and Resource Management". McGraw-Hill, Toronto (1968).
3.   Scott H. Gordon, "The Economic Theory of a Common-Property Resource: The Fishery", J.P.E., LXII, 124-142, (1954) and A.D. Scott, "The Fishery: The Objectives of Sole Ownership", J.P.E., LXII, 116-124, (1955).
4.   Parzival, Copes, "Factor Rents, Sole Ownership and the Optimum

        Level of Fisheries Exploitation". The Manchester School.
        (June) pp. 145-163, (1972).
5. V.L. Smith, "Economics of Production from Natural Resources".
        A.E.R., LVIII No. 3, 4090431, (1968) and Colin Clark,
        "Mathematical Bioeconomics". John Wiley and Sons. Toronto,
        (1976).
6.a.James A. Crutchfield and Guilo Pontecorvo, "The Pacific Salmon
        Fishery: A Study of Irrational Conservation". Resources for
        the Future. Baltimore (1969).
  b.Richard F. Fullenbaum and F.W. Bell, "A Simple Bio-Economic
        Fishery Model: A case study of the American Lobster Fishery".
        Fish.Bull. Vol.72, No.1, (1974).
  c.John M. Gates and Virgil J. Norton, "The Benefits of Fisheries
        Regulations: A case study of the Yellowtail Flounder Fishery".
        Univ.of Rhode Island Marine Technical Report. No.21, (1974).
7. Canada, "Policy for Canadian Commercial Fisheries". Dept. of
        Fisheries and Oceans (1976).
8. See Appendix I. "Policy Objectives for Fishery Management and
        Development".
9. C.L. Mitchell and N.G.F. Sancho, "Optional Fishing Effort of
        Canada's Offshore Fisheries -- An Application of Economic
        Optimisation Techniques". Mathematical Biosciences.34 (1977)
10. P.H. Pearse, "Rationalization of Canada. West Coast Salmon
        Fishery", in Economic Aspects of Fish Production. O.E.C.D.
        Paris, (1972).
11. Canada, "Benefit-Cost Analysis Guide" Planning Branch. Treasury
        Board Secretariat. Ottawa, (1976).
12. Canada, "Annual Report: Salmonid Enhancement Program - 1977".
        Dept. of Fisheries and Oceans. Vancouver, (1978).
13. Richard M. Bird, "Charging for Public Services: A New Look at
        an Old Idea". Canadian Tax Foundation. Toronto, (1976).
14. Canada, "Discussion Paper - Cost-Recovery and S.E.P.". Dept. of
        Fisheries and Oceans. Vancouver, (1978).
15. Sol Sinclair, "A Licensing and Fee System for the Coastal
        Fisheries of British Columbia". Dept. of Fisheries and Oceans.
        Vancouver, (1979).

PLANNING MODEL FOR SMALL-SCALE FISHERIES DEVELOPMENT

G.E. Lierens

Van de Bunt, Management Consultants

Amsterdam

INTRODUCTION

This paper records a study of the application of modelling in small scale fisheries development planning.

A computerized planning model has been designed and tested. The paper describes the basic structure and principles of the model and comments on the results after having applied the model to the Sri Lanka fisheries. Consequently advantages and limitations of the applied model are discussed and some recommendations made on how to apply planning-tools in general and in this model in particular in the field of small-scale fisheries development.

GENERAL

The study, upon which this paper is based, was an activity of the regional FAO/UNDP project "Development of Small-Scale Fisheries in Southwest Asia". One of the project objectives was to develop tools for policy and programme planning for small-scale fisheries development. This work can be seen as a logical follow-up, at field level, of conceptual work and modelling earlier undertaken at FAO headquarters; an Expert Consultation on Quantitative Analysis in Fishery Industries Development was held in Rome in 1975 (FAO Fisheries Report no 167) and, in this connection, a first Small-Scale Fisheries Development Model was prepared (FAO Fisheries Circular no 337). Consequently the main objective of the undertaken study was to determine whether an improved "complete" planning model for small-scale fisheries could be developed and, especially, how and to what extent such a model could be used beneficially,

especially in relation and addition to other (to be developed)
planning-tools.   The model should also be applied for analysis and
preparation of policies to support the implementation of development
plans.   The following sub objectives of a to be developed model
could be specified:

- examination of possible future development trends per fishing
  sector and for the fisheries in general;
- assessment of effects of alternative policy measures concerning
  subsidies, fish prices, exports/imports, etc.;
- determination of input requirements for development, such as
  investments and foreign exchange for alternative development
  plans;
- investigation into the relative importance of the major variables
  influencing the system;
- provision of a quantitative and comprehensive description of the
  sector;
- to facilitate the collection of an improved data base;
- possibilities for combination of optimal developments in each
  fishery sector towards a national fisheries development plan,
  which satisfies as much as possible national development
  priorities.

Whether the model design and study has been considere
successful is discussed in section "conclusions" of this paper.
The model is coded in Fortran IV G and can be implemented on
medium size computers.

The model is documented in working paper no 19 of the Project
"Small-Scale Fisheries Promotion in South Asia" (RAS/77/044),
Colombo, October 1978.

MODEL STRUCTURE AND PHILOSOPHIES

General

The model is divided into a number of sub-models:

1.   fishery resources;
2.   catching units;
3.   handling/distribution of landings;
4.   processing and exports;
5.   costs and foreign exchange;

Model sub-sectors 1 - 4 indicate a division of the sector
fisheries along the usual sub-sectors. Model sub-sector 5 covers
the costs structure of all other sub-models.

Relations between the various model variables describe the existing situation in the fisheries (basis year) and by further assigning yearly growth factors to these variables the development, year by year, over the chosen planning period is simulated. The model is deterministic, all the variables take predetermined (input) values. Consequently the probabilities of the fishing sector's "reality" is not reflected in the output information. This may make the planning rather rigid, since predetermined input values only suggest accuracy, while nothing can be said about the probabilities that the calculated estimates for development in the sector will materialize. However, it is believed that the determination of probability distributions for the various variables may prove to be too difficult a task.

The planning horizon of the model can be chosen freely, thus enabling the planner to use the model for both medium as well as long term planning. The model is not an optimizing model, i.e. it does not necessarily reflect a most efficient or an optimum plan. Only by applying the model with different values of the input variables can one get an indication in which direction an optimum plan should be formulated.

The only built-in decision criterion is related to the economics of catching units. When, in a certain year, the earnings of the owners become marginal, no expansion or replacement of catching units will take place and boats will be withdrawn if and when they become uneconomical for the operators (see Catching units)

Fishery Resources

The respective quantities of pelagic and demersal resources are fixed. The model calculates the grade of exploitation of both the demersal and pelagic maximum sustainable yields. However, no restrictions are imposed by the model in case the maximum sustainable yield is exceeded.

Catching units

There are a limited number of groups of catching units. Each type of unit is a rigid combination of a particular craft and fishing gear and is considered to catch either demersal or pelagic fish. Crew size, number of fishing days per year and "theoretical" catch per unit per day (i.e. possible catch effort) are determined by the given values in the basis year and the assumed growth factors. The total theoretical demersal as well as total pelagic production effort is the sum of the joint efforts of the respective (demersal or pelagic) catching units. Total "real" catch, either demersal or pelagic, is obtained by adjusting the total "theoretical" production effort by means of the catch–effort relation (i.e. yield curve), which is in general:

$$\text{catch} = c_1 \,(\text{effort}) - c_2 \,(\text{effort})^2;$$

c_1, c_2 constants (which depend also on max. sustainable yields)

The net earnings of the boat owners are calculated as the balance between sales proceeds and the sum of variable and fixed costs and crew share. Gross earnings are the net earnings plus capital costs (depreciations and interests).

The crew share, divided amongst the crew, is the share of the sales after subtraction of variable costs. In case a fisherman owns his boat, it is assumed that he is entitled to his part of the crew share as well.

The mechanism of the decision criterion which determines the number of catching units in a group, is as follows.

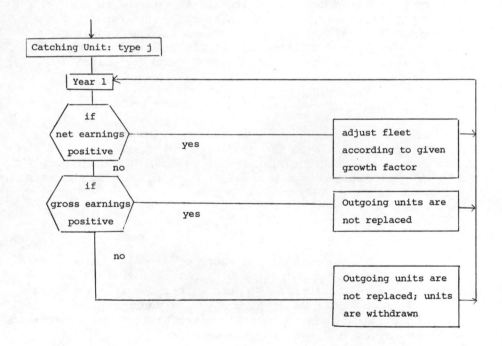

Handling/Distribution of the landings

The handling/distribution function covers the transport of fish from the landing site to the first market (private, retail or wholesale).   Handling of fish comprises of the unloading from the craft, (re)icing, boxing and loading in vans; manpower is the major cost determining factor.   Costs of distribution are mainly determined by the average distance of transport.   In the handling/ distribution sector the model can only consider one type of unit (most probably this will be a transport van).

The producer's fish price is determined by a Quantity-Price relation (see graph).

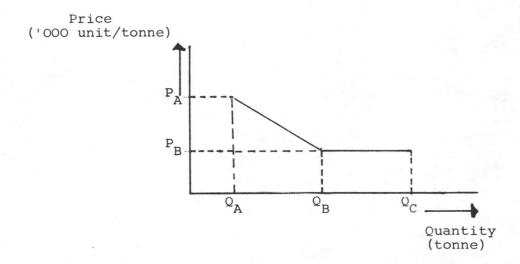

Figure 1.

It is assumed that the landings of each catching-unit group are homogenous and that they have their own relation between quantity and price.   The relation is made linear for quantities smaller than Q_B (see graph).   Beyond Q_B, glut landings between B and C are supposed to fetch a constant price (P_B).   Quantities beyond Q_C are not marketable for neither fresh consumption or processing and are consequently dumped.   The function ABC is estimated by the user on basis of actual statistics of producer's prices.   In principle, any linear function, derived at by linear regression analysis could be used within the model.   This price-quantity curve is believed to be rather crucial for the(output) results. However very often it occurs that data on these curves are not readily available.

It is believed that the here chosen relation is the most

realistic possible compromise between a theoretical and a practical
price-quantity relation. The perishable nature of fish was one
of the main reasons to decide upon a price-quantity curve, instead
of a price-demand curve.

Processing and Export

It is assumed that fixed portions of the fish landed by the
various "catching unit groups" are reserved for the different
processing industries (e.g. freezing, drying). These portions may
vary for "glut" and "normal" landings. Export of both fresh and
processed fish is considered in the model. The processing
industries and export agents pay the same price for their raw
material as the wet fish traders do.

Costs and Foreign Exchange

For the sub-sectors used in the model, i.e. production by the
catching units, handling/distribution and processing, the costs are
divided into:

- variable costs (costs as function of size/intensity of production)
- capital costs (depreciation, interest incurred by the usage of
  capital assets)
- fixed costs other than capital costs.

When determining the annual depreciation of the units, the
depreciation method of the decreasing percentage of the replacement
value has been applied. The rate of depreciation which is decided
by the user, may reflect the decreasing economic performance of
the asset due to obsolescence. The real replacement value of the
asset at the moment of depreciation is used. Interest is calculated
on the capital still invested in the asset (that part, which is not
yet depreciated).

The foreign exchange needed as well as the foreign exchange
earnings in the sector are calculated. The extra charges to be
paid by the importers as well as the bonuses to be received by the
foreign exchange earners are taken into account (if applicable in
particular country).

INPUT

The input required for implementation of the model can be
divided into three categories:

1.   country data, e.g. (working) population, growth of population,

2.   <u>fisheries sector data describing existing situation,</u>
e.g. maximum sustainable yields, manpower employed in various
sectors, costs structures in various sectors, number of craft for
the various year groups, catch per unit in basis year, crew share
system, number of fishing days in basis year, price-quantity
relations for fish, sector capacities in basis year, fractions of
landings to be processed in basis year.

3.   <u>fisheries planning data, determined by policies and develop-
      ment goals</u> e.g. number of planning years, number of year groups,
      increases in catch productivity, handling efficiency.

The input data of categories 1 and 2 form factual information,
while data under 3 represent ambitions, assumptions, policies etc.
determined by views of the Government and its development planners.

The user is free to decide upon the number of catching units
and processing groups. There is a temptation to include many groups
in order to describe the sector as realistically as possible. However, a large number makes the model implementation cumbersome,
requires more computer capacity, while it is often not possible to
obtain detailed input data for many separate groups. The
possibility of combining a number of real existing catching-units
(or to a much lesser extent processing groups) into one suitable
theoretical unit should be considered. One should not forget that
the model is to be used for (overall) development planning.

The quantity of input required is substantial and there may be
problems to handle the input volume. Very often no real reliable
fisheries statistics are available.

However, on the other hand, the systematic collection of such
data required as input for the model can be a very useful exercise
and spark off a "consciousness" for statistics.

OUTPUT

Output results (per planning year) could be arranged according
to the existing (model) sub-sector, as is indicated in the following table. It will be of course obvious that some results are
relevant for more than one sub-sector. The model supplies the
drafters of a national (fisheries) development plan with the following key-figures: (per year).

- total manpower involved in fisheries;
- total landings;
- per capita fish consumption;
- needed investments;
- needed foreign exchange;
- foreign exchange earnings;

PLANNING MODEL APPLIED FOR SRI LANKA FISHERIES

The following remarks are based on runs with Sri Lanka
Fisheries' data as input.   It should be emphasized that the aim
of these runs was primarily to test the model and no (alternative)
development planning was undertaken.   Therefore these are only
comments on magnitudes and trends.

The planning horizon was 10 years and 7 (pelagic) catching
groups were specified.   The annual growth for the number of motorized boats was fixed at 9-10% and for the traditional motorized
sector this percentage amounted to 5%.   No further growth was
anticipated in the (traditional) unmotorized sector.   In reality
these "growth" figures are likely to be less.   As a result the
landings increased ower the 10 years with about 75%.   However the
per capita consumption, which is a national objective of high
priority, increased over these 10 years by 50%.   The utilization
of the pelagic sustainable yield increases from a "safe" percentage
in the basis year (year zero) to a more critical grade, which is
to be expected with this large growth.   The built in "decision-
criterion" guards, to a certain extent, against uneconomical
fishing and so sustains the growth.   The supply-price curve seemed
to have worked, but once again the collection of reliable statistics
seemed critical for the entire model.   The price remains rather
stable, in spite of the increase in landings.   Probably this is
partly due to the big demand for fish and the readiness to pay a
relative high price.   Presently there is a shortage of fish in Sri
Lanka.

Motorized boats turn out to be profitable.   It occurs that
there is hardly any bright economical future for the unmotorized
craft.

Modernization, reflected by high growth rates for the motorized sector, has a negative effect on employment.   However this
effect is partly neutralized by the fact that within the industry
additional employment is created due to bigger landings.

However the same modernization requires a large investment
and moreover the needed foreign exchange rises substantially.

CONCLUSIONS

The model could be a tool for policy formulation and evaluation
at national level; however, lack of proper data and experience of
modelling restricts useful application of the model.

Generally it seems that the application of an optimization
model is less appropriate.   The formulation of a criterion function

for an entire National Fisheries development plan is, at least, awkward or even impossible.   Small optimizing models may support the decision-making process within the fisheries' sub-sectors; features of such models should be: small size, easy to handle, low quantity of input data, modest required computer capacity.   In this context it is believed that the use of linear programming models for sub-sectors could ne useful.

On the other hand it should be stated that in developing economies the central governments play a major role in the development of the fisheries.   The governments not only stimulate further development of the fisheries industry by means of subsidies, but very often also initiate as well as organize this development by starting investment projects as well as supplying infra structural and institutional facilities.   Therefore there is certainly scope and interest to have this tool to draft an overall national fisheries development plan.   Besides big private industries, if they exist at all, are not yet in the opportunity to apply modelling techniques.

This model was designed to simulate and cover all aspects of a small-scale fishery industry, i.e. resources, production, distribution, processing, investments, prices, incomes.   This approach makes the model rather complex, intransparent and more difficult to handle.

The division of the production into catch unit groups, which is maintained throughout the entire model structure, turned out to be somewhat artificial and difficult to manage.   The structure of the model should perhaps be more focussed on a division of sectors along the kind of products of the fisheries, i.e. fish species, wet fish consumed, processed fish.

Another disadvantage, as already stated before, is the required quantity of input.   It cannot be stressed enough that the model can only be applied meaningfully in case the input-data is realistic and reliable.   On the other hand the model can be used as an "excuse" and tool to set up a proper data collection system.

There is the danger that the model drafts a national fisheries development plan, which cannot be implemented, although necessary, due to lack of funds and administrational possibilities.   In the end the gap between "what is really possible" and development plan can become too wide and interest in the modelling may be lost.

Another constraint is the size of the required computer installation, which is not always available.

One of the biggest advantages of model design is the structural insight acquired by designers and users into the problems and possibilities of the entire small-scale fisheries sector.   It is

TABLE. OUTPUT OF MODEL (PER PLANNING YEAR)

Subject	Fisheries resources	Catching units
Capacity (in operation		number of units per catching unit type
Manpower		in sub-sector
Landings	grade of exploitation of max. sust. yields	average landings per day per catching unit type; and total per catching group; and total in sector
Utilizations of landings: - dumped - waste - consumed fresh - processed		per catching unit type and total in sector
- exported		fresh fish per catching unit type
Price (beach price)		per catching unit
Income		per fisherman per catching unit type
(Net) Earnings		per unit per catching unit type
Investment		in sub-sector
Number of new units		per catching unit type
Value added		per catching unit type
Foreign earnings		costs

TABLE (CONTINUED)

Handling/ Distribution	Processing/ Exports	National (Fisheries) Development
number of units	number of units per processing unit type	
in sub-sector	in sub-sector	in sector
		total demersal/pelagic landings
total sub-sector	per processing type unit	
	total export	
per unit	per unit per processing unit type	
in sub-sector	in sub-sector	total in sector
in sub-sector	per processing unit type	
in sub-sector	in sub-sector	
costs	costs; earnings	balance costs- earnings

felt by the staff of the project under which this work was undertaken, that discussions and research concerning model structure
and variables were very valuable.

In conclusion, it is believed that continuous development of
quantitative planning and decision-making tools is useful. However, it is also important not to over-estimate the possibilities
of quantitative methods, which should only be considered as helpful
additional tools for the fisheries development planner.

MODELING AND SIMULATION OF INTERDEPENDENT FISHERIES, AND OPTIMAL EFFORT ALLOCATION USING MATHEMATICAL PROGRAMMING

Lee G. Anderson[1], Adi Ben-Israel[2] Gary Custis[3] and Charles C. Sarabun[1]

1. College of Marine Studies & Department of Economics, University of Delaware, Newark, Delaware,19711,USA.
2. Department of Mathematical Science, University of Delaware, Newark, Delaware 19711, USA.
3. Gerber Scientific Instruments, Inc., South Windsor, Connecticut, 06074, USA.

INTRODUCTION

Recent technological developments (intensified fishing by modern fleets, sea pollution) and trends in international law (expanded territorial limits) underscore the role of fisheries (taken here to mean the stocks of fish and enterprises that exploit them) as an important resource to be optimally exploited and conserved. In the U.S.A., optimal management of fisheries is required by law, the Fishery Conservation and Management Act of 1976 (the "200 mile law"), and there, as elsewhere, the trend is toward more regulations and conservation measures.

Two approaches to the analysis of fisheries management are discussed here: simulation and mathematical programming. The simulation approach is necessitated by the complexity of inter-dependent fisheries, with their intricate and hard-to-quantify interactions. Of the tools presently available to the fisheries manager, simulation seems the best way to realistically evaluate and compare alternative management policies.

Mathematical programming can be used, under somewhat simplifying assumptions, to determine optimal harvesting and optimal effort allocation in fisheries, subject to relevant constraints on resources (e.g., fishing vessels, shore facilities, minimal employment). We propose using dynamic programming to determine the optimal harvest (or standardized effort) and then allocate it

421

optimally, using linear programming, among the various vessels.

Computer programs of both the simulation model (Sarabun[1]) and the optimal effort-allocation (Custis[2]) are available from the authors.

MODELING AND SIMULATION OF INTERDEPENDENT FISHERIES

The problem of interdependent fisheries is one of the most complex facing modern fisheries management agencies. Fisheries can be interdependent for technological and biological reasons. They are biologically interdependent if the different stocks of commercial fish compete in any way for food, habitat, etc., or if they have a predator-prey relationship. Fisheries are technologically inter-dependent when the harvest of one stock can lead to the harvest, intentional or not, of a by-catch of another stock. The purpose of this simulation model is to allow for an analysis of these issues in a systematic way. The model is based upon earlier simulation models of independent fisheries by Huppert et al.[3] and Devanney et al.[4] but it is different in that it can handle both types of interdependencies. It is also different from other inte-grated simulation models such as in Andersen and Ursin[5] in that specific attention is focused on the activities of the harvesting sector. Due to constraints on space, the description of the model and its applications is necessarily limited. It will only be possible to sketch out the basic components of the model and show one application.

The model is divided into a biological sector and an economic sector with the link between them being the amount of fishing days produced by the economic sector. This is described schematically in Figure 1. The economic sector consists of different fleets which direct their effort at particular species of fish. Each fleet can be composed of more than one vessel type. Profit is the driving force in the economic sector. In trying to maximize profits, the various vessels produce fishing effort which, when applied to the fish stock, obtains certain amounts of catch. The exact amount caught will depend upon the catchability coefficient of the vessel and the size of the stock at that particular time. The amount of fish sold determines the profit (or loss) for the fishermen and consumer surplus for the purchasers of the fish.

The biological sector contains the stocks of fish which are exploited. Stocks are measured in numbers of fish and are broken down into age groups or cohorts. Recruitment, aging, and natural mortality are described in this sector.

The link between the two sectors is effort, which is produced in the economic sector, and impacts the biological sector through

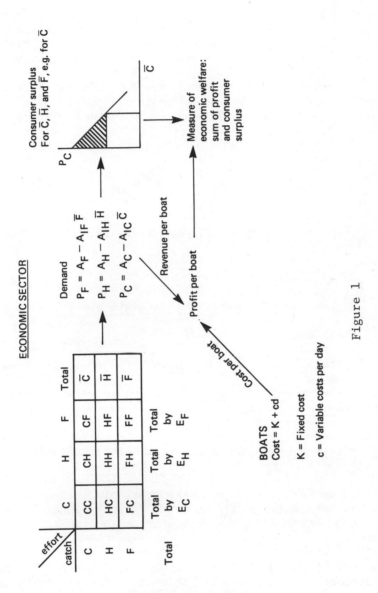

Figure 1

fishing mortality.   The number of fish leaving the biological
sector as a result of this mortality, is converted to weight as it
enters the economic sector.

One of the main problems with this model is to set appropriate
values for the parameters. Particularly difficult to specify are
the catchability coefficient and the parameters which provide for
a comparison of the relative power of the various vessels.

The biological sector of the model is described by a recruitment function and by the following system of differential equations
describing the interactions between the various species/cohorts and
the effects of fishing mortality.

$$\frac{dN_i}{dT} = \sum_{j=1}^{m} A_{ij} N_j \qquad\qquad i = 1,\ldots,m$$

Here

m the total number of cohorts (of all species combined)

$$m = \sum_{k=1}^{s} m_k$$

m_k the number of cohorts in the kth species, $k = 1,\ldots,s$

s the number of species

N_i the size of the ith cohort (in numbers of fish), $i=1,\ldots,m$

A_{ij} the coefficient of interaction between the jth and ith
           cohorts ($A_{ij}N_j$ is the net change in the ith cohort due
           to the jth cohort), in particular,

$A_{ii} = -(F_i + M_i)$ the coefficient of mortality (fishing and
                      natural) for the ith cohort, $i=1,\ldots,m$

Because we were unable to obtain reliable estimates of the
off-diagonal interactions A_{ij} ($i \neq j$), default values of zero are
currently used pending further research or additional biological
information.

The recruitment options are Ricker, Beverton-Holt, linear
density dependent and bi-modal. Each has a deterministic and
stochastic mode. See Barnes[6].

Figure 2 provides a more detailed breakdown of the economic
sector and its link to the biological sector. The basic elements
in the economic sector are the fleets each directing its effort at
particular species/cohorts. Each fleet can be broken down into

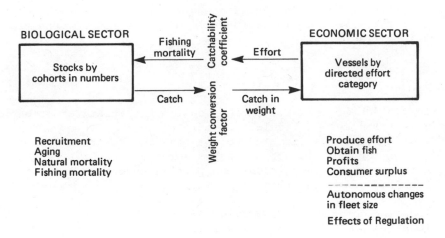

Figure 2.

specific vessel types, say by the size of the boat, its ability to
handle the expected weather conditions and the expected maintenance
days required.   The numbers of days fished, d, is assumed equal
for all vessels of the same type. The effort per vessel, e, is
λd, where λ is a relative fishing power coefficient. The total
amount of effort produced by each vessel type is ne where n is the
number of vessels. Using the relative power coefficients is
necessary to express effort of the various types of vessels in
common terms so that they may be added to obtain the total directed
effort of the fleet. These are represented in the diagram as
capital E's.

    In order to consider technological interdependence between the
stocks, fishing mortality for each species/cohort is determined as
a function of the sums of efforts (directed and nondirected), of
all fleets.   For example, the fishing mortality on the ith cohort
of the stock, F_i, is a linear function of the total efforts from
the various fleets,

$$F_i = \Sigma \; \phi_{ij} E_j .$$

The coefficient of each E_i (ϕ_{ij}) is called the relative catchability
coefficient for the particular jth effort and the ith cohort. For
practical purposes we have had to use the same relative catchability
coefficient for each cohort of each species although it is possible
to consider partial recruitment and partial susceptibility if we
can obtain adequate information on it.

    Figure 3 contains another view of the economic sector which
describes the profit and social welfare effects. As the effort
is applied to the stocks, catches from the various species are
obtained. The total catch from any species can be broken down into
directed catch by the effort of the fleet involved and bycatch of
other fleets. This is specified in the matrix in the left hand
corner of the figure.   The diagonal elements represent the directed
catches of cod, haddock, and flounder, respectively, while the off
diagonals are bycatch.   For example in the first row, the first
element is a directed catch of cod by the cod fishery, the second
element is the bycatch of cod in the haddock fishery, and the third
element is the bycatch of cod in the flounder fishery. The final
element is the total amount of cod harvested.

    The price of fish of each species is determined by a demand
curve which shows the relationship between price and the amount of
fish landed. It is possible that the price can differ depending
upon whether the fish is landed as a directed catch or as a bycatch
and the program can handle this if necessary. Given the amount
of total catch, and hence the price, of each species, revenue per
boat is expressed as price times the share of total catch attribut-

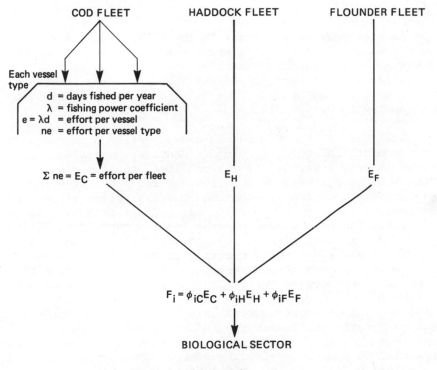

Figure 3

able to each vessel. Given a cost function which takes into account
fixed and variable cost per day fishing, it is possible to determine
the profit level per boat. Clearly, different types of vessels will
in general have different profit levels.

Given the demand curve, it is also possible to determine
consumer surplus which is the area between the demand curve and the
price line. A correct measure of the social value obtained from
fishing in terms of pure economic efficiency is the sum of consumer
surplus and the profits (Anderson[7] p.65-72). Our simulation model
provides a measure of this sum. By determining how it varies with
different regulation techniques or with different numbers of boats,
etc. it is possible to compare various management programs.

It is also possible to simulate the following fisheries management options: quotas for the fishery as a whole or for particular
fleet/vessel types, landing taxes, or taxes on days fished;
restrictions on the number of vessels, their fishing power or the
number of days fished by each. It is also possible to simulate
entries and exits of vessels.

In such a short paper a complete description of a run of the
simulation model cannot be provided. (Sample runs are available
from the authors, however.) It is possible, nonetheless, to
describe some sample outputs and Figure 4 contains three important
tables from a study of a hypothetical cod and yellowtail flounder
interdependent fishery.   The years covered are from 1978 to 1992,
however these tables are for the year 1981.   The first one shows
the type and number of vessels that direct effort at cod and
yellowtail flounder during each of the years. It also shows the
revenue, cost, and profit per vessel and the average price of fish
(determined by the demand curve) for the given year.

The next table shows current value of the social welfare
indicators (profit and consumer surplus) as well as the sum of the
present values up to that time. That is, the 1978 present value of
total social surplus as of 1982 is $20,102,000 and the current
value in 1981 is $16,219,000. (See also Figure 5.)

The third table shows the directed catch, by-catch, and total
catch of each of the species by each of the vessel classifications.
For this hypothetical example, the smallest yellowtail vessel
catches 97 metric tons of yellowtail and 24 metric tons of cod.

OPTIMAL EFFORT ALLOCATION USING MATHEMATICAL PROGRAMMING

The problem here is to determine management policies (e.g.
harvests, effort allocations) that are optimal in the sense of
maximizing a prescribed criterion (e.g. fishermen net income plus

ECONOMIC SUMMARY FOR 1981

DIRECTED SPECIES	VESSEL TYPE	VESSEL NUMBER	REVENUE	COST	PROFIT	CREW SHARE
			(DOLLAR FIGURES IN THOUSANDS OF DOLLARS)			
YELLOWTAIL	0-60 TONS	70	$109.	$70.	$39.	$0.
YELLOWTAIL	61-125 TONS	31	$131.	$110.	$21.	$0.
YELLOWTAIL	125+ TONS	17	$187.	$144.	$43.	$0.
TOTAL YELLOWTAIL	AVERAGE PRICE WAS $1.02 PER KG		$14,869.	$10,745.	$4,124.	$0.
COD	0-60 TONS	26	$99.	$70.	$30.	$0.
COD	61-125 TONS	14	$119.	$110.	$9.	$0.
COD	125+ TONS	6	$170.	$144.	$27.	$0.
TOTAL COD	AVERAGE PRICE WAS $0.36 PER KG		$5,278.	$4,220.	$1,058.	$0.

SOCIAL WELFARE SUMMARY FOR 1981
(DOLLAR FIGURES IN THOUSANDS OF DOLLARS)

SPECIES	PROFIT		CONSUMER SURPLUS		SUM	
	CURRENT	PRESENT VALUE	CURRENT	PRESENT VALUE	CURRENT	PRESENT VALUE
YELLOWTAIL	$4,124.	$8,600.	$6,323.	$27,865.	$10,447.	$36,465.
COD	$1,058.	$2,562.	$4,714.	$11,792.	$5,772.	$14,353.
TOTAL	$5,182.	$11,162.	$11,037.	$8,940.	$16,219.	$20,102.

FISHING CATCH-BYCATCH BREAKDOWN FOR 1981

DIRECTED SPECIES	VESSEL TYPE	NUMBER OF VESSELS	CATCH-BYCATCH		TOTAL CATCH
			YELLOWTAIL	COD	
YELLOWTAIL	0-60 TONS	70	9.7597E+01	2.4955E+01	1.2255E+02
YELLOWTAIL	61-125 TONS	31	1.1712E+02	2.9946E+01	1.4706E+02
YELLOWTAIL	125+ TONS	17	1.6716E+02	4.2741E+01	2.0990E+02
TOTAL CATCH BY YELLOWTAIL EFFORT			1.3304E+04	3.4018E+03	
COD	0-60 TONS	26	1.8848E+00	2.6783E+02	2.6971E+02
COD	61-125 TONS	14	2.2618E+00	3.2139E+02	3.2365E+02
COD	125+ TONS	6	3.2282E+00	4.5872E+02	4.6194E+02
TOTAL CATCH BY COD EFFORT			1.0004E+02	1.4215E+04	

Figure 4

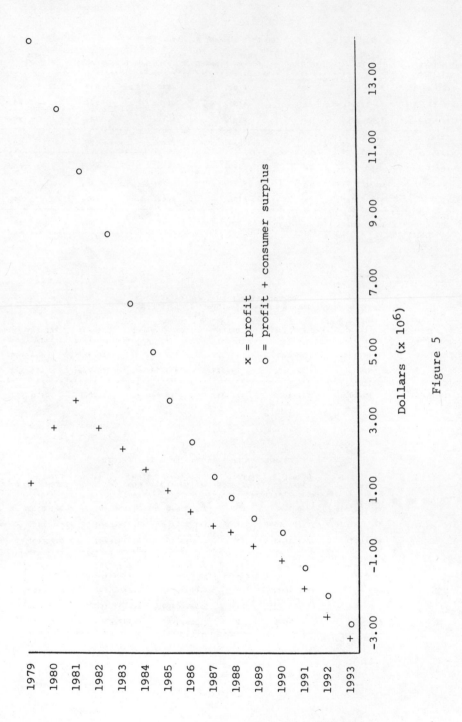

Figure 5

consumer surplus) subject to relevant constraints on resources (e.g.
fishing vessels, shore facilities, minimum employment).

   While simplified fishery models can be analyzed using modern
optimal control theory, e.g. Clark[8]; Cliff and Vincent[9]; Quirk and
Smith[10]; a less elegant and more tractable approach is required by
the realistic fishery model considered here.

   Here, <u>fishing effort</u> measured in fishing days per standard
<u>period</u> (e.g., month, year) is allocated to each vessel according
to its type, so as to maximize the discounted present value of the
fishery, subject to constraints of biological (conservation,
technological (capacity) and political (e.g., employment) type.
We denote by:

i   the <u>number of periods remaining in the planning horizon</u>.
        A reasonable approximation to the optimal policy and value
        in an infinite-horizon model is indicated by the converg-
        ence, with say i = 10 years to go, of optimal policies and
        values in an finite-horizon model (say 20-30 years).

S   the <u>fish stock</u> (in biomass or numerical count), a scalar
        (for a single species) or a vector (for several species).
        S is discretized, so that actual values of S are rounded
        to the nearest values S_k from among finitely many stock
        levels $(S_1, S_2, \ldots, S_K)$.

AS   the <u>set of admissible stock levels</u>, a subset of $(S_1, S_2,$
        $\ldots, S_K)$ obtained by excluding levels which are too low
        (at which the species would be endangered) or too high
        (beyond the carrying capacity of the environment). The
        set AS would have to be specified by biologists. The set
        AS is assumed, for simplicity, to be constant in time.
        With minor programming changes, time-dependent admissible
        stock levels, AS(i), may be incorporated.

V   the <u>number of different vessel types</u>

N_v the <u>number of vessels</u> (assumed identical) of <u>type v = 1,</u>
        $\ldots$, V.

D_v the <u>number of fishing days allocated to each vessel of</u>
        <u>type v</u>

E   the <u>fishing effort</u> for the whole fishery per period, given
        in standard units as

$$E = \sum_v N_v D_v \lambda_v$$

where
$\quad\quad \lambda_v \quad$ the fishing power <u>rating of vessel type</u> v.

The effort E is also discretized and limited to one of the values
$(E_1, E_2, \ldots, E_J)$. Further we denote by

$\quad\quad T_j(S_k)$ the <u>stock in any period resulting from applying effort</u>
$\quad\quad\quad E_j$ <u>to stock S_k in the previous period</u>. (Again this is
$\quad\quad\quad$ assumed stationary in time, but can be made time-depend-
$\quad\quad\quad$ ent, say $T_j(S_k, i)$, with only minor changes).

$\quad\quad E(S_k)$ the <u>set of admissible efforts at stock level</u> S_k, i.e.,
$\quad\quad\quad E(S_k) = (E_j : T_j(S_k) \epsilon \ AS).$

$\quad\quad R(S_k, E_j)$ the <u>return</u>, in each period of applying effort E_j
$\quad\quad\quad$ to stock S_k. This return consists of current profit
$\quad\quad\quad$ to fishermen, including depreciation of vessels, as well
$\quad\quad\quad$ as benefit to society in the form of consumer surplus.
$\quad\quad\quad$ This is the same measure of social value used in the
$\quad\quad\quad$ simulation model.

$\quad\quad \alpha \quad$ the <u>interest rate</u> in each period

$\quad\quad OPV_i(S_k)$ the <u>optimal (discounted) present value</u> of the fishery
$\quad\quad\quad$ with stock level S_k and i periods to go.

Using Bellman's optimality principle[11] the optimal present
value can be shown to satisfy the functional equation

$$(*) \quad OPV_{i+1}(S_k) = \max_{E_j \epsilon E(S_k)} \left(R(S_k, E_j) + \frac{1}{1+\alpha} OPV_i(T_j(S_k)) \right)$$

The dynamic programming solution of (*) yields simultaneously
the OPV and the optimal (standardized) effort E for each stock
level S_k and time i. The optimal return $R(S_k, E_j)$ in (*) is found
by the optimal allocation of the effort among the various vessel
types, determined here by solving the linear programs:

(LP) Find (D_v) so as to

$$\text{minimize} \quad \sum_{v=1}^{V} C_v N_v D_v$$

(1) $$\sum_{v=1}^{V} N_v D_v \lambda_v = E_j$$

(2) $$L_m \leq \sum_{v=1}^{V} A_{mv} D_v \leq U_m, \quad m = 1, \ldots, M$$

$$D_v \geq 0, \quad v = 1, \ldots, V$$

where

C_v - the <u>cost per fishing day in vessel type v,</u>

the constraint (1) sets the total fishing effort at E_j, and constraints
(2) express the technological and physical limitations of the fleet
and shore facilities, as well as political requirements for minimal
employment, etc. A linear program (LP) has to be solved for each
admissible effort E_j encountered in the solution of the dynamic
program (*). These linear programs are small (M+1 constraints,
V variables) and their solution requires a small part of the total
computational effort, while achieving the aggregation of the various
vessel type - efforts ($N_v D_v$: v = 1, ..., V) into one decision
variable, the (standardized) effort, E. (Linear programming is thus
used here to overcome the "curse of dimensionality" of dynamic
programming, which would otherwise restrict the number V of decision
variables to 3-4.)

Alternatively, quadratic programming is used if costs are
assumed to be quadratic functions, a generalization which requires
only a minor additional computational effort, since the methods of
linear programming (e.g., the Simplex Algorithm) are easily adapted
to solve quadratic programs, e.g., Gass[12].

The optimal present value calculated above is the maximal
present value of the net addition of goods produced (fish) over
time. The corresponding optimal effort allocation over time
represents an ideal policy given that economic efficiency is the
main criterion of management, subject to the constraints as noted
above.

Our approach differs from other applications of dynamic
programming in fisheries management (e.g., Clark[13]; Mendelssohn
and Sobel[14]; Mendelssohn[15]; Sancho and Mitchell[16]; Smith and
Silvert[17] in the consideration of more than one species, and several
vessel types whose efforts ($N_v D_v$: v = 1, ..., V), the decision
variables in this optimal allocation problem, are aggregated by
using the auxilliary linear programs (LP), or quadratic programs.
Another novelty is that the stock transforms ($T_j(S_k)$ are determined,
for each pair of population-effort levels ((S_k, E_j): k = 1, ..., K,
j = 1, ..., J)), by simulation, allowing for the incorporation of
realistic assumptions regarding the dynamics and recruitment of the
populations in question, and the economic sector.

The optimal effort allocation program DYFISH consists of three
FORTRAN programs:

    (i)   DYNPR2   a dynamic program for maximizing the optimal
                               present value of the fishery

    (ii)   FISHES   a single step simulator

(iii) LP or QP   auxilliary programs for minimizing a
linear or a quadratic cost, respectively, subject to
linear constraints.

To illustrate how DYFISH works, consider a two-species fishery
with two-fleets each consisting of vessels of three types. Taking
10 admissible population levels for each species and 6 effort levels
for each fleet, the fishery is described by two grids, the 10 x 10
population nodes (numbered 1,..., 100) and the 6 x 6 effort nodes
(numbered 1, ..., 36).

From a given population pair S_k, (k = 1, ..., 100), one can
reach, in one step, any S_ℓ in T_j (S_k), (j = 1, ..., 36), where in
general more than one effort pair E_j connects S_k and S_ℓ. For each
S_k (k = 1, ..., 100) and S_ℓ (which can be reached from S_k in one
step) the simulator FISHES determines the optimal effort E_j
(taking S_k into S_ℓ), denoted by

$$(k \to \ell) = j$$

in the following output of FISHES:

EFFORT BETWEEN NODES

(1 → 1) = 28   (1 → 2) = 27 . . . . . (1 → 22) = 3
(1 → 23) = 1 . . . . . (2 → 1) = 30 . . . .
. . . .
. . . .
(100→85) = 12   (100→86) = 11 . . . . .(100→97) = 3
(100→98) = 2   (100→99) = 1

Thus from S_1 only S_1, S_2, ..., S_{23} can be reached in one step, and
E_{27} is the optimal effort taking S_1 into S_2.

For each effort combination (E_j: j = 1, ..., 36), the auxilliary
program LP or QP determines the optimal allocation of fishing days,
to each of the 6 vessel types (three types in each fleet). A sample
result for the effort level E_4 = (8841.124, 16980.382) is

TOTAL EFFORT BY SPECIES 1 = 8841.124
TOTAL EFFORT BY SPECIES 2 = 16980.382

VESSEL TYPE	NUMBER	TIME PER VESSEL	TOTAL FIXED COST	TOTAL VAR COST	TOTAL COST	FRACTION
1	55	59,2932	570130	1457506	2037636	.3689
2	47	50.0000	930036	1612100	2542136	.3190
3	40	50.0000	880000	1532000	2412000	.3122
4	55	150.0000	570130	3712500	4282630	.4859
5	47	50.0000	930036	1612100	2542136	.1661
6	40	107.0721	880000	3280690	4160690	.3481

THE TOTAL COST FOR SPECIES 1 is 6991772
THE TOTAL COST FOR SPECIES 2 is 10885456

Where species 1 and 2 are harvested by vessel types (1, 2, 3) and
(4, 5, 6) respectively. Here 47 vessels of type 5 are each allocated 50 fishing days per year, at a total cost of $2,542,136. The
effort E_4 is optimally allocated at a minimal cost of $6,991,772 +
$10,985,456. Following this preparatory work, the dynamic program
DYNPR2 finally yields, for a given planning horizon (here 25 years)
and interest rate here(10%), the optimal present value $OPV_{25}(S_k)$
for each k = 1, ..., 100, summarized in the following table.

TIME HORIZON IS 25 YEARS. DISCOUNT FACTOR 0.10.

FROM NODE	PRESENT VALUE
1	56,631,744
2	58,340,396
-	
-	
-	
-	
-	
99	123,708,130
100	125,349,730

The solution is represented in terms of the optimal population
paths

OPTIMAL POPULATION LEVELS FOR THE GIVEN REPORTING PERIOD. THE
BEGINNING NODE IN LEFT MOST COLUMN

0	1	2	3	4	5	6	7	8	9	10
1	23	45	56	67	77	77	77	77	77	77
2	24	45	56	67	77	77	77	77	77	77
.	.	.	.	.	.	.	.	.	.	.
.	.	.	.	.	.	.	.	.	.	.
99	98	98	98	98	98	98	98	98	98	98
100	99	98	98	98	98	98	98	98	98	98

showing rapid convergence (5 years) to one of six optimal equilibrium levels (77, 78, 87, 88, 97, 98). These optimal paths are
represented graphically in Fig. 6.

Thus, for example, from population node 1 (the left most node
in Fig. 6) the optimal path is

$$1 \rightarrow 23 \rightarrow 45 \rightarrow 56 \rightarrow 67 \rightarrow 77$$

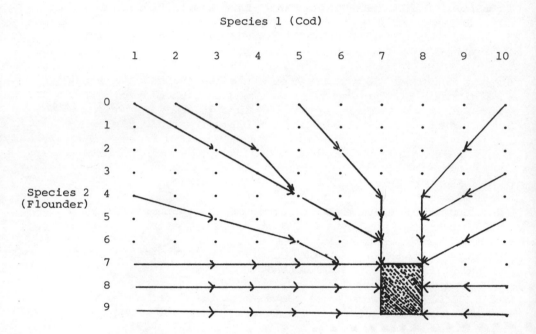

Figure 6.   Optimal Population Paths

The six equilibria (77, 78, 87, 88, 97, 98) bound a region
(the shaded area in Fig. 6) which is likely to contain the "true"
equilibrium point. By concentrating on a small part of the
population grid (containing the shaded area), discretizing it and
applying DYFISH again, the solution is "fine-tuned", obtaining a
better approximation of the optimal equilibrium population.

REFERENCES

1. C.C. Sarabun, FISH2: Program and Users Guide, University of
    Delaware, (1979).
2. G. Custis, DYFISH: Program and Users Guide, University of
    Delaware, (1979).
3. D. Huppert, D. Brown and V. Hongskul, NORSIM II: An Expanded
    Simulation Model for the Study of Multispecies Exploitation,
    NORFISH Technical Report 40, University of Washington,(1973)
4. J.W. Devanney , H. Simpson and Y. Geisler, The MIT Single-
    Species Fishery Simulator, Massachusetts Institute of
    Technology, (1977).
5. K.P. Andersen and E. Ursin, A Multispecies Extension to the
    Beverton and Holt Theory of Fishing, with Accounts of
    Phosphorus Circulation and Primary Production, The Danish
    Institute for Fishery and Marine Research, 1977.
6. V. Barnes, Recruitment Modes for FISH2 (University of Delaware,
    1979).
7. L.G. Anderson, Economics of Fisheries Management, (Johns
    Hopkins University Press, Baltimore, 1977).
8. C.W. Clark, Mathematical Bio-economics. (John Wiley and Sons,
    New York, 1976a).
9. E.M. Cliff and T.L. Vincent, An Optimal Policy for a Fish
    Harvest, J. Optimiz. Th. Appl., 12, (1973), 485-496.
10. J.P. Quirk and V.L. Smith, Dynamic Economic Models of Fishing,
    Economics of Fisheries Management, University of British
    Columbia, (1970), 3-32.
11. R.E. Bellman, Dynamic Programming, (Princeton University Press,
    1957).
12. S. Gass, Linear Programming: Methods and Applications, (4th
    edition, McGraw-Hill, 1975).
13. C.W. Clark, A Delayed-Recruitment Model of Population Dynamics,
    with an Application to Baleen Whale Populations, J. Math.
    Biology, 3, (1976b), 381-391.
14. R. Mendelssohn and M.J. Sobel, Capital Accumulation and the
    Optimization of Renewable Resource Models, (to be printed).
15. R. Mendelssohn, Optimal Harvesting Strategies for Stochastic
    Single-Species, Multiage Class Models, Math. Biosciences,
    (to be printed).
16. N.G.F. Sancho and C. Mitchell, Economic optimization in controll-
    ed fisheries". Math. Biosciences, 27, (1975), 1-7.

17. W.R. Smith and W. Silvert, Dynamic Programming Techniques in
        Fisheries Management, Fisheries Management Project Volume 2,
        Institute for Environmental Studies, Dalhousie University,
        Halifax, N.S. Canada (no date).

ACKNOWLEDGEMENTS

     Research supported by Sea Grant 04-6-158-44025 to University
of Delaware.

ADDITIONAL PAPERS

THE VALUE OF CATCH-STATISTICS-BASED MANAGEMENT TECHNIQUES FOR

HEAVILY FISHED PELAGIC STOCKS WITH SPECIAL REFERENCE TO THE RECENT

DECLINE OF THE SOUTHWEST AFRICAN PILCHARD STOCK

D.S. Butterworth

Department of Applied Mathematics,
University of Cape Town,
Rondebosch, 7700, South Africa.

INTRODUCTION

The relatively short life-span of pelagic species, coupled with substantial year-class size fluctuations, makes it unlikely that management on the basis of models hypothesising a steady-state situation will prove adequate. Harvesting in excess of very conservative levels of exploitation necessitates monitoring of the current biomass of the stock.

Catch statistics, that is catch and effort data, and allied length and age distribution data from the catch samples, can provide measures and indicators of biomass trends. These statistics have the advantage of being relatively cheap to collect, compared to data for direct-survey-type assessments such as egg/larvae, acoustic, or aerial methods.

Catch-statistics-based biomass estimates are however known to be subject to error or ambiguous interpretation. The essential test is whether they can nevertheless provide adequate accuracy for successful management.

The problems inherent in catch-statistics biomass indicators are briefly reviewed; then the case of the Southwest African Pilchard which has recently exhibited "a classic case of collapse" (ICSEAF Scientific Advisory Committee 1978) is considered to establish whether the catch-statistics-based indicators exhibited adequate advance warning of this.

CATCH–STATISTICS–BASED BIOMASS INDICATORS

Catch per Unit Effort

One may define the fishing mortality F by the equation

$$\frac{dC}{dt} = FN \quad \ldots \quad (1)$$

(C = number of fish captured
(
(N = number of fish in population

In principle F is a function of the fishing effort E, the number
of fish in the population N (or the biomass), and the other factors
usually classed as "environmental".

The use of Catch per Unit Effort (CPUE) as a biomass index is
based on the assumption that F is linearly proportional to E. However in pelagic context, the concept of measuring effort becomes
problematic. Also the F ∝ E hypothesis is open to question (see
Ulltang[10] and references therein), for example due to stock
concentration in relatively few shoal-groups in specific areas.

The possible existence of "Paloheimo-Dickie"[7] effects for
pelagic stocks, whereby the ^{F}/E ratio rises as the biomass falls,
is significant as it means the CPUE index may not fully reflect a
stock decline

Virtual Population Analysis

VPA is most readily illustrated by Pope's[8] approximation
(corresponding to pulse-fishing at mid-year) which is reasonable
for small values F and M. The initial size of a year-class is
given by

$$N_0 = C_0 e^{M/2} + C_1 e^{3M/2} + C_2 e^{5M/2} + \ldots + C_n e^{(\frac{2n+1}{2})M} +$$

$$N_{n+1} e^{(n+1)M} \quad \ldots \quad (2)$$

where if the year-class initiates in year 0,
N_i = number of fish of age "i" present at start of year "i"

C_i = number of fish of age "i" captured in year "i".

As $N_k \sim e^{-k(F+M)} N_0$, the series converges with the first few
terms usually the only ones to make significant contributions.

The method therefore gives good estimates of stock size   in previous years, but for more recent years the portion of the series for which data is available has not yet converged. One has usually to turn again to effort measures to estimate F and thence the biomass for the current year. VPA estimates are thus increasingly less accurate for more recent years.

Mean (Modal) Lengths

In the steady-state situation, heavier fishing mortality will result in a shorter average life-time (and hence length) for the fish. Thus a drop in the adult modal length would be an indicator of stock decline, and conversely.

On its own however, this index is ambiguous for pelagic species due to large fluctuation in year class size. A drop in the adult modal length may in fact be a reflection of entry of a particularly strong incoming year-class, rather than an increasing exploitation level, for example.

A BRIEF HISTORY OF THE SOUTHWEST AFRICAN PILCHARD FISHERY

The Southwest African Pilchard stock has been harvested predominantly by a local industry based in Walvis Bay. Fishing was initially in the immediate vicinity of Walvis Bay, but later expanded to cover most of the inshore region of ICSEAF Division 1.4 and the northern portion of Division 1.5 (see Figure 1). Recently fishing has extended into Division 1.3 as the stock has declined.

Commercial exploitation of the Southwest African Pilchard commenced shortly after the war, and during the 1950's quotas were set at a conservative level (see Figure 2). However during the 1960's succeeding quota increases were permitted until, after reaching a record annual catch of 1390 thousand tonnes in 1968, landings dropped sharply to only 510 thousand tonnes in 1970.

Stricter controls and an intensive research program (the "Cape Cross program") were instituted towards the end of 1970. Stock improvement was detected, predominantly on the basis of aerial survey techniques utilised from 1972 to 1974, and quotas rose from 330 to 560 thousand tonnes.

Midway through the 1976 season reported scarcity of pilchard led to a 20% quota reduction. Aerial surveys recommended early in 1977 and failed to reveal any major pilchard concentration, following which the quota was reduced to 250 thousand tonnes, and later during that season further to 200 thousand tonnes.

A 1978 quota of 125 thousand tonnes proved arbitrary, as

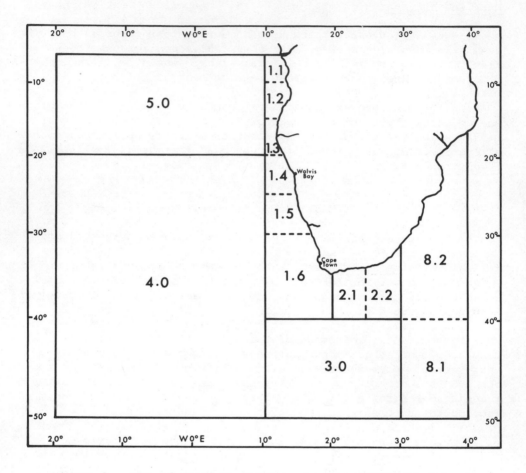

Figure 1:   Map of Southern Africa showing ICSEAF Divisions.

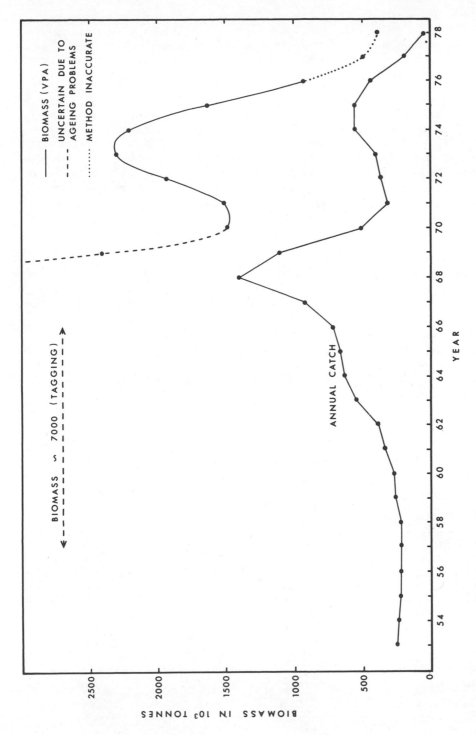

Figure 2: Annual landings of Pilchard by the local Southwest African industry from 1953 to 1978, together with VPA Biomass estimates from 1969 to 1978.

directed pilchard fishing was curtailed in mid-season with the
catch at only 46 thousand tonnes – the "classic case of collapse".
The 1979 season has confirmed this bleak picture.

ASSESSMENT OF THE SOUTHWEST AFRICAN PILCHARD STOCK

The Biomass estimates for the stock subsequent to 1970 have
been calculated by VPA (see Figures 2 and 4). Unfortunately age
data prior to this is inadequate to permit confident extension of
this method to earlier years. Tagging returns yield an estimate of
the order of 7 million tonnes over the period 1957-67 (Newman[6]),
but this figure is very sensitive to the value assumed for initial
tagging mortality.

Technical details of the VPA calculation used are elaborated
elsewhere (Butterworth[1], Butterworth and le Clus[2]). Fishing mortalities for the current year (1978) have been set in proportion to
values in previous years according to effort ratios (achieved
iteratively):

$$F_i^{78} = E^{78} \frac{1}{6} \sum_{y=72}^{77} (F_i^y/E^y) \quad \ldots \quad \ldots \quad \ldots \quad \ldots \quad (3)$$

where

$$F_i^y = \text{fishing mortality or "i" year old fish in year "y".}$$

$$E^y = \text{fishing effort in year "y"}$$

The effort index used is vessel-hours at sea, taking only
pilchard directed fishing into account, and weighted by vessel
hold-capacity. Directed effort has been (semi-arbitrarily) defined
as trips where the vessel's landing comprised more than 50% of the
species concerned.

The projections shown in Figure 4 have been calculated using
the average recruit/adult stock ratio over previous years to estimate
the incoming year-class. The relative biomass increase shown in
this figure for 1979 may well be spurious, for reasons given in
the Section following entitled CPUE.

The CPUE index shown in Figure 6 is for directed catch divided
by directed effort weighted for hold-capacity. It has been normalised to the mean biomass given by VPA over the 71-78 period for
comparative purposes.

The adult-stock recruit plot shown in Figure 3 clearly indicates
the reasons behind the stock's initial increase then catastrophic
decline in the 1970's. Improvements over the 71-74 period were

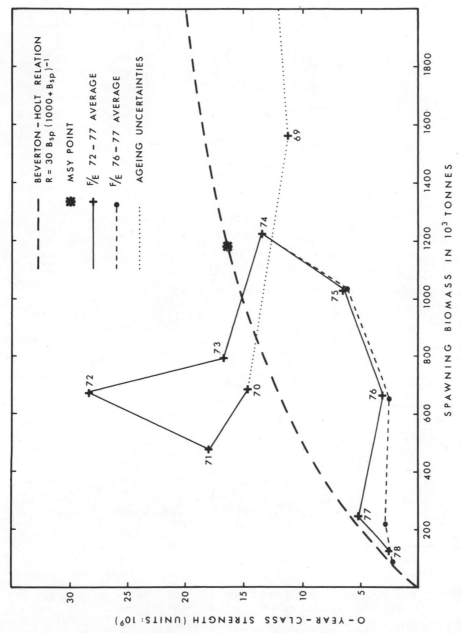

Figure 3: Spawning stock – recruit plot for the Southwest African Pilchard from 1969 to 1978; a Beverton-Holt relation eyefit to the data is shown.

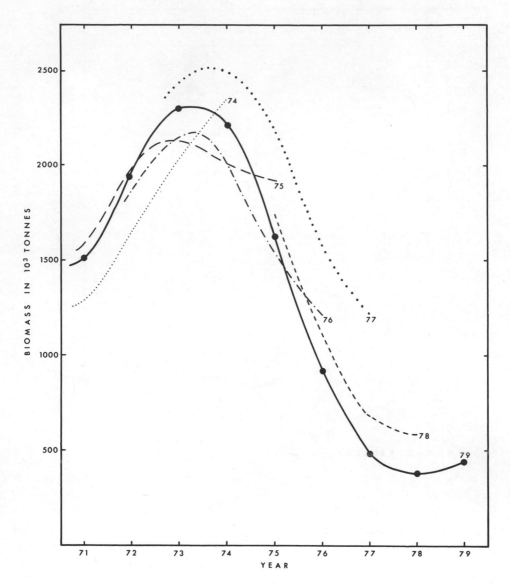

Figure 4: VPA Biomass estimates and projections as calculable at
          the start of successive years 1974–1979.

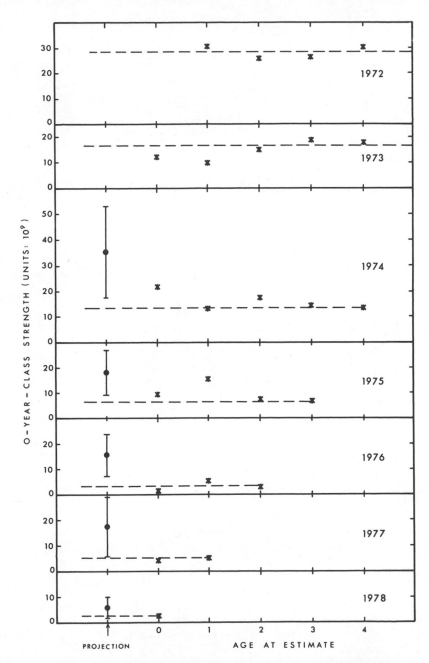

Figure 5: The accuracy of the O-year-class strengths as determined by VPA successive years after their entry to the fishery: 1972-1978.

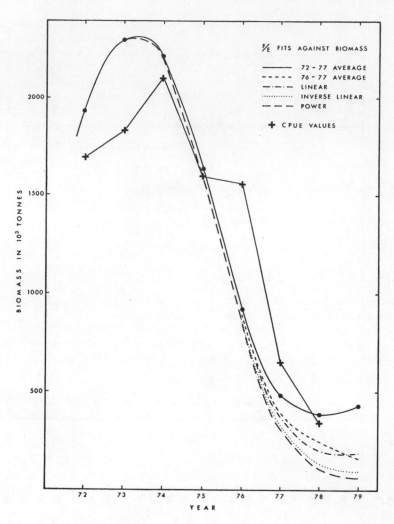

Figure 6:   VPA Biomass estimates from 1972 to 1978 and projections
            for various prescriptions for the determination of F

            values for the current year. Plots are for $\bar{F}/E$ the

            the average of values over previous years, and for $\bar{F}/E$
            negatively correlated with biomass (Paloheimo-Dickie
            effect) for parametrizations given by equation (5).
            Comparison with CPUE values refers to directed catch per
            directed hours at sea weighted for hold capacity and
            normalised to the mean VPA estimate over the period.

predominantly the result of a particularly strong 1972 year-class; fishing pressure on the stock left inadequate reserves to offset poor year-classes in 1975 and particularly in 1976.   Thereafter the biomass had fallen to levels at which depensatory effects may well play additional damaging roles.

Such an analysis is easily stated well after the event when it is probably too late to take management action that will lead to significant improvements. The relevant question is whether the catch-statistics-based biomass indicators could have given warning of the decline in time to take remedial action.

THE DETECTABILITY OF THE BIOMASS DECLINE

VPA

Figure 4 shows VPA Biomass estimates for the stock that would have resulted from calculations on data available at the commencement time of previous seasons.

It is noticeable that at the start of 1974, VPA projects a continuing stock increase whereas the stock had in fact slightly declined. A year later in the presence of a strong stock decline, VPA gives only a weak indication of such (in fact without taking hold-capacity weighting into account in the effort index, VPA still projects a slight increase).

Only at the start of 1976 does VPA give a clear indication of stock decline, by which time the stock had already dropped from $2\frac{1}{4}$ to less than 1 million tonnes.

With what success did VPA indicate the weakness of the 1975 and 1976 year-classes, whose role was crucial in the collapse? Figure 5 shows that only by the start of 1978 was the weakness of the 1975 year-class confirmed. The 1976 year-class weakness was indicated by 1977, but by then the stock had dropped to $\frac{1}{2}$ million tonnes.

Adult Modal Length

Table 1 shows that the adult modal length rose steadily from 1970 to 1976, broadly indicative of continuing health of the stock. In fact the rise over the latter part of this period was not a reflection of a low exploitation rate at all, but of the last stages of, in particular, the strong 1972 year-class being followed by successively weaker year-classes.

Only during 1977 did this index first reflect directly the fall in the stock size.

Table 1. Adult Pilchard Modal Length Size in the
          Commercial Catch, compared to subsequent VPA
          Biomass estimates

YEAR	Modal Length (cm)	Biomass (10^3 tonnes)
1970	19.2	1500
1971	20.2	1510
1972	20.2	1930
1973	20.2	2300
1974	20.7	2220
1975	20.7	1640
1976	21.2	920
1977	20.2	490
1978	19.2	380

CPUE

Figure 5 shows the CPUE values for each year. At the start
of 1975 an increasing stock was still indicated (really a reflection
mainly of full recruitment of the 1972 year class the previous
year).   The drop for the 1975 season would not have seemed too
serious as CPUE was still at a level comparable to 1972 and 1973 –
surprisingly this level maintained for 1976 as well in spite of
severe stock decline.

Only during 1977 did CPUE values clearly reflect the major
stock decline.

The maintenance of the CPUE value during 1976 merits closer
attention. Figure 7 plots the $\bar{F}/E$ ratio against biomass. $\bar{F}$ is a
weighted average of the fishing mortalities VPA calculates for the
various age-classes in a given year.

The large (compared to the stock size) CPUE value for 1976 is
reflected by a $\bar{F}/E$ value in Figure 7A considerably greater than the

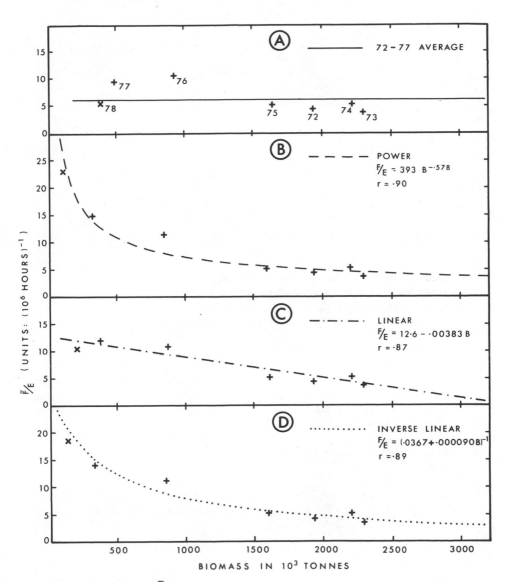

Figure 7: Plots of $\bar{F}/E$ against Biomass for the various prescript-
ions used for the Biomass estimates of Figure 6.

1972–75 average. This larger value maintains for 1977, and in
the presence of decreasing biomass, is strongly suggestive of a
"Paloheimo–Dickie" effect. Attempts have been made to incorporate
this into the VPA calculations by replacing the F/E = constant
hypothesis by

$$F/E = f(B) = \begin{cases} \alpha B^m & \text{Power} \\ \alpha + mB & \text{Linear} \\ (\alpha + mB)^{-1} & \text{Inverse Linear} \end{cases} \quad \dots \quad \dots \quad (4)$$

where at each repetition of the VPA analysis in the overall iteration, the regression parameters α and m are calculated, then the
iteration continued using $E_{effective}$ in equation (3) where

$$E_{effective} = E \cdot f(B) \quad \dots \quad \dots \quad \dots \quad \dots \quad (5)$$

   The results are shown in Figures 6 and 7. Stock projections
for 1979 using equation (4) are considerably lower: between 100
and 200 thousand tonnes compared to 450 thousand tonnes, and show
both the potential importance of this effect in current biomass
assessment, and more generally the increasing lack of certainty in
VPA estimates for more recent years.

   However, attempts to repeat this technique for data available
at the commencement time of previous seasons (to see whether the
biomass decline could be more satisfactorily detected) meet with
iteration convergence problems, essentially because too few data
points are available at low biomass levels to determine the
parameters of $f(B)$ adequately.

   A summary of the conclusions of this section on the detectability of the biomass decline is given in Table 4. A more complete
analysis, however, still requires estimates of what management
options were (in retrospect) available to avert the decline

MSY ESTIMATES

   Before proceeding to simulate alternative histories of the
stock behaviour under different harvesting strategies, it is useful
to try to get a measure of a MSY, to have an order of magnitude
indication of the harvesting potential of the stock.

   Figure 8A shows a plot of catch against effort for 1972–78.

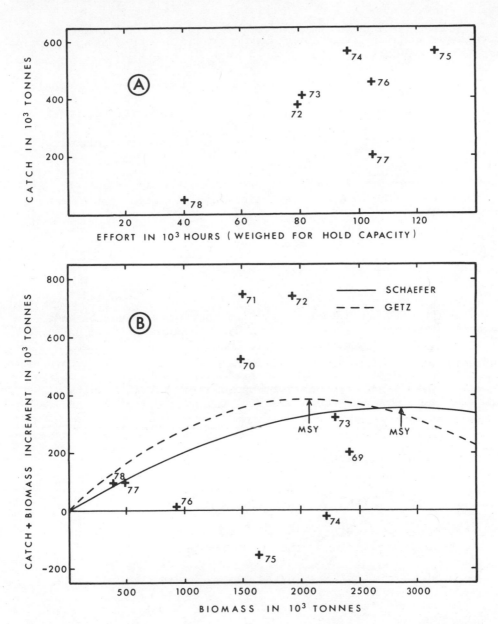

Figure 8:   Diagram A shows a plot of yield against (directed)
            effort for the Pilchard fishery from 1972 to 1978.
            Diagram B shows a plot of (Catch + Biomass increment)
            against Biomass (from VPA) for this period; fits for
            Schaefer and Getz[5] sustainable-yield-type models are
            shown, with the implied MSY's indicated.

Clearly a production curve fit to this data would provide only irrelevant results – not surprisingly as the steady-state assumption is scarcely justified.

In figure 8B, data is shown fitted to a Schaefer-type model:

$$\Delta B = \alpha B (B_\infty - B) - C \quad \dots \quad \dots \quad \dots \quad \dots \quad \dots \quad \dots \quad (6)$$

where ΔB = Biomass increment from year considered to following year.

Calculations give the following, where the error quoted corresponds to one standard deviation, and illustrates the large uncertainty:

C_{MSY} = 356 ± 428 thousand tonnes

B_{MSY} = 2870 ± 4460 thousand tonnes.

An alternative model by Getz[4] incorporates a Beverton and Holt stock-recruit relation into a dynamic-pool-analysis using an age-structured Beverton and Holt model.

Figure 3 shows the (spawning-) stock recruit curve used (an eye-fit) and the point on it corresponding to MSY.   Results of calculations using the model are also shown in Figure 8B and yield:

C_{MSY} = 386 thousand tonnes

B_{MSY} = 2070 thousand tonnes.

SIMULATION UNDER DIFFERENT HARVESTING STRATEGIES

Any detailed simulation necessitates assumption of some form of stock-recruit relation.   The model used here is:

$$R = R(\text{stock-recruit}) \times \text{Environmental Factor} \dots \quad \dots \quad (7)$$

$R(\text{stock-recruit})$ is taken from the Beverton-Holt form shown in Figure 3:

$$R(\text{stock-recruit}) = \frac{30 B_{sp}}{1000 + B_{sp}} \quad \dots \quad \dots \quad \dots \quad \dots \quad (8)$$

Units: B_{sp} : thousand tonnes
        R : 10^9

The "Environmental Factor" for each year is obtained by comparing the historic R for each year with R(stock-recruit), and is listed in Table 2. Values for recent years are less well determined because of the VPA increasing inaccuracies for this period.

Figure 9A compares the behaviour of the stock under fixed quota levels to the historic situation from 1971 to 1978. The definition of a "sustainable yield" over this period in the presence of fluctuations is semi-arbitrary, but we shall utilise the prescription of commencing the 1978 season with a biomass no lower than the 1971 minimum.

Calculations indicate such a sustainable yield to be 228 thousand tonnes per annum, which is considerably lower than the steady-state type models of Schaefer/Getz shown in Figure 8 indicate: respectively 356/386 for MSY, or 275/357 predicted for a biomass level of 1500 thousand tonnes corresponding to the minimum we permit.

The simulation also permits comparison of different harvesting strategies, all to result in a 1979 biomass the same as that in 1971. The results are shown in Table 3 and also in part in Figure 9B.

In attempting to assess what remedial courses of action were possible at various times to prevent the collapse of the stock, it is again necessary to specify semi-arbitrarily a current stock level which, even if not large, can at least be considered as tolerable, and indicative of "non-complete collapse" (or alternatively, above a level at which depensation mechanisms may come into play).

A value of 1000 thousand tonnes has been selected for this purpose. Results, which have been expressed in terms of constant reduced quotas imposed at a certain time and kept fixed thereafter, are shown in Table 4 and Figure 10.

DIRECT SURVEY ESTIMATES

Figure 11 compares direct survey estimates of the stock with the subsequent VPA. Aerial estimates (Cram and Agenbag[3]) have been compared to a biomass plot not including the contribution of the O-year class, as surveys were carried out early in the year before any significant proportion of these fish would have joined the shoals. Estimates have been renormalised to the mean VPA estimates for 1972-74 for comparative purposes; the multiplicative packing density parameter required for absolute aerial estimates was in any case poorly determined. Errors on the estimates have

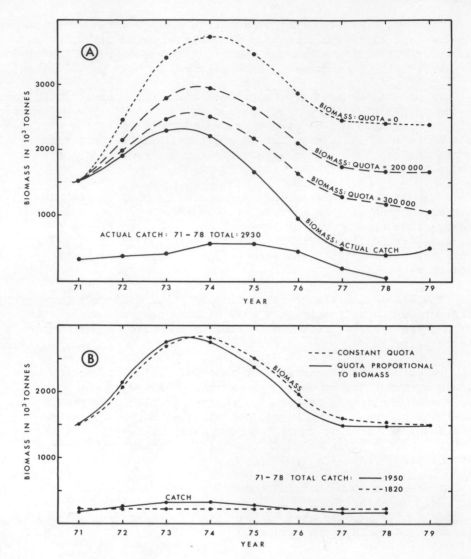

Figure 9:   Diagram A shows Simulation Model results for the stock
            biomass under harvesting strategies corresponding to
            constant quotas from 1971 to 1978 of various magnitudes,
            compared to the VPA estimate of the actual biomass trend
            under the historic catches.   Diagram B compares Biomass
            and Catch profiles corresponding to constant quota and
            quota proportional to current biomass strategies, that
            would have left the 1979 biomass equal to that at the
            start of 1971.

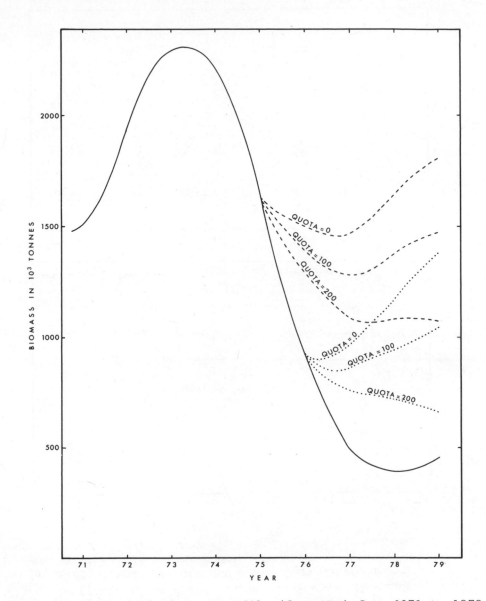

Figure 10: The actual Biomass profile (from VPA) from 1971 to 1979 is compared to the simulation results of imposing constant quotas of various levels from 1975 onwards and from 1976 onwards.

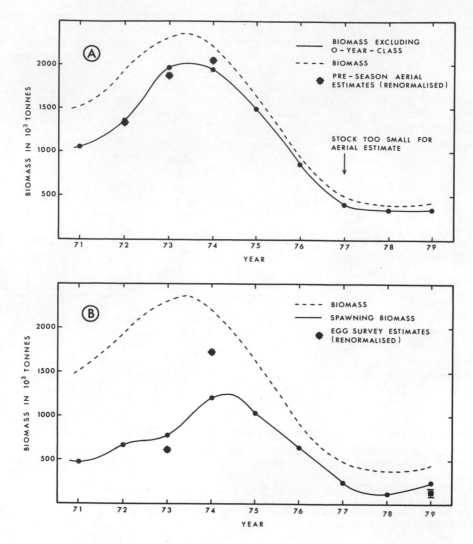

Figure 11: Diagram A compares VPA estimates of stock (excluding the
O-year-class) with pre-season aerial survey estimates
(renormalised to the VPA average over the 1972-1974
period).   Diagram B shows similarly renormalised
egg-survey estimates compared to VPA spawning stock
estimates.   The error shown for the 1979 egg survey
estimate reflects only the normalisation uncertainty.

Table 2.   "Environmental" Factors used to multiply Stock-
           Recruit Relation in Simulation.

Year	Environmental Factor
1971	1.85
1972	2.36
1973	1.27
1974	0.81
1975	0.42
1976	0.26
1977	0.89
1978	0.72
1979	1.00 (assumed)

Table 3.   Comparison of Simulated Total Yields from Stock
           1971-78 under different Harvesting Strategies

Strategy for Quota	Percentage	Range of quota fluctuation about mean	Comparison Total Yield With Strategy 1
1. Constant quota	(228 thousand tonnes)	-	-
2. Proportional current year's biomass	12%	35%	+6.7%
3. Proportional current year's spawning biomass	21%	62%	+5.4%
4. Proportional previous year's biomass	11%	40%	+3.3%
5. Proportional biomass 2 years previously	10%	34%	-3.5%

not been shown, but were crudely reckoned at   50%.

Figure 11B compares VPA spawning stock estimates with estimates
from egg surveys. Estimates have been normalised by the mean VPA

Table 4. Summary of stock assessment indications at various times compared to actual stock situation. The convention used is (+2/+1/0/-1/-2) corresponds to (strongly increasing/ increasing/steady/ decreasing/strongly decreasing)

Beginning of Year		1974	1975	1976	1977	1978
(VPA)		+2	0	-2	-1	-2
(CPUE)	Catch-statistics based.	+1	+2	-1	0	-2
(Modal length)		0	+1	0	+1	-2
(Aerial)	Direct	+1	-	-	-2	-
(Egg)		+2	-	-	-	-
(Year-class strength) 1975		-	(0)	-1	0	-2
(compared to average) 1976		-	-	(0)	-2	-2
Actual Biomass		0	-1	-2	-2	-2
(Historic quota) pre-season		+20%	+0%	+0%	-47%	-38%
(decision) mid-season				-20%	-58%	-77%
(compared to previous season)						
Quota reduction required for 1000 thousand tonne 1979 biomass		-42%	-61%	-78%	Not poss-ible	Not poss-ible

estimate for 1973-1974; in any case recent research on repeated spawning (Smith[9]) places doubt on the possibility of independent absolute estimates via the integration-type method used.

Data shown constitute a preliminary evaluation incorporating a revaluation of King's[5] estimates; comparability of the surveys requires further attention before final estimates are made. The

error shown for the 1979 estimates reflects the normalisation uncertainty only; overall error estimates have yet to be evaluated.

SUMMARY AND CONCLUSIONS

     The results of previous sections of the paper are summarized
in Table 4.   In brief catch-statistics-based indicators only
confirmed the severe stock decline with absolute certainty by the
beginning of 1978; preliminary indications of a decline were
apparent from these methods at the beginning of 1976, but by this
time the stock had already declined substantially, and a quota cut
of the order of 80% (scarcely a practical possibility at that time)
would have been required for stabilisation.

     Compared to the indications of the catch-statistics indices
(not all of which were actually available at the time) the actual
management decisions taken do not appear unreasonable, despite being
proved inadequate by subsequent events.

     The crude empirical moral to be drawn from this case seems to
be that with short-lived fluctuating pelagic species, not only is
it too late to wait for certain confirmation of a decline before
acting, but that:

(i)   stock increases should be apparent for longer than one year
      before quota increases are allowed.
(ii) a substantial drop in the rate of biomass increase constitutes
      sufficient evidence for a quota reduction, even though no
      actual biomass decrease may be apparent.

     Part of the problem for the catch-statistics-based estimates
is the Paloheimo-Dickie-type effect, which appears to have played
a role in this case. More subtly, this can mean that the fishermen themselves are less likely to be aware of a fall in the stock
size, thus being the more difficult to convince of the necessity
of imposing stricter management measures.

     Direct survey methods are clearly desirable in spite of their
greater cost; their performance in this case is promising, but it
is most unfortunate none took place in the critical 1975-76 period
which would have provided a more complete test. It is unlikely
though that direct survey methods on their own can in general
provide adequate information; catch statistics remain necessary
to test whether a biomass increase is a reflection perhaps of only
a single strong year class, for example.

     The simulation used suggests that steady-state techniques for
MSY estimates may be over-optimistic; however, this result may
depend on the specific case studied and debatable assumptions of

the model, so is not necessarily a general feature of the effect
of large year-class fluctuations.

Overall the results indicate, once again, the need for
conservatism in pelagic stock management.

ACKNOWLEDGEMENTS

Data used in this paper is derived from diligent collection
over the period by the Sea Fisheries Branch, Department of
Industries, South Africa.

The author acknowledges financial support for travel purposes
from the University of Cape Town, and Council for Scientific and
Industrial Research, South Africa. The full version of this paper
appears in Colln scient. Pap. Int. Commn S.E.Atl. Fish. 7 (1980).

REFERENCES

1. D.S. Butterworth, A preliminary alternative assessment based
       upon a revised method of pilchard ageing. Addendum to: An
       Assessment of the Southeast Atlantic pilchard population in
       ICSEAF Divisions 1.4 and 1.5, 1953-1977. Colln scient. Pap.
       Int. Commn S.E. Atl. Fish. 5 (1978) 45-52.
2. D.S. Butterworth and F le Clus, Assessment of the anchovy
       (Engraulis capensis) in ICEASF Divisions 1.4 and 1.5, 1968-78.
       Colln scient.Pap. Int. Commn S.E. Atl. Fish. 6 (1979)171-182.
3. D.L. Cram and J.J. Agenbag, Airborne Shoal Tonnage Estimation,
       in the Cape Cross Program, Phase IV Report, Int. Rep. Sea
       Fish. Branch, S. Afr. (1974).
4. W.M. Getz, Sustainable yield strategies from an age-structured
       two-season Beverton-Holt type model, TWISK32, NRIMS, CSIR
       (S.Afr.) (1978).
5. D.P.F. King, Pilchard stock estimation by means of egg and
       larval surveys, in the Cape Cross Program, Phase IV Report.
       Int. Rep. Sea Fish. Branch, S. Afr. (1974).
6. G. Newman, A stock assessment of the pilchard (Sardinops
       ocellata) at Walvis Bay in SWA. Investl Rep. Div. Sea Fish.
       S. Afr. 85 (1970) 13pp.
7. J.E. Paloheimo and L.M. Dickie, Abundance and fishing success.
       Rapp.P.-V. Reun.Cons. int. Explor. Mer, 155 (1964) 152-163.
8. J.G. Pope, An investigation of the accuracy of virtual
       population analysis using cohort analysis. Int.Commn Northwest
       Atl. Fish. Res. Bull. 9 (1972) 65-74.
9. P.E. Smith, Pers. Commn (1979).
10. Ø. Ulltang, Factors of pelagic fish stocks which affect their
       reaction to exploitation and require a new approach to their
       assessment and management. ICES Symposium on the biological
       basis of pelagic fish stock management 34. (1978).

A BIOLOGICAL PREDICTOR MODEL DEVELOPED IN SUPPORT OF AN OPERATIONS

RESEARCH APPROACH TO THE MANAGEMENT OF THE NEW ENGLAND

GROUNDFISH FISHERY

Emma M. Henderson and Guy D. Marchesseault

Northeast Fisheries Center, New England Fishery
Woods Hole, Management Council Staff,
Massachusetts, Peabody, Massachusetts,
U.S.A. U.S.A.

THE GROUNDFISH MODELING PROGRAM

The modeling project

The staffs of the Northeast Fisheries Center and the New England Fishery Management Council have been deeply involved in developing a comprehensive groundfish management plan using an operations research approach. Approximately one year ago a list of recommended objectives for groundfish management was compiled and a research program initiated to analyze and evaluate feasible alternatives. The program called for bioeconomic modeling and the co-authors were assigned to the biology sector of the modeling task. The following subset of objectives essentially defined the biological modeling requirements.

> The overall objective of the plan shall be to generate over the period of the plan the greatest possible joint economic and social net benefits from the harvesting and utilization of the groundfish resource, ensuring that by the end of the period the relevant groundfish stocks shall be in conditions which will produce enhanced and relatively stable yields from the groundfish fishery in future years.

The overall objective recognizes the following relevant dimensions of management goals.

(a) The management unit for the groundfish plan shall
    consist of cod, haddock, and yellowtail flounder
    as well as of other species in the mixed trawl
    fishery that the Council considers necessary to
    bring under regulation.

(b) Over the plan period expected total removals will
    be established on a yearly basis consistent with
    the overall objectives as constrained by (1) an
    acceptable probability of achieving the biological
    stock conditions by the end of the plan period, and
    (2) a minimum spawning stock level for each species
    which ensures an acceptable probability of continued
    recruitment.

(c) A multiple year planning period so as to make a
    determination within the framework of the plan of
    how alternative harvesting levels in the immediate
    future affect the options available for later years.

To summarise, the model should predict stock variables over
a multi-year period with prescribed annual removals. The
variables should include stock, spawners and recruits. The
variables should be random variables. The model should also
establish the relationship between annual removals and final
stock size, between spawners and recruits, and between removals
and spawners. Finally, the model should calculate the probability
distribution of each variable at the end of the planning period.

Because of time constraints and the lack of certain data,
modeling would not deal with species interactions, seasonal
variability, or size-selective gear. The model had to be
designed for the existing data base not only to be operational in
the near future, but also to maintain continuity with past
analyses.

The authors have developed a general simulation model encompassing
the above needs and not restricted to any specific application.

A quadratic programming optimization (QPO) model

One such application under way was an economic model in which
the set of annual quotas would be obtained by maximizing the
present value of gross ex-vessel earnings over a 5-year period.
Optimization would be subject to a set of constraints (linear
inequalities involving the quotas and the other stock variables)
to ensure satisfying the biological management goals. A review
of the equations being developed for the simulation model showed
that such relationships were nonlinear. In addition, a survey

of software indicates that stochastic software would be difficult to locate or to use, while nonlinear constraints either would not be allowed, or would not guarantee an optimum solution. These interfacing problems were eventually solved as follows. Constraints could be formulated using a linearized version of the model equations developed by introducing extra assumptions and approximations. A conditional set of the required numerical values could be obtained from preliminary studies using the simulation model in a deterministic mode. The resulting linear constraints and the quadratic economic objective function would be the input for the QPO software. The proposed set of optimum quotas output by the QPO model would then become input for the simulation model in the stochastic mode for assessing the probability that all desired management goals were met, but some iteration with revised constraints will be required to obtain a final set of quotas.

DEVELOPMENT OF THE MODEL

The conceptual model

The first step in designing the model was to judge the suitability of treating the parameters and variables as constant or time dependent or density dependent, perhaps with time lags. Variables were also judged to be deterministic or random variables. The following decisions were made. The model would calculate numbers and biomass of stock, spawners, and recruits. Classical population dynamics equations would be used for growth and mortality. Realistic variability in stock size and recruitment could be introduced by modeling variable year-class strength and partial recruitment processes. Pre-recruit mortality appeared to be the only candidate for a random variable and could be modeled using historical data. Thus year-class strength would be multiplied by a log-normal survival factor. This random variable would induce stochastic variation in the other stock variables, providing the desired probability distributions. Among the numerous options for modeling a spawner-recruit relation first priority went to developing a spawner-recruit equation, with "recruitment independent of parent stock" as second choice. It appeared feasible to include the above complexity without using a complex model. In addition, the lack of data mentioned earlier implied the pilot model would be a single-species model with annual time steps, and with fishing mortality independent of size.

The model is previewed in this section by outlining the corresponding conceptual model. The next section contains the model equations and definitions. Parameter estimation is not usually considered part of a model, but the new approaches required for predictive modeling led to devising estimation procedures as well as deriving equations. Since these derivations formed

the major share of the modeling problems, a discussion is included.
Later sections treat partial recruitment, the spawner-recruit
equation, data requirements, an abbreviated derivation of the
linear constraints for the QPO model and a brief review of the
computer programs. Further details can be found in (1).

The model assumes knife-edge recruitment at weight w_1.
Consider a set of K+1 weights, w_1, w_2,..., w_K, $w_{K+1} = W_\infty$,
spaced off by one year's growth. Fish in the "kth weight class"
grow from weight w_k to w_{k+1} during the year. Recruits occupy
the first weight class their first year in the fishery and
the kth class their kth year in the fishery. The final class
contains all fish whose weight exceeds w_K. In the initial year,
t, fisheries data is used to divide the recruited stock into
weight classes. The number of fish assigned to the kth class
is N_{kt}. The total recruited stock is obtained by summing N_{kt},
the biomass by summing the products $N_{kt} w_k$. The annual yield
is the exploitation rate times the effective biomass. The
effective biomass is calculated using $\bar{w}_k$, the average weights
during the year. To calculate spawners one must reduce numbers
of fish by the losses up to spawning time, use weights at
spawning time, and sum only over classes containing mature fish.
These steps provide all the required quantities for the first
year.

To advance one year we note that fish formerly in class k
will now be in class k+1, while the number of fish has been
reduced by the survival factor. Thus, new values can be
calculated except for k=1. $N_{1',t+1}$ is calculated as follows. Each
year class begins recruiting at age r and is fully recruited at
age R. We show that the number recruiting at any age can be
expressed as a fraction of year-class abundance at age r.
Abundance at age r is calculated from a spawner-recruit equation
with lag r. $N_{1',t+1}$ is then obtained by summing new recruits from
the appropriate age groups. Thus it is possible to calculate
forward one year at a time.

The model equations

The model is based on the following ten equations. Definitions
and comments are given below.

$$N_t = \sum_{k=1}^{K} N_{kt} \tag{1}$$

$$\bar{B}_t = \sum_{k=1}^{K} N_{kt} \bar{w}_k \tag{2}$$

$$\mu_t = \frac{Y_t}{\bar{B}_t} \tag{3}$$

$$\dot{s}_t = \alpha + \beta \mu_t \tag{4}$$

$$NS_t = \sum_{k_\alpha}^{K} N_{kt} s_t^{\tau} \tag{5}$$

$$BS_t = \sum_{k=k_\alpha}^{K} N_{kt} w_{\tau k} s_t^{\tau} \tag{6}$$

$$N_{k,t+1} = N_{k-1,t} s_t \qquad\qquad k=2,\ldots,K-1 \tag{7}$$

$$N_{K,t+1} = (N_{K-1,t} + N_{Kt}) s_t \tag{8}$$

$$A_{rt} = (C_1 NS_{t-r} + C_2 BS_{t-r}) e^{-\varepsilon_t} \tag{9}$$

$$N_{1,t} = \sum_{i=0}^{R-r} \Pi_i A_{r,t-i} \tag{10}$$

N_{kt} = number of fish in the kth weight class at the beginning of year t

N_t = total number of recruited fish at the beginning of year t

$\bar{B}_t$ = effective biomass of recruited fish during year t

   $\bar{B}_t$ and $\bar{w}_k$ have been defined to satisfy equation (3)

$\bar{w}_k$ is estimated as the weight at mid-year

μ_t = annual exploitation rate on recruited fish in year t

Y_t = annual yield or quota in year t

    Y_t is input for each year in the planning period.

s_t = annual survival rate of recruited fish in year t

    The approximation (5) was derived to avoid iterative solutions.

Observed values" for the regression were obtained by calculating pairs of values of s and μ over a range of F's. ($\hat{\alpha}$ = .81745, $\hat{\beta}$ = -.88953, r^2 = .99997 for M = .2 and .3<s<$e^{-.2}$)

NS_t = number of spawners in year t

BS_t = biomass of spawners in year t

τ = spawning date expressed as a fraction of a year

k_α = designates the first class containing mature fish

$w_{\tau k}$ = spawning weight of fish in class k

A_{rt} = abundance of r-group fish in year t

    Equation (9), C_1, C_2 and ε are discussed later. In the deterministic model $\varepsilon_t \equiv 0$. In the stochastic model ε_t is generated from random normal variable software with $\mu = \bar{\varepsilon}$ and $\sigma = s_\varepsilon$.

    N_{1t} = the number of new recruits entering the fishery in year t

    Equation (10) and estimation of coefficients, Π_i, are now discussed.

Partial recruitment

    Projected partial recruitment follows the average pattern established by past data. The derivation is similar to that given in Allen (2).

    Calculation of Π_i requires the intermediate quantities Π_{it} and R_{it}. R_{it} is the number of i-group fish recruiting in year t. C_{it} is the number of i-group fish landed in year t. Let μ_t be the common exploitation rate on fully recruited fish. from

          total recruits = new recruits plus escapement

and

$$\text{recruits} = \frac{\text{catch}}{\text{exploitation rate}}$$

it follows that

$$R_{rt} = \frac{C_{rt}}{\mu_t} \tag{11}$$

$$R_{it} = \frac{C_{it}}{\mu_t} - \frac{C_{i-1,t-1}}{\mu_{t-1}} s_{t-1} \qquad i=r+1,\ldots,R$$

Allen's formulas can be adapted to give A_{rt} and Π_{it}.

$$A_{rt} = R_{rt} + R_{r+1,t+1}e^M + \ldots + R_{R,t-R}e^{(R-r)M} \tag{12}$$

$$\Pi_{it} = \frac{R_{r+i,t}}{A_{rt}} \qquad i=0,\ldots,R-r \tag{13}$$

Since the model uses the Π's to project into the future, it is preferable to eliminate the dependence on t. Therefore, we average Π_{it} over the set of t's represented in the available data.

$$\hat{\Pi}_i = \bar{\Pi}_{it} \qquad i=0,\ldots R-r, \text{ t in data set}$$

The quantity M in (12) should be M_u, the annual mortality rate on unrecruited fish, but has been set equal to M for lack of an estimator.

For those familiar with VPA we point out that the variable N_{rt} obtained by VPA calculations corresponds to the variable A_{rt} in (12). However, the VPA partial recruitment coefficients do not correspond to the Π's in any simple way.

Spawner-recruit equations

Projected deterministic year-class abundance follows the average pattern established in prior years. Projected deviations in abundance are simulated using the parameters of the historical probability distribution. Equation (9)   is derived beginning with the premise that fecundity is linear in weight. Thus

$$\text{eggs per } i^{th} \; Q = \phi \, (w_i - w_\alpha)$$

$$\text{total eggs spawned} = \sum_{\text{spawning } \female\text{'s}} \text{eggs per } \female$$

$$= \phi x \text{ (biomass of spawning } \female\text{'s)} - \phi w_\alpha x$$
$$\text{(number of spawning } \female\text{'s)}$$

$$A_{rt} = \text{(total eggs spawned in year t-r)} x \\ \text{(survival from t-r to t)}$$

If survival to age r in year t is a log-normal random variable with exponent

$$M_{rt} = \bar{M}_r + \varepsilon_t$$

then A_{rt} takes the form

$$A_{rt} = (C_1 NS_{t-r} + C_2 BS_{t-r}) e^{-\varepsilon_t}$$

Since the coefficients C_1 and C_2 have absorbed $e^{-\bar{M}_r}$, ϕ, w_α and the $\male \female$ ratio, these quantities need not be estimated individually. $\hat{C}_1$ and $\hat{C}_2$ are estimated from the bilinear regression

$$A_{rt} = C_1 NS_{t-r} + C_2 BS_{t-r} \qquad\qquad (14)$$

using available data. We then calculate

$$\hat{A}_{rt} = \hat{C}_1 NS_{t-r} + \hat{C}_2 BS_{t-r} \qquad\qquad \text{t in data set}$$

and

$$\varepsilon_t = \ln \frac{\hat{A}_{rt}}{A_{rt}} \qquad\qquad \text{t in data set}$$

This set of ε_t yields $\bar{\varepsilon}$ and s_ε.

If it is impossible to estimate $\hat{C}_1$ and $\hat{C}_2$ because of lack of suitable data or lack of fit of the model, equation (9) is altered to

$$A_{rt} = C_0 e^{-\varepsilon_t}$$

The choice $C_0 = 0$ or $C_1 = C_2 = 0$ is made in the input to the program. Since w_α is approximated by $\dfrac{-C_1}{C_2}$ and C_1 should be negative, C_2 positive, we have some criteria for rejection of the model.

The data base

The models described in this paper have been designed specifically for studying New England groundfish stocks. Parameters have been estimated using length-frequency data from commerical samples for several years, length-age and weight-length equations, commercial catch-at-age data for consecutive years plus an initial $\hat{F}$, and VPA results for spawners, recruits and F's for consecutive years. The initial data for the model comes from commercial length-frequencies, total catch, and $\hat{F}$ for the year preceding the planning period. Although the data bases are not uniform for all species, the estimation procedures described here can be modified as needed. Deficiencies in data can be supplemented by estimation procedures discussed in the assessment documents. Clark and Overholtz (3), McBride and Sissenwine (4), Serchuk et. al. (5).

QPO constraints

To obtain linear constraints for use with the QPO model it was necessary to adapt some equations, refit some parameters and introduce another parameter, which was estimated by tuning to the simulation model. Recursion relations provided the two equations

$$B_t = a_{ot} + \sum_{i=1}^{t-1} a_{it} Y_i \qquad (15)$$

$$BR_t = b_{ot} + \sum_{i=1}^{t-2} b_{it} Y_i \qquad (16)$$

B_t and BR_t are the biomasses of total recruited stock and of new recruits at the beginning of year t. The coefficients a_{it} and b_{it} are complicated expressions involving recruitment parameters and initial data, so are specific to the planning period as well as to the stock. Omitted details will be in (1). Equations (15) and (16) allow writing constraints only of the type

$$B_t \geq BMIN$$

$$BR_t \geq BRMIN$$

The computer program

The current programs were written for debugging and experimenting Programs are in FORTRAN-IV for use on the Xerox Sigma-7 computer used by NEFC. The authors expect to improve the input file system, add computer graphics and provide documentation in the near future. Separate programs were written for the deterministic and stochastic models becuase of differences in usage.

The deterministic program, DET, outputs N_t, $\bar{B}_t$, NS_t, A_{rt}, and N_{lt}, for each year in the planning period. One of the quantities Y_t, F_t, Z_t, μ_t, or s_t is chosen as input. The others are calculated and listed as output.

The stochastic program, STO, generates random normal variables, ε_t, and multiples A_{rt} by $e^{-\varepsilon}t$. The stock variables are only considered at the end of the planning period. The program runs a large number of complete periods (say 300) and cumulates the probability of occurrence of a set of values for each variable. The set of values is determined by frequency classes. The number of classes and the width of the class intervals are selected to suit the range of the variable and the desired accuracy. The probability distributions are output, along with $E(x)$, $V(x)$, $max(x)$ and $min (x)$ for each variable, x. Current output variables are N_T, NS_T, and N_{lT}.

Both programs print out the input data and parameters for the run, including program control variables selected by the operator.

DISCUSSION

The model was primarily developed as a tool for the operations research studies demanded by the objectives. It can now be used to explore concepts such as "minimum spawning stock" or "acceptable probability," to establish possible ranges of each variable and to evaluate their importance.

The model equations show that simple concepts will not suffice to predict the future or to describe a stock. For example, in the equation

$$N_t = N_{lt} + \sum_{k=2}^{K} N_{kt}$$

"stock" is the sum of incoming recruits and escapement. The first factor is the sum of random variables and requires prediction, while the second is evaluated from past values and harvesting rates. The first factor reflects the effect of distant harvests and stock levels, the second reflects the effect of more immediate harvests and stock levels. "Biomass" is subject to the same comments. "Recruits" is a group of fish containing fractions of several year classes. Conversely, the effect of one year class on the fishery is not seen in a single year. For "spawners" the composition is important and neither numbers nor biomass can be neglected.

This type of model should provide useful management advice in the interim period between the 1-year ahead assessment projections and full-scale multispecies models. The model does not replace assessments, but is based on their results.

The model will not be validated until the predictions are tested with new data, not used in fitting the parameters. At this time parameter estimates have not been completed, nor has the QPO software been selected. To illustrate some possible uses of the output it is necessary to use some of the debugging material, rather than the final product.

Test runs used F_t as the input parameter with F_t constant over a 5-year planning period. Figure 1 shows how the natural rate of increase of a stock is decreased by fishing. We can conclude that the value of F has an effect on recruitment, but F alone is not controlling recruitment. Other factors could be the composition of the stock and the initial stock level. It appears that there is an equilibrium F at about .4 or .5, while F = .6 is detrimental to both stock and recruits. The annual values have been connected with straight lines only for convenience. Intermediate values should not be read from these graphs.

The above results are from the deterministic model. We can also examine the probability that the values in the final year are those shown. Figure 2 contains histograms for the probability distributions of recruits and total stock. The distributions are not identical, but are of the same one-sided, left-skewed type. The shape of the curves depends in part on the assumed underlying log-normal distribution. The height of the curves depends in part on the choice of class intervals. A plot is shown for F = O. The curves for other F's are similar, but the mode moves to the left as F increases. The deterministic value, X, falls between the mode, M, the mean, μ (obtained from computer output). Results are preliminary as some experimenting remains to be done in choosing appropriate class intervals, parameters in the random number generator, and the number of trials to use to produce these curves.

The upper graph in Figure 3 is a plot of yield as a function of F and demonstrates that yield decreases annually as F increases beyond the equilibrium value. The equilibrium value of F shows clearly. The lower graph illustrates that the average yield is inadequate for decision making. The average yield is a maximum for F = .6, but the other graphs show why that is an undesirable level.

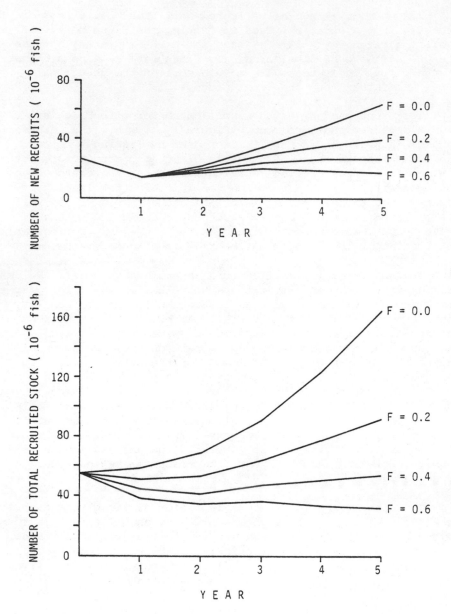

Figure 1. Predicted annual status of stock and recruits for a
hypothetical species over a 5-year planning period
under various constant fishing mortality rates.

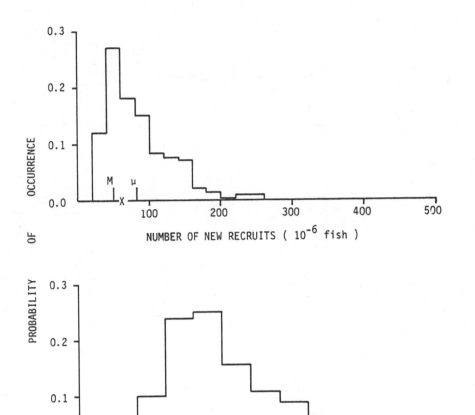

Figure 2. Probable status of stock and recruits for a hypothetical
species at the end of a 5-year perod with no fishing(See
text for details.)

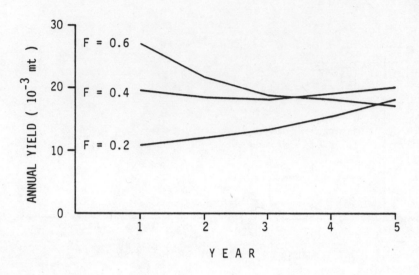

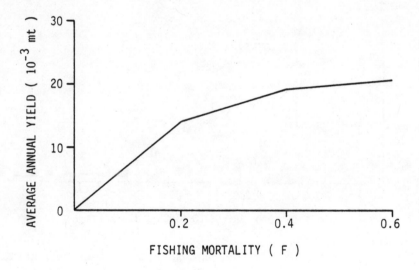

Figure 3. Predicted annual yield and average annual yield from
a hypothetical species over a 5-year planning period
under various constant fishing mortality rates.

References

(1) E.M. Henderson, Biological models for the New England groundfish fishery, NEFC Lab. Ref.No. 79-47 (1979).

(2) K.R. Allen, Methods for estimating exploited populations, J. Fish. Res. Bd. Can. 23 (10), 1553-1574 (1966).

(3) S.H. Clark and W.J. Overholtz, Review and assessment of the Georges Bank and Gulf of Maine haddock fishery, NEFC Lab. Ref. No 79-05 (1979).

(4) M.M. McBride and M.P. Sissenwine, Yellowtail flounder (Limanda ferruginea): status of the stock, February 1979, NEFC Lab. Ref. No. 79-06 (1979).

(5) F.M. Serchuk, P.W. Wood, R. Lewis, J.A. Penttila, and B.E. Brown, Status of the Georges Bank and Gulf of Maine cod stocks February 1979, NEFC Lab. Ref. No. 79-10 (1979).

Acknowledgements

We wish to thank B. Brown, F. Serchuk, M. Sissenwine, S.Clark, J. Mueller, and J. Kirkley of NEFC, and L. Vidaeus, H. Russell, and S. Wang of the NEFMC staff for discussions contributing to the success of this project.

FROM THE POINT OF VIEW OF A CIVIL SERVANT

Johán H. Williams

Counsellor, Department of Economic Planning
The Norwegian Ministry of Fisheries

I will make some comments upon what I feel are the main
problems for model-making as well as all-levels-decision-making
in the fishery industry.   It is a pity but still a fact, that
today it seems impossible to get reliable fish-stock-estimates which
can be used for economic planning for a longer period than the next
week.   It's understandable that this causes some problems when you
are going to suggest a problem-solution to the decision-making
politicians.   From my point of view, this is also the main problem
in model-building, as it strictly limits the use of the different
models for long-term planning of the fishery-industry.   (When I
say long-term planning of the fisheries, that is what in ordinary
economic planning is called short-term planning, 1 to 4 years).

What I want to say further is that from my point of view, as
a civil servant, not a research-scientist, it is more desirable
that some of the existing models would be developed in a way that
it could be used in our planning and decision-suggestion.   This
would be preferable instead of developing a lot of new and more
advanced models, of which none can be used for any practical purpose.

As long as all model-building-scientists stress that their
main goal is to help society to a better future, it is a failure
to search for optimization in models with a narrow approach and
over-simplified surroundings.   What the decision-makers want is an
answer to the humble question of what direction is the right one to
go from the present situation.   We have realized that we will
never reach Nirvana.

A problem referring to the relations between the scientists
and the administrators concerns what is happening when we have to

say we can't use the models being presented to us.   When the
scientist is told that the model will remain just a model he gets
angry and tells us that we are a bunch of bloody ignorants, not
understanding what is best for us, which is to buy his model.

The next step for the scientist, an economist, a biologist
or anyone else, is to use his model for a case study.   With the
model he will find a solution for a certain problem in the fishing
industry: for instance, size of the fleet related to certain size
of the fish-stocks.   The result will then be presented to the
administration and the politicians.

When, for example, an economist presents his model-solution
to the decision-makers they know the way to go from the narrow
point of view of the economist's theories.   But when the politicians
and administrators don't exactly do what the scientist's model tells
them to, the scientist gets mad again, and deeply offended. He then
often uses the press or other fora, like this one, to accuse the
decision-makers for stupidity. But it must be understood, the real
world is not as simple as the world of the model. The bureaucracy
and the politicians do have to take into account factors that compli-
cate the picture a lot compared with the model, and as a result this
leads to a "not-optimal" solution from the scientist's point of view.

What is needed to avoid this, I think, is a greater effort to
have model-making and case solution reflect the society as it is.
To obtain this it will be necessary to have cooperation between
different kinds of scientists, economists, technologists and
biologists.   Furthermore, the administrators with training in taking
political considerations should also be participating in this work.

The above discussion implies that no one should think of his
specific field as superior to whatever anyone else is doing. -
Unfortunately, this attitude is not so very uncommon today among
scientists as well as among administrators.

FORUM - "BACK TO COMPREHENSIVE MODELLING IN FISHERIES"

Report by C. Curr

Very little discord entered the final discussion session of
the conference. Perhaps the excellent organisation and hospitality
of our hosts had conspired to reduce the probability of disagree-
ment between participants in this forum.   Whilst the title of the
session bade us return to the theme of comprehensive modelling in
fisheries, and the opening speakers addressed their remarks to it,
the ensuing discussion drifted immediately from a strict adherence
to the theme, and remained aloof, but not distant from it.   Rather
than extend deliberation on comprehensive versus piecemeal,
sequential modelling techniques, the conference drifted towards the
framework within which models are conceived, devised and used.
This was not a bad thing, as it gave contributors the opportunity to
discuss those comprehensive aspects of modelling which immediately
concerned them. Had the forum been able to continue longer, it is
likely that the discussion would have reverted to the relationship
between the extent to which a single model should attempt to
represent a possibly wide problem area, and the research framework.
How much do factors such as objectives, data and clients restrict
modelling techniques and limit the models' explicit range over the
problem area?

If the outcome of the forum can be expressed in terms of a
single theme, the most appropriate might be the wide perspective
into which fisheries modelling must fit.   Most contributors
alluded to this, without suggesting that models themselves should
explicitly represent a global situation. Rather they should become
more acceptable within the operational framework, as a means
towards eventually achieving better-defined objectives, making more
efficient use of data.   A scientist without the gift of wide
vision, interpreting an objective in his own way, might well use an

operations research technique to produce a pure, but sterile
optimum solution to his own view of the problem.   Worse still, he
might develop an esoteric technique, unintelligible even to fellow
scientists.   Such situations, extreme or otherwise, were tacitly
acknowledged by the forum, which agreed that the problem-solver
must himself be obliged at the outset to contemplate the wider
perspective into which his efforts must blend.

    First, the objective or objectives (here, a note of discord
on the concept of multiple objectives).   Those participants
experienced in consultancy projects were quick to point out the
difference in requirements between different sorts of client.   The
objectives of a commercial client, aggressively exploiting a
fishery through his markets, might prove easy to define compared
to those of a government.   In its role of regulating a national
economy, rather than allowing full rein to free market forces,
a government may favour specialist group interests from time to time,
apparently at the expense of the general public.   Whether such
motivation favours the fishing industry or not, and whether or
not the operational researcher sympathises with a particular policy,
he must consider how to maximise the usefulness of his results to
a government client.   Only cooperation, and humble cooperation
between scientists, economists and civil servant administrators can
produce models useful in planning fisheries development.   Otherwise
models will remain just models, never to be extended back into
the real world they strive to represent.   It was thus agreed that
there is a general need to clarify objectives, and to consider them
in the widest perspective necessary.   There is little merit in
building a model of fish resources in an African river, however
excellent it might be, if the researcher remains oblivious of the
imminent construction of a dam, which will alter most of his
assumptions about the future.   The final objective of any modelling
project must be not to propose a unique optimal solution to the
problem, but rather to indicate the direction in which the client
should move in time towards achieving his goal.   In an extreme
case, the outcome of a modelling exercise, if carried out
immediately, might have no value at all to society.   An example
quoted was a sudden and extended ban on fishing, causing first the
collapse of a local industry, and later, on resumption of fishing,
an over-supply of cheap fish.

    Both the objectives that should be set, and the modelling
techniques that should be used, depend greatly on the data
available, or potentially available.   Much discussion centred on
the problems of data.   It was argued that we tend to lump data
into a single concept, whereas they fall into different types,
available at different costs.   It is important to define what
precision is required, before collecting copious data which may
not only exceed their usefulness, but also divert resources from
seeking out other important facts.   Since uncertainty will remain

no matter how extensive the data, it is important to consider the
cost of this uncertainty before deciding what further data might be
worth having.   It is generally true that we should make better
use of our data: this was stressed in the field of fish stock
assessment, where better statistical analysis and organisation of
data would be advantageous. Even so, the central problem of
assessment, to estimate the number and age structure of fish
present in a stock now, remains unsolved by what might be called
traditional methods. More recent research using egg or accoustic
surveys, or direct count methods shows that fishery resource
analysts and their masters appreciate the high cost to the
industry of uncertainty in this area.   Even so, for certain
species, such as pelagic fish, shrimp and, worst of all, squid, we
should not be too hopeful of being able to get any lead time on
stock abundance estimates.   Perhaps the great variation in
opinion noted amongst fishery resource biologists, from
pessimistic to optimistic, is the final expression of uncertainty
in this area.

It seems that fisheries operations research workers may profit
from further examining methods used in other fields of application,
not necessarily within the food industry.   For instance, inventory
was suggested as a possible analogy for fish stocks.

There was general agreement that the choice of modelling
techniques depends on both the data available and the feasible
range of end results.   It is often true that the client's final
choice of action does not lie within a continuum of options.
Vessels, ports and processing machinery are available in discrete
sizes; fishing seasons come and go.   Together with present and
future uncertainty, these factors combine to make reality hard to
define, and harder still to represent in a global model.   The
more a modeller tries to approach reality, the less likely he is
to be justified, if only because of the expense incurred.   There
is much to be said for the sequential use of simple models
(possibly with sequential sampling from data) within the
perspective framework already advocated.

Because of the difficulty, and undesirability of attempting
realism in modelling, it is unlikely that a practical model will
itself answer all the questions required by the objective.   What
it can do is often overlooked.   A feature of all models, whether
mathematical or physical, is their power to illuminate the problem
area and to educate the user, often to his surprise. Just as there
is a strong case for producing results which can be used in
adaptive management of the real system, so there is an equally
powerful argument for O.R. workers to use their models in an
adaptive way.   They should learn by observing how models work,
and adapt accordingly, or change to another model if necessary.
This argues very much in favour of using simple partial models

adaptively and sequentially, but in no way conflicts with the
necessity of maintaining a perspective framework through consultation
and cooperation, which is ultimately the only route towards
comprehensive modelling.

AUTHOR INDEX

SUBJECT INDEX